Digital Logic Design

Digital Logic Design

B. Holdsworth, BSc(Eng), MSc, FIEE
Senior Lecturer in Electronics,
Chelsea College, University of London

Butterworths
London Boston Sydney Wellington Durban Toronto

All rights reserved. No part of this publication may be reproduced or transmitted in any form or by any means, including photocopying and recording without the written permission of the copyright holder, application for which should be addressed to the Publishers. Such written permission must also be obtained before any part of this publication is stored in a retrieval system of any nature.

This book is sold subject to the Standard Conditions of Sale of Net Books and may not be resold in the UK below the net price given by the Publishers in their current price list.

First published 1982

© Butterworth & Co (Publishers) Ltd, 1982

British Library Cataloguing in Publication Data

Holdsworth, Brian
   Digital logic design.
   1. Digital electronics 2. Electronic circuit design
   I. Title
   621.3815'3   TK7868.D5   80–40833

ISBN 0-408-00404-5
      0-408-00566 1 Pbk

Typeset by Scribe Design, Gillingham, Kent

Preface

This book has been developed from the material given to undergraduates in the Electronics Department at Chelsea College, and to practising engineers and scientists who have attended the one week *Digital Design* courses held at regular intervals in the same institution. Essentially, the book is an undergraduate text, but it should be useful to all practising engineers who, having had no formal education in digital techniques, have found themselves thrust into the digital cauldron at their place of work. Alternatively, sixth formers who are pursuing electronics as a hobby, prior to entering a university, and also students who are doing a Master's degree in digital techniques, and not having previously studied the subject, should find the contents of this book helpful.

Digital design techniques have become increasingly important over the last decade and now form part of the curriculum at most universities and polytechnics. Initially, the design techniques employed discrete components, but as the decade progressed, SSI circuits and MSI circuits were introduced and became widely available to the design engineer so that the design process became in some cases a matter of assembling a group of ICs to form a digital system. More recently, LSI chips have also become available, and this has led to the development of the microprocessor. As a consequence of this development, there are now two alternative methods of design available to the digital engineer. The design problem may be solved using hard wired logic, or, alternatively, it can be solved by using a microprocessor. The first technique requires the methods of design dealt with in this book, whilst the second method is a software solution of the same problem. In the final analysis, the criterion governing the choice of design method must be its cost effectiveness.

In practice, the solution to many problems requires a mixture of the two techniques. It is therefore essential that the digital engineer should have a good working knowledge of Boolean algebra and its application to problems in digital design. Additionally, he should have a sound knowledge of programming techniques. It is the author's opinion that a student who has a good grounding in the design of hard wired logic circuits of both the combinational and sequential types will find the transition to designing with a microprocessor a less difficult step.

The book is divided into twelve chapters. The first three of these chapters cover the introductory work necessary for solving the combinational problems presented in Chapter 4. Chapter 1 outlines the essential laws of Boolean algebra and Chapter 2 extends this work to K-map plotting techniques and the simplication of Boolean functions. Although simplification is, to some extent, becoming a dead art as a result of the development of MSI circuits, the author feels that it is more than useful for the reader to develop his or her manipulative skills by employing these techniques. Chapter 3 is devoted to NAND and NOR logic. Most present day SSI based circuits are designed with NAND and NOR gates rather than AND, OR and NOT gates, and this chapter deals with the implementation of Boolean functions using these types of gates. The final chapter of the section, Chapter 4, examines and solves a number of commonly occurring combinational problems.

The next five chapters are concerned with sequential circuit design. Chapter 5 presents the properties and develops the characteristic equations of a number of different types of flip-flop and the work in this chapter in conjunction with the contents of the earlier chapters is used to develop the sequential circuit techniques used in Chapters 6 to 9. Chapters 5 and 6 deal with the design of synchronous and asynchronous counters using either discrete flip-flops or shift registers. In Chapter 8, a synchronous sequential circuit is analysed and a design procedure is developed which is the reverse process of the analytical technique. Chapter 9 is devoted to the design and implementation of event driven logic circuits using the NAND sequential equation.

Chapter 10 examines the properties of a number of MSI circuits and shows how they can be used in combinational and sequential circuit design, whilst Chapter 11 looks at the production of static and dynamic hazards in combinational circuits and essential hazards in sequential circuits. Finally, Chapter 12 is designed to show that a microprocessor is nothing more than a rather complicated synchronous sequential machine.

The author wishes to thank a number of people for the assistance they have given to the production of this book. First, I wish to thank

Preface

Professor D. Zissos of the University of Calgary who has had a considerable influence on my thinking and first stimulated my interest in the subject. Next, I would like to thank the students who have passed through my hands, for it was they who, through the years, helped me to refine my initial hesitant efforts from which the contents of this volume have been developed. Finally, I would like to thank my wife for typing the manuscript and also the whole of my family, for this book could never have been written without their tolerance.

Contents

	Preface	V
1	Boolean algebra	
1.1	Introduction	1
1.2	The logic of a switch	2
1.3	The AND function	2
1.4	The OR function	4
1.5	The inversion function	5
1.6	Implementation of Boolean equations using switches or electronic gates	6
1.7	The idempotency theorem	7
1.8	The theorems of union and intersection	8
1.9	The redundancy or absorption theorem	9
1.10	The determination of the complementary function	10
1.11	Theorems on commutation, association and distribution	12
1.12	The formation of optional products	13
	Problems	14
2	Karnaugh maps and function simplification	
2.1	Introduction	16
2.2	Product and sum terms	16
2.3	Canonical forms	18
2.4	Boolean functions of two variables	18
2.5	The Karnaugh map	20
2.6	Plotting Boolean functions on a Karnaugh map	22
2.7	Simplification of Boolean functions	25
2.8	The inverse function	27
2.9	'Don't-care' terms	28
2.10	The plotting and simplification of P-of-S expressions	30
	Problems	32

Contents

3 NAND and NOR logic
- 3.1 Introduction — 35
- 3.2 The NAND function — 35
- 3.3 The implementation of AND and OR functions using NAND gates — 37
- 3.4 The implementation of S-of-P expressions using NAND gates — 38
- 3.5 The NOR function — 41
- 3.6 The implementation of OR and AND functions using NOR gates — 42
- 3.7 The implementation of P-of-S expressions using NOR gates — 43
- 3.8 The implementation of S-of-P expressions using NOR gates — 43
- 3.9 Gate expansion — 45
- 3.10 Miscellaneous gates — 46
- 3.11 The tri-state gate — 49
- 3.12 The exclusive-OR gate — 49
- Problems — 54

4 Combinational logic design
- 4.1 Introduction — 56
- 4.2 The half-adder — 57
- 4.3 The full adder — 58
- 4.4 The four-bit parallel adder — 60
- 4.5 The carry look-ahead adder — 60
- 4.6 The full subtractor — 64
- 4.7 The 2's complement — 66
- 4.8 The 1's complement — 67
- 4.9 Binary representations for arithmetic units — 67
- 4.10 Addition and subtraction using 2's complement arithmetic — 68
- 4.11 Binary multiplication — 70
- 4.12 Code conversion — 71
- 4.13 Binary to Gray code converter — 74
- 4.14 Interrupt sorters — 76
- 4.15 Daisy-chaining — 79
- Problems — 80

5 Single-bit memory elements
- 5.1 Introduction — 83
- 5.2 The T flip-flop — 83
- 5.3 The SR flip-flop — 87
- 5.4 The JK flip-flop — 91
- 5.5 The D flip-flop — 95
- 5.6 The latching action of a flip-flop — 96
- Problems — 98

Contents

6	Counters		
	6.1	Introduction	101
	6.2	Scale-of-two up-counter	101
	6.3	Scale-of-four up-counter	103
	6.4	Scale-of-eight up-counter	103
	6.5	Scale-of-2^n up-counter	104
	6.6	Series and parallel connection of counters	105
	6.7	Synchronous down-counters	106
	6.8	Scale-of-five up-counter	106
	6.9	Decade binary up-counter	112
	6.10	Decade binary down-counter	113
	6.11	Decade Gray code 'up' counter	113
	6.12	Scale-of-16 up/down counter	118
	6.13	Asynchronous binary counters	119
	6.14	Scale-of-ten asynchronous up-counter	122
	6.15	Asynchronous resettable counters	123
	6.16	Integrated-circuit counters	124
	6.17	Cascading of IC counter chips	127
	Problems		128
7	Shift register counters and generators		
	7.1	Introduction	131
	7.2	The four-bit shift register with parallel loading	133
	7.3	The four-bit shift-left, shift-right register	134
	7.4	The use of shift registers as counters	134
	7.5	The universal state diagram for shift registers	135
	7.6	The design of a decade counter	137
	7.7	Shift register sequence generators	140
	7.8	The ring counter	143
	7.9	The twisted ring or Johnson counter	147
	7.10	Shift registers with exclusive-OR feedback	150
	Problems		156
8	Clock-driven sequential circuits		
	8.1	Introduction	158
	8.2	Analysis of a clocked sequential circuit	160
	8.3	The design procedure for clocked sequential circuits	163
	8.4	The design of a sequence generator	171
	8.5	Moore and Mealy state machines	174
	8.6	Pulsed synchronous circuits	175
	8.7	State reduction	179
	8.8	State assignment	184
	Problems		189

9 Event-driven circuits
- 9.1 Introduction — 195
- 9.2 The museum problem — 195
- 9.3 Races and cycles — 199
- 9.4 Race-free assignment for a three-state machine — 202
- 9.5 The pump problem — 203
- 9.6 Race-free assignment for a four-state machine — 206
- 9.7 A sequence detector — 209
- Problems — 215

10 Digital design with MSI
- 10.1 Introduction — 220
- 10.2 Data selector or multiplexer — 221
- 10.3 The multiplexer as a logic function generator — 222
- 10.4 Decoders and demultiplexers — 231
- 10.5 Decoder applications — 232
- 10.6 Read-only memories (ROMs) — 237
- 10.7 Addressing techniques for ROMs — 239
- 10.8 Design of sequential circuits using ROMs — 241
- 10.9 Programmable logic arrays (PLAs) — 246
- 10.10 Design of sequential circuits using PLAs — 249
- 10.11 Arithmetic using MSI chips — 252
- 10.12 Decimal addition with MSI adders — 254
- Problems — 257

11 Hazards
- 11.1 Introduction — 261
- 11.2 Gate delays — 261
- 11.3 The generation of spikes — 262
- 11.4 The production of static hazards in combinational networks — 264
- 11.5 The elimination of static hazards — 266
- 11.6 Design of hazard-free combinational hazards — 269
- 11.7 Detection of hazards in an existing network — 272
- 11.8 Dynamic hazards — 274
- 11.9 Essential hazards — 277
- Problems — 278

12 An introduction to microprocessors
- 12.1 Introduction — 280
- 12.2 Binary multiplication — 281
- 12.3 The hardware requirements of a binary multiplier — 282
- 12.4 The binary multiplier — 282

Contents

12.5	Flow charts	285
12.6	The programme-controlled binary multiplier	288
12.7	Word length	289
12.8	The programme counter	290
12.9	Instructions and the instruction register	291
12.10	The hexadecimal number system	293
12.11	Comparison of the simple machine with a practical microprocessor	295
12.12	A general block diagram for a microprocessor system	296
12.13	Programming a microprocessor	299

Answers to problems 301

Bibliography 331

Index 333

1
Boolean algebra

1.1 Introduction

In a digital system, the electrical signals that are used have two voltage levels which may, for example, be 5 volts and 0 volts. The electrical devices used in such systems can generally exist in one of these two possible voltage states indefinitely, providing the power supply is maintained. For example, a bipolar transistor that is non-conducting in a 5 volt system will have approximately 5 volts between collector and emitter. However, when the transistor is turned on and is conducting, it can be arranged, with a suitable choice of load, that the voltage between collector and emitter is approximately zero. Hence, a digital system can be described as a binary system and the two voltage levels used can arbitrarily be assigned the binary values 0 and 1. The two states defined in this way can have a logical significance: they can indicate the presence of a particular condition, or alternatively, its absence.

There is an algebra, developed in the last century, by the Reverend George Boole, an English clergyman, which is well suited for representing the situation described above. This branch of mathematics is called *Boolean algebra*. It is a discrete algebra in which the variables can have one of two values, either 0 or 1 and, as might be imagined, there is a set of rules and a number of theorems associated with the algebra which allow the manipulation of Boolean equations.

Boolean algebra is the foundation of all digital design work, and this chapter will be concerned with the development of the rules and the statement of the theorems associated with the algebra. A clear understanding of the principles of the algebra will lead to a facility for dealing with the algebraic part of design problems, and is therefore extremely useful to anyone wishing to practise digital design.

1.2 The logic of a switch

Consider the switch shown in figure 1.1(*a*) connected between two points P and Q. The status of the switch is expressed in terms of a Boolean variable A which can have two values, 0 and 1. If the switch x is open, $A = 0$, and if the switch is closed, $A = 1$. The status of the

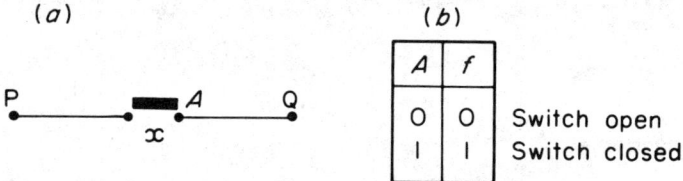

Figure 1.1. (a) Single switch contact circuit. (b) Truth table for a single switch contact

connection PQ can also be expressed in terms of a Boolean variable f which has two possible values, 0 or 1. If there is no connection between P and Q then $f = 0$, and if there is a connection between P and Q, $f = 1$. These results are tabulated in figure 1.1(*b*). This kind of table is called a *truth table* and it is clear from an inspection of the table that

$$f = A$$

The above equation is a *Boolean equation,* and the two variables f and A are called *Boolean variables* or, alternatively, *binary variables,* since they can only have two values, 0 or 1.

1.3 The AND function

Two switches x and y are connected in series between the points P and Q as shown in figure 1.2(*a*). The status of switch x is A and the status of switch y is B. Both A and B are Boolean variables. If switch x is open, $A = 0$ and if it is closed $A = 1$. Similarly, $B = 0$ or 1 depending on whether y is open or closed. As in the case of the single switch contact, the status of the connection PQ is expressed in terms of the Boolean variable f, the value of which depends upon the absence or presence of the connection between the two points P and Q.

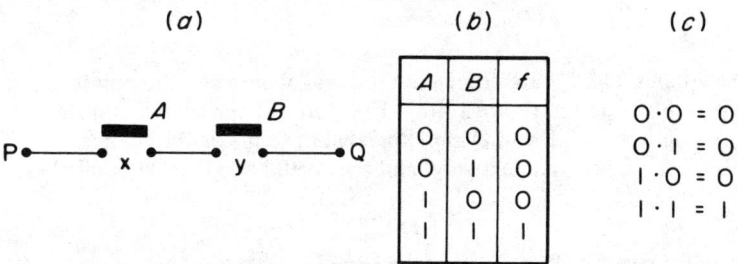

Figure 1.2. (a) Switch contact circuit which implements the AND function. (b) Truth table for the AND function. (c) Rules of binary and Boolean multiplication.

There are four possible combinations of the variables A and B, and these are tabulated in the truth table shown in figure 1.2(b). For example, if x is open and y is open, then $A = 0$ and $B = 0$, and there is no connection between the points P and Q; consequently $f = 0$. If, however, x and y are both closed, then $A = 1$ and $B = 1$, and it follows that $f = 1$.

The truth table shown in figure 1.2(b) is that of the AND function, sometimes referred to as the *Boolean multiplication function,* and it is written in algebraic form as

$$f = A \cdot B$$

where the 'dot' is interpreted as AND. In practice, the dot is usually omitted and the equation is written as

$$f = AB$$

The rules of Boolean multiplication are identical to those of binary multiplication and they are summarised in the table shown in figure 1.2(c).

In the digital world the AND function is implemented electronically by an electronic gate called an AND gate. The generally accepted

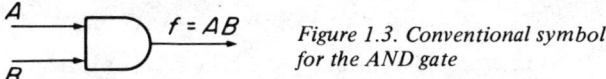

Figure 1.3. Conventional symbol for the AND gate

symbol for the AND gate is shown in figure 1.3. The output of this gate is logical "1" if, and only if, both of the inputs A and B are logical "1". For any other combination of the inputs the output f is logical "0".

Boolean algebra

1.4 The OR function

Two switches x and y are connected in parallel between the points P and Q as illustrated in figure 1.4(a). The status of the switch and the connection between P and Q are expressed in terms of the Boolean variables A, B and f, respectively, and the truth table for the parallel

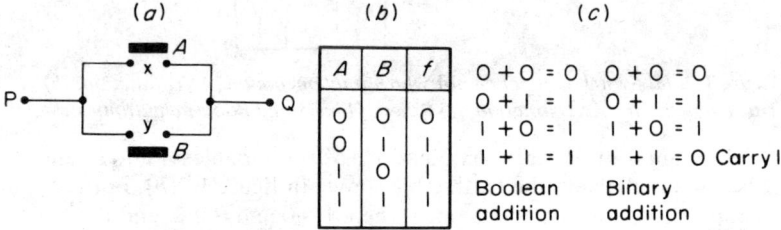

Figure 1.4. (a) Switch contact circuit which implements the OR function. (b) Truth table for the OR function. (c) Rules of Boolean and binary addition

circuit is shown in figure 1.4(b). If both switches x and y are open, $A = 0$ and $B = 0$; it is then clear that there is no connection between P and Q, and consequently $f = 0$. On the other hand, if switch x is closed and switch y is open, $A = 1$ and $B = 0$, and a connection exists between P and Q through the closed contact of x; consequently $f = 1$.

The truth table of figure 1.4(b) is that of the OR function which is sometimes referred to as the *Boolean addition function*. Examination of the truth table shows that a connection exists between points P and Q if x is closed OR y is closed OR if both x and y are closed. Strictly speaking, this function should be referred to as the *inclusive-OR function*. The reason for doing this is simply because $f = 1$ when $A = 1$ and $B = 1$; in other words it includes this condition. When the exclusive-OR function is defined later, it will be seen that it excludes this condition.

The OR function is expressed in algebraic terms by the equation

$$f = A + B$$

where the + should be interpreted as meaning OR. The rules of logical addition are tabulated in figure 1.4(c) alongside the rules of binary addition, and it will be noticed that they differ in one respect. In binary addition $1 + 1 = 10$, where 0 is the sum digit and 1 is the carry digit to be added in at the next most significant stage of the addition, whereas in Boolean addition $1 + 1 = 1$.

In digital systems the OR function is implemented by an electronic gate and the conventional symbol for this gate is shown in figure 1.5. In this circuit the output of the gate will be logical "1" if input A is logical "1" or input B is logical "1" or if both inputs A and B are

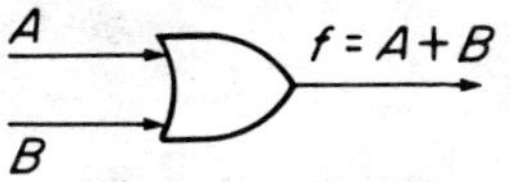

![OR gate symbol with inputs A, B and output f = A+B]

Figure 1.5. Conventional symbol for the OR gate

simultaneously logical "1". For any other condition of the inputs the output is logical "0".

1.5 The inversion function

The operations of addition and multiplication are available in Boolean algebra, but the operations of division and subtraction do not exist. However, there is one other fundamental operation that does exist which is referred to as *inversion* or, alternatively, as *complementation*. Consider the switch shown in figure 1.6, connected between points P

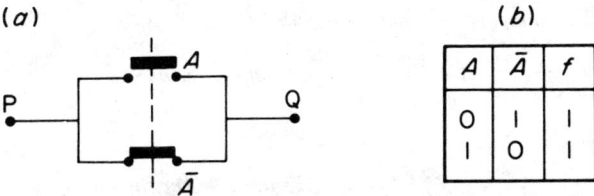

Figure 1.6 (a) Switch circuit, and (b) truth table, for $f = A + \bar{A}$.

and Q, which has a pair of ganged contacts, one of which is normally open and the other normally closed. The normally open contact is represented by the Boolean variable A, whilst the normally closed contact is represented by the Boolean variable $\bar{A}$. For the upper contact $A = 0$ when it is open and clearly for the lower contact, which is closed when the upper contact is open, $\bar{A} = 1$. $\bar{A}$ is said to be the inverse of A, or alternatively it is *not A*. Hence the bar over the Boolean variable A is used to signify the inversion of that variable.

The truth table for the circuit is shown in figure 1.6(b). When $A = 0$, $\bar{A} = 1$ and $f = 1$; similarly, when $A = 1$, $\bar{A} = 0$ and $f = 1$. This means that there is always connection between P and Q no matter which of the two possible positions the switch contacts are in. Hence the equation of this circuit is

$$A + \bar{A} = 1$$

and it is an algebraic statement of the complementation theorem.

6 Boolean algebra

The dual of the above equation, or of any Boolean equation for that matter, is obtained by replacing each + with a · or vice versa, and by changing all the 1's to 0's or vice versa. In the above case the dualisation process leads to the equation

$$A \cdot \bar{A} = 0$$

The switch contact circuit corresponding to this equation is shown in figure 1.7(a) and the truth table for the circuit is shown in figure 1.7(b).

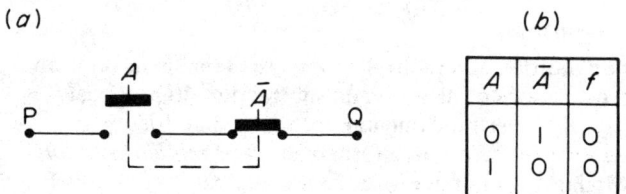

Figure 1.7. (a) Switch contact circuit, and (b) truth table, for $f = A \cdot \bar{A}$

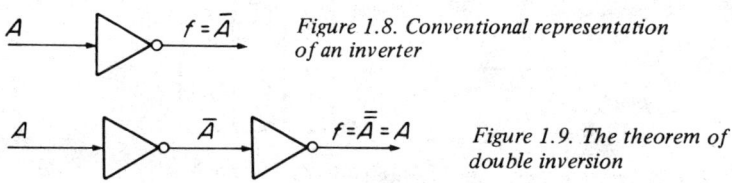

Figure 1.8. Conventional representation of an inverter

Figure 1.9. The theorem of double inversion

An inverter can be implemented electronically and is conventionally represented as shown in figure 1.8. In this circuit, if the input A is logical "0", then $f = \bar{A}$ at the output is logical "1", and vice versa.

If the two inverters are connected in cascade, as shown in figure 1.9, then the input to the second inverter is $\bar{A}$ and its output is the inverse of $\bar{A}$ which can be written as $\bar{\bar{A}}$. Clearly however, a double inversion infers that the output of the second gate is the same as the input to the first gate; hence

$$A = \bar{\bar{A}}$$

1.6 Implementation of Boolean equations using switches or electronic gates

It is now possible to implement a Boolean equation using switches or, alternatively, electronic gates. In the switch contact circuit a + in a

Boolean algebra

Boolean equation is interpreted as a pair of parallel branches, a · is interpreted as a connection of switches in series, and a ¯ is interpreted as a normally closed switch contact. For example, the function

$$f = AB + \bar{B}C$$

would be implemented as shown in figure 1.10(a) or in terms of electronic gates as shown in figure 1.10(b).

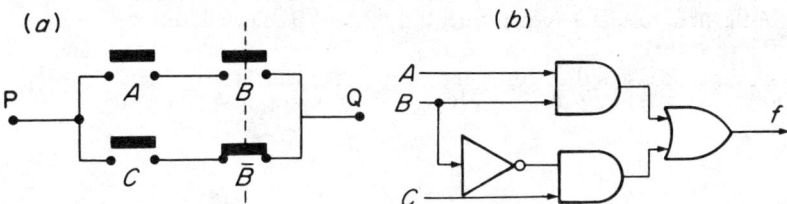

Figure 1.10. $f = AB + \bar{B}C$, implemented (a) with switches, and (b) with gates

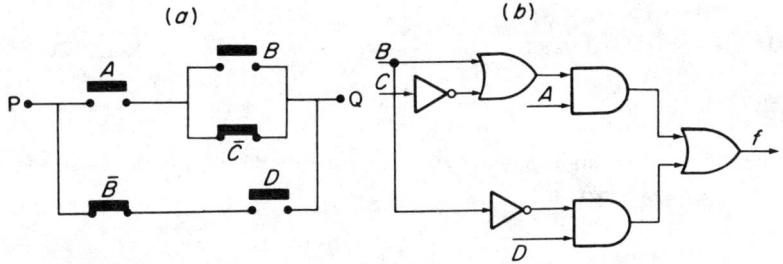

Figure 1.11. $f = A(B + \bar{C}) + BD$ implemented (a) with switches, and (b) with gates

A more complicated Boolean equation such as

$$f = A(B + \bar{C}) + \bar{B}D$$

would be implemented with switches as shown in figure 1.11(a) and with electronic gates as shown in figure 1.11(b).

1.7 The idempotency theorem

This theorem states that

$$A + A = A$$

Clearly, if $A = 1$ the equation reduces to

$$1 + 1 = 1$$

or alternatively if $A = 0$ the equation reduces to

$$0 + 0 = 0$$

both these results agreeing with the rules of Boolean addition given in figure 1.4(c).

There is also a dual form of this theorem which is obtained by using the rules described above. In this case the dual function is

$$A \cdot A = A$$

If $A = 1$ this expression reduces to

$$1 \cdot 1 = 1$$

and if $A = 0$ it reduces to

$$0 \cdot 0 = 0$$

these last two results agreeing with the rules of Boolean multiplication given in figure 1.2(c).

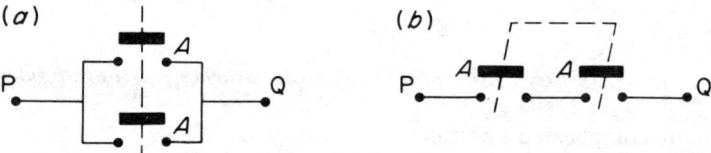

Figure 1.12. (a) Implementation of $f = A + A$. (b) Implementation of $f = A \cdot A$

The theorem is illustrated by the two switch contact circuits shown in figure 1.12.

1.8 The theorems of union and intersection

These theorems state that

$$(a) \left. \begin{array}{l} A + 0 = A \\ A \cdot 1 = A \end{array} \right\} \text{ and } (b) \left. \begin{array}{l} A + 1 = 1 \\ A \cdot 0 = 0 \end{array} \right\}$$

and it is left to the reader to prove the validity of these theorems.

1.9 The redundancy or absorption theorem

The theorem states that

$$A + AB = A$$

and it may be proved as follows:

$$
\begin{aligned}
A + AB &= A \cdot 1 + AB \\
&= A(B + \bar{B}) + AB \quad \text{Complementation theorem} \\
&= AB + A\bar{B} + AB \\
&= AB + A\bar{B} \quad \text{Idempotency theorem} \\
&= A(B + \bar{B}) \\
&= A \quad \text{Complementation theorem}
\end{aligned}
$$

The original equation $f = A \cdot 1 + AB$ is said to be expressed in the sum-of-products form (S-of-P). For example, the term AB is the product of two Boolean variables A and B, and is consequently called a *product term* (P-term).

In other words, the theorem states that in any Boolean equation that is expressed in S-of-P form, a product that contains all the factors of another product is redundant. As a consequence it allows the elimination of redundant products in a sum-of-products expression. For example, in the equation

$$f = AD + A\bar{B}D + ACD$$

the product terms $A\bar{B}D$ and ACD can be eliminated because each contains all the factors present in AD, and hence the expression can be reduced to

$$f = AD$$

The theorem is described in terms of the switch contact circuit in figure 1.13. Clearly, the diagram shows that if switch contact A is made

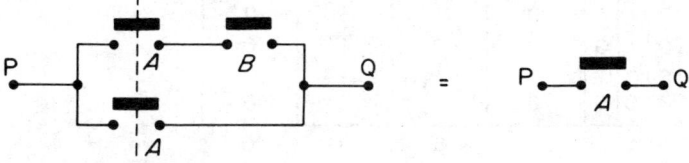

Figure 1.13. Switch contact circuits illustrating the absorption theorem $A = A + AB$

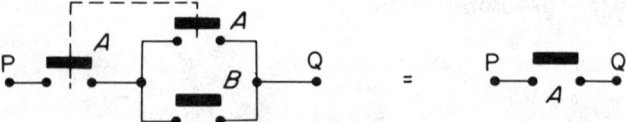

Figure 1.14. *Switch contact circuits illustrating the dual form of the absorption theorem*

then a connection exists between P and Q irrespective of whether switch contact B is closed or open, and hence contact B is redundant.

The dual of the equation $A = A + AB$ is obtained in the usual way and is given by

$$A = A(A + B)$$

and the corresponding switch contact circuit is shown in figure 1.14.

1.10 The determination of the complementary function

The complement of a Boolean equation can be obtained by replacing each variable by its complement in the corresponding dual equation. For example, the dual of $f = A + \bar{B}C$ is

$$f_d = A(\bar{B} + C)$$

and the inverse or complementary function is given by

$$\bar{f} = \bar{A}(B + \bar{C})$$

That this is so can be confirmed with the aid of the truth table shown in figure 1.15. Examination of columns 9 and 10 of this table shows

A	B	C	Ā	B̄	C̄	B̄C	B+C̄	A+B̄C	Ā(B+C̄)
0	0	0	1	1	1	0	1	0	1
0	0	1	1	1	0	1	0	1	0
0	1	0	1	0	1	0	1	0	1
0	1	1	1	0	0	0	1	0	1
1	0	0	0	1	1	0	1	1	0
1	0	1	0	1	0	1	0	1	0
1	1	0	0	0	1	0	1	1	0
1	1	1	0	0	0	0	1	1	0
1	2	3	4	5	6	7	8	9	10

Figure 1.15. *Truth table for confirming that $\bar{f} = \bar{A}(B + \bar{C})$ is the inverse of $f = A + \bar{B}C$*

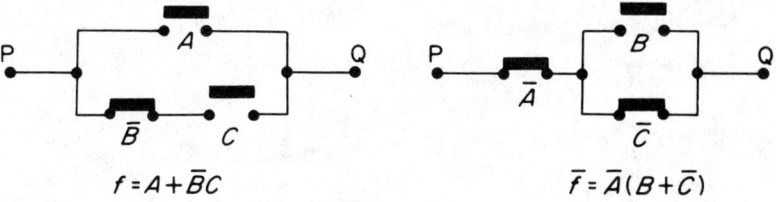

Figure 1.16. Complementary circuits

that $\overline{A}(B+\overline{C})$ is the complement of $A+\overline{B}C$. The two complementary switch contact circuits are shown in figure 1.16.

Looking now at the basic AND and OR functions, the dual of the AND function $f = AB$ is given by

$$f_d = A + B$$

and the inverse function is obtained by inverting the variables in the dual equation so that

$$\overline{f} = \overline{A} + \overline{B}$$

Similarly, the dual of the OR function $f = A + B$ is given by

$$f_d = AB$$

and the inverse function by

$$\overline{f} = \overline{A}\,\overline{B}$$

Thus the inverse of the AND function is given by the OR function of the inverted variables in the AND equation, and the inverse of the OR function is given by the AND function of the inverted variables in the OR equation. The process of complementation applied to the AND and OR functions is illustrated with gate circuits in figure 1.17.

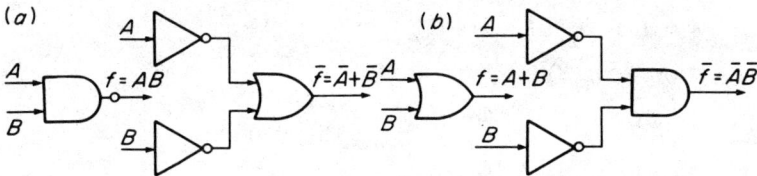

Figure 1.17. (a) Inversion of the AND function. (b) Inversion of the OR function

Boolean algebra

Now the complement of $f = AB$ is given by $\bar{f} = \bar{A} + \bar{B}$, but

$$\bar{f} = \overline{AB}$$

and hence

$$\overline{AB} = \bar{A} + \bar{B}.$$

Similarly, the complement of $f = A + B$ is given by $\bar{f} = \bar{A}\,\bar{B}$, but

$$\bar{f} = \overline{A + B}$$

and hence

$$\overline{A + B} = \bar{A}\,\bar{B}$$

These two equations are the two forms in which de Morgan's theorem is normally expressed. In their most general form they may be written as

$$\overline{AB\ldots N} = \bar{A} + \bar{B} + \ldots \bar{N}$$

and

$$\overline{A + B + \ldots N} = \bar{A}\cdot\bar{B}\ldots\bar{N}.$$

1.11 Theorems on commutation, association and distribution

Besides the theorems discussed previously, there are also theorems on *commutation, association* and *distribution*. The statement of these theorems is given below:

$$\left.\begin{array}{l} A + B = B + A \\ AB = BA \end{array}\right\} \text{Commutation}$$

$$\left.\begin{array}{l} A + (B + C) = (A + B) + C \\ A(BC) = (AB)C \end{array}\right\} \text{Association}$$

$$\left.\begin{array}{l} A + BC = (A + B)(A + C) \\ A(B + C) = AB + AC \end{array}\right\} \text{Distribution}$$

Perhaps the most useful of these theorems is the first form of the

Boolean algebra

distribution theorem, particularly when there is a need to factorise Boolean equations.

1.12 The consensus theorem

If a function such as $f = AC + B\bar{C}$ contains the variable C in one term and its complement $\bar{C}$ in another term, then an optional product or consensus term can be formed by taking the Boolean multiplication of the remaining variables, in this case A and B. Furthermore, the optional product can be added to the original function f to give a new function $f' = AC + B\bar{C} + AB$ whose value is identical to f for all values of the variables A, B and C.

A	B	C	AC	B$\bar{C}$	f	AB	f'
0	0	0	0	0	0	0	0
0	0	1	0	0	0	0	0
0	1	0	0	1	1	0	1
0	1	1	0	0	0	0	0
1	0	0	0	0	0	0	0
1	0	1	1	0	1	0	1
1	1	0	0	1	1	1	1
1	1	1	1	0	1	1	1

Figure 1.18. Proof of the consensus theorem by truth table

All possible combinations of the variables A, B and C are tabulated in the truth table shown in figure 1.18. An examination of the columns headed f and f' show that these two functions are identical for all possible values of A, B and C.

This can also be proved algebraically, as follows:

$$f' = AC + B\bar{C} + AB = AC + B\bar{C} + AB(C + \bar{C})$$
$$= AC + B\bar{C} + ABC + AB\bar{C}$$
$$= AC(1 + B) + B\bar{C}(1 + A)$$

but $1 + A = 1 + B = 1$

$$\therefore f' = AC + B\bar{C} = f$$

The consensus term may therefore be defined as one whose presence in a Boolean equation does not alter its value. Hence, the optional product in the Boolean equation $f' = AC + B\overline{C} + AB$ is redundant. However, consensus terms have their uses, since when they are formed and introduced into a Boolean equation they may eliminate other terms and as a consequence simplify the original equation. For example,

$$f = C + AB\overline{C}$$

The consensus term here is AB. This can be added to the original equation to give

$$f = C + AB\overline{C} + AB$$

and by the use of the redundancy theorem it can be seen that the term $AB\overline{C}$ is now redundant and can be eliminated so that

$$f = C + AB$$

It should be noticed that with the elimination of one of the original terms $AB\overline{C}$, AB is no longer an optional term in the equation, it is now essential.

This technique of forming consensus terms and adding them to an equation without altering its value is very useful to the digital designer. It will be seen in a later chapter of this book that the technique can be used for eliminating static hazards in combinational circuits.

Problems

1.1 Find the complement of each of the following expressions and do not simplify the result:

(a) $f_1 = A + \overline{B}C$
(b) $f_2 = A(B + C) + B\overline{D}(\overline{A} + C)$
(c) $f_3 = [A\overline{B} + C(\overline{A} + DE)][\overline{B} + AC(\overline{E} + \overline{B}\overline{D})]$.

1.2 Simplify each of the following expressions using the method of optional products:

(a) $f_1 = A\overline{C} + B\overline{C}D + A\overline{B}C + ACD$
(b) $f_2 = B + \overline{A}\overline{B} + ACD + A\overline{C}$
(c) $f_3 = B\overline{C}\overline{D} + \overline{A}B\overline{D} + AB\overline{C} + A\overline{B}D + A\overline{C}D$.

1.3 Construct a truth table for the following functions and from the truth table obtain an expression for the inverse function:

(a) $f_1 = A + B\bar{C}$
(b) $f_2 = AC + BC + AB$
(c) $f_3 = (A + \bar{B})(\bar{A} + \bar{B} + C)$
(d) $f_4 = BD + \bar{A}C + \bar{B}\bar{D}$.

1.4 Expand the following expressions using de Morgan's theorem:

(a) $f_1 = \overline{A + B \cdot \overline{ABC} \cdot \overline{AC}}$
(b) $f_2 = \overline{(AB + \bar{B}C) + (\overline{BC} + \bar{A}B)}$
(c) $f_3 = \overline{(AB + \bar{B}C)(AC + \bar{A}\bar{C})}$.

1.5 Draw (i) the switch contact circuits and (ii) the AND/OR implementations for the following Boolean functions:

(a) $f_1 = \bar{A} + B(\bar{C} + \bar{D})$
(b) $f_2 = (\bar{A} + B)(\bar{B} + C) + (AB + C)$
(c) $f_3 = (A + B + C)(\bar{A} + D) + B\bar{C} + A(B + D)(\bar{C} + D)$.

1.6 Prove the following identities:

(a) $\bar{A}B + A\bar{B} = (\bar{A} + \bar{B})(A + B)$
(b) $(AB + C)B = AB\bar{C} + \bar{A}BC + ABC$
(c) $BC + AD = (B + A)(B + D)(A + C)(C + D)$
(d) $AB + ABC + A\bar{B} = A$.

2

Karnaugh maps and function simplification

2.1 Introduction

One of the objectives of the digital designer when using discrete gates is to keep the number of gates to a minimum when implementing a Boolean function. The smaller the number of gates used the lower the cost of the circuit. Simplification can be achieved by a purely algebraic process, but this can be tedious, and the designer is not always sure that the simplest solution has been produced at the end of the process.

A much easier method of simplification is to plot the function on a Karnaugh map and with the aid of a number of simple rules to reduce the Boolean function to its minimal form. This particular method is very straightforward up to and including six variables; above six it is better to go to a tabulation method like Quine–McCluskey, or possibly to use an iterative method of reduction which involves the formation of optional products.

Function simplification is not so important in these days of MSI (medium-scale integration) and LSI (large-scale integration) as it was when discrete gates were used for implementing Boolean functions, although it gives an insight into the methods used for manipulating the algebra, and as such it is worthy of the reader's attention.

2.2 Product and sum terms

A product term of n variables is the logical product of all n variables, where any of the n variables may be represented by the variable itself or

Karnaugh maps and function simplification

its complement. In the case of two variables A and B there are four possible combinations of the variables and these are tabulated in figure 2.1. Corresponding to these four combinations of the variables there are four possible P-terms which can be obtained as follows. In the first row of the table $A = 0$ and $B = 0$, hence $\bar{A}\bar{B} = 1$. The P-term is formed on

A	B	P-terms	S-terms
0	0	$P_0 = \bar{A}\bar{B}$	$S_0 = A+B$
0	1	$P_1 = \bar{A}B$	$S_1 = A+\bar{B}$
1	0	$P_2 = A\bar{B}$	$S_2 = \bar{A}+B$
1	1	$P_3 = AB$	$S_3 = \bar{A}+\bar{B}$

Figure 2.1. The P- and S-terms of two variables

the values of the variables which make the value of the P-term equal to one, hence $P_0 = \bar{A}\bar{B}$. The other three P-terms are obtained in the same way.

A sum term of n variables is the logical sum of all n variables where any one of the variables may be represented by its true or complemented form. The S-terms are formed on the values of the variables which make the value of the S-term equal to zero.
Now $P_0 = \bar{A}\bar{B} = 1$ for $A = 0$ and $B = 0$
and $\bar{P_0} = \overline{\bar{A}\bar{B}} = 0$
giving $\bar{P_0} = S_0 = A + B = 0$ for $A = 0$ and $B = 0$
and the other three S-terms can be obtained by the same method.

For the three variables A, B and C there are eight possible combinations of the variables and consequently there are eight P-terms and eight S-terms. If there are n variables there are 2^n possible combinations of those variables and this leads to 2^n P-terms and 2^n S-terms. It is clear that the number of P-terms and S-terms rises sharply with n.

One important property of P-terms is that the sum of all 2^n P-terms of n variables is equal to one, i.e.

$$\sum_{i=0}^{2^n-1} P_i = 1,$$

where the Boolean sum is being referred to in the above expression. The dual of this equation is

$$\prod_{i=0}^{2^n-1} S_i = 0$$

where the product referred to in the above equation is the Boolean product.

In the case of two variables the logical sum of all the P-terms is given by the expression

$$S = \bar{A}\bar{B} + \bar{A}B + A\bar{B} + AB$$
$$= \bar{A}(\bar{B} + B) + A(\bar{B} + B)$$
$$= \bar{A} + A$$
$$= 1$$

Taking the dual of the expression for the sum gives

$$(\bar{A} + \bar{B})(\bar{A} + B)(A + \bar{B})(A + B) = 0$$

and this represents the product of all the sum terms of two variables.

2.3 Canonical forms

If a Boolean function is written as a sum-of-product terms or as a product-of-sum terms it is said to be in *canonical form*. For example, the following equation is written in canonical S-of-P form

$$f = AB\bar{C} + A\bar{B}C + \bar{A}\bar{B}C$$

whilst the next equation is written in the canonical P-of-S form

$$f = (\bar{A} + \bar{B} + \bar{C})(A + B + \bar{C})(A + \bar{B} + C)$$

2.4 Boolean functions of two variables

There are a limited number of Boolean functions of two variables. Each Boolean function will consist of a certain number of the P-terms; for example,

$$f = \bar{A}B + A\bar{B}$$

Karnaugh maps and function simplification

is a Boolean function of two variables and contains two of the four available P-terms. The total number of Boolean functions of two variables can be obtained in the following way.

Let the presence of a P-term in a two-variable function be indicated by a 1 and its absence be indicated by a 0. For example, if the P-term $\bar{A}\bar{B}$ is present in the expression its presence will be represented by a 1, if absent, its absence will be indicated by a 0. If all four P-terms are absent this will be indicated by a column of four 0's as shown in the table in figure 2.2 and it follows that this Boolean function will be $f_0 = 0$.

P-terms	f_0	f_1	f_2	f_3	f_4	f_5	f_6	f_7	f_8	f_9	f_{10}	f_{11}	f_{12}	f_{13}	f_{14}	f_{15}
$P_0 = \bar{A}\bar{B}$	0	0	0	0	0	0	0	0	1	1	1	1	1	1	1	1
$P_1 = \bar{A}B$	0	0	0	0	1	1	1	1	0	0	0	0	1	1	1	1
$P_2 = A\bar{B}$	0	0	1	1	0	0	1	1	0	0	1	1	0	0	1	1
$P_3 = AB$	0	1	0	1	0	1	0	1	0	1	0	1	0	1	0	1

Figure 2.2. Table for determining all Boolean functions of two variables

There are two ways the entry in the first row can be allocated: it can be either 0 or 1. There are also two ways the entry in the second row can be allocated. When combined with the first row allocation, this leads to four ways that the first two rows can be allocated with 0's and 1's. For four rows it follows that there are $2^4 = 16$ ways in which the 0's and 1's can be allocated. These allocations are shown in figure 2.2 and the 16 Boolean functions of two variables can be written down immediately from this table, as shown in figure 2.3.

$f_0 = 0$	Zero	$f_5 = B$	Identity	$f_{10} = \bar{B}$	NOT
$f_1 = AB$	AND	$f_6 = \bar{A}B + A\bar{B}$	Exclusive OR	$f_{11} = A + \bar{B}$	OR (not B)
$f_2 = A\bar{B}$	AND (not B)	$f_7 = A + B$	OR	$f_{12} = \bar{A}$	NOT
$f_3 = A$	Identity	$f_8 = \bar{A}\bar{B} = \overline{A+B}$	NOR	$f_{13} = \bar{A} + B$	OR (not A)
$f_4 = \bar{A}B$	AND (not A)	$f_9 = \bar{A}\bar{B} + AB$	Coincidence	$f_{14} = \bar{A} + \bar{B} = \overline{AB}$	NAND
				$f_{15} = 1$	One

Figure 2.3. The 16 Boolean functions of two variables

As the number of variables increases the number of Boolean functions that can be formed increases rapidly. For three Boolean variables there are $2^8 = 256$ possible Boolean functions, for four variables there are $2^{16} = 65\,536$ possible Boolean functions and for n variables there are $2^{(2^n)}$ possible Boolean functions.

2.5 The Karnaugh map

For two variables there are four P-terms and these can be conveniently plotted on a map as shown in figure 2.4. The map consists of a square which is divided into four cells, one for each of the P-terms. The variable A is allocated to the rows of the map whilst the variable B

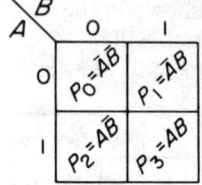

Figure 2.4. The map for two Boolean variables

is allocated to the columns of the map. Hence the top left-hand cell represents the P-term which is the conjunction of $\bar{A}$ and $\bar{B}$, namely $\bar{A}\bar{B}$; and the bottom right-hand cell represents the P-term which is the conjunction of A and B, namely P_3. This kind of map is called a *Karnaugh map*.

Karnaugh maps can be marked in a variety of ways. Each cell can be numbered with the subscript of the P-term which occupies the cell; for example, the top left-hand cell is marked with a 0, the subscript of the P-term that occupies that cell, as shown in figure 2.5(a). Alternatively, the cells can be marked with the binary representation of the subscript, as shown in figure 2.5(b).

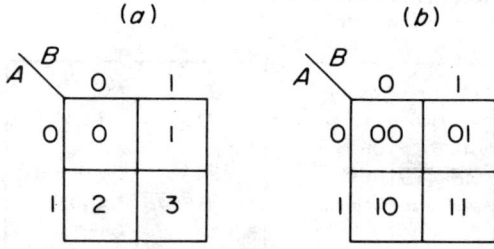

Figure 2.5. (a), (b) Alternative methods for marking a Karnaugh map

For three variables the map contains eight cells, one for each of the P-terms as shown in figure 2.6(a). The variable A is allocated to the two rows of the map, whilst the variables B and C are allocated to the four columns. There are four combinations of these two variables, and each combination is allocated to a column of the map.

Karnaugh maps and function simplification

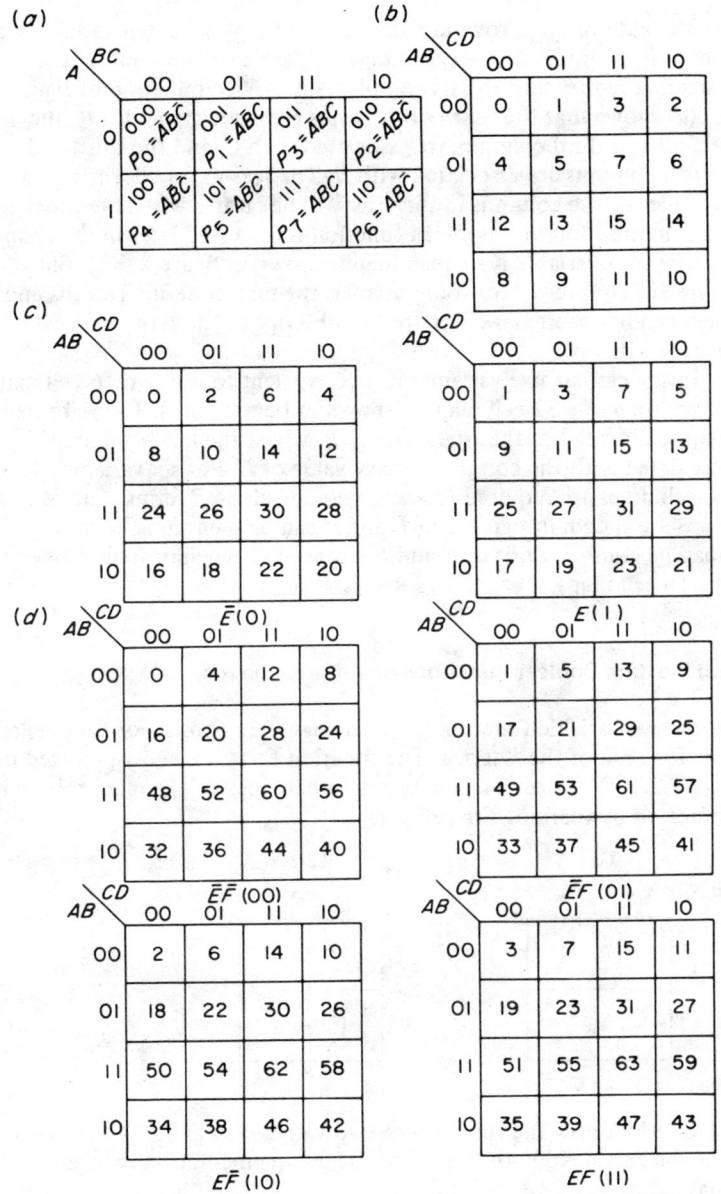

Figure 2.6. Karnaugh maps for (a) three variables; (b) four variables, (c) five variables, and (d) six variables

The columns and rows are allocated in the way shown so that two adjacent columns are always associated with the true value of the variable or, alternatively, its complement. An examination of figure 2.6(a) shows that the first two columns are associated with $\bar{B}$, the second and third columns are associated with C, and the third and fourth columns are associated with B. The reason for allocating the variables to the columns in this way will be clearer when the procedure for minimisation of a Boolean function is discussed later in the chapter.

The four-variable Karnaugh map is shown in figure 2.6(b). Since there are 16 P-terms for four variables the map contains 16 cells and each cell has been marked with the subscript of the P-term which occupies it.

In the case of five variables it is convenient to use two 16-cell maps rather than one 32-cell map, as shown in figure 2.6(c). The right-hand map is allocated to the true value of E, whilst the left-hand map is associated with the complementary value of E. For six variables, four 16-cell maps are required to accommodate all 64 P-terms. The four maps are shown in figure 2.6(d) and it can be seen that the four possible combinations of E and F have each been allocated to one of the 16-cell maps.

2.6 Plotting Boolean functions on a Karnaugh map

For a two-variable function the Karnaugh map consists of four cells, one for each of the P-terms. The function $f = \bar{A}B$ is shown plotted in figure 2.7(a). It occupies the top right-hand cell of the map, this being indicated by marking the cell with a '1'.

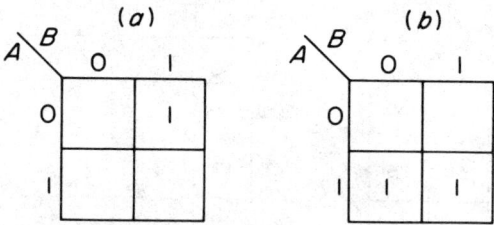

Figure 2.7. (a) $f = \bar{A}B$. (b) $f = AB + A\bar{B} = A$

Consider now the function of two variables A and B, $f = A$. This may be expanded with the aid of the complementation theorem such that

$$f = A(B + \bar{B}) = AB + A\bar{B}$$

This function is plotted in figure 2.7(b) and occupies the two cells on the bottom row of the map. Hence a two-variable function from which one of the variables has been eliminated occupies two adjacent cells when plotted on a two-variable Karnaugh map.

Since there are eight P-terms for three variables, a three-variable function $f(A, B, C)$ requires an eight-cell Karnaugh map as shown in figure 2.8(a). The function $f = ABC + \bar{A}\bar{B}\bar{C}$ is shown plotted on this map. The marked cells in this case are not adjacent and this is an indication that the two terms which make up this function cannot be combined to form a simpler function.

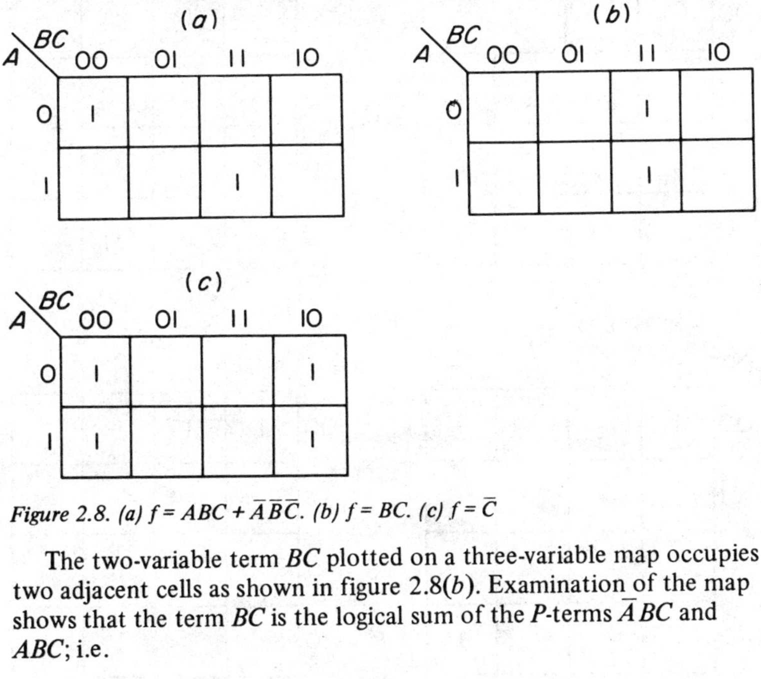

Figure 2.8. (a) $f = ABC + \bar{A}\bar{B}\bar{C}$. (b) $f = BC$. (c) $f = \bar{C}$

The two-variable term BC plotted on a three-variable map occupies two adjacent cells as shown in figure 2.8(b). Examination of the map shows that the term BC is the logical sum of the P-terms $\bar{A}BC$ and ABC; i.e.

$f = \bar{A}BC + ABC$

$= (\bar{A} + A) BC$

$f = BC$

A one-variable term plotted on a three-variable map occupies four adjacent cells. For example, the term $f = \bar{C}$ is shown plotted on figure 2.8(c). An inspection of this map seems to indicate that the four cells are not adjacent, however, if the two ends of the map are folded over to form a vertical cylinder, it can be seen that the first and last columns of the map may be regarded as adjacent.

On a four-variable map a *P*-term occupies one cell as shown in figure 2.9(*a*). Similarly, three-variable, two-variable and one-variable terms when plotted on a four-variable map will occupy two, four and

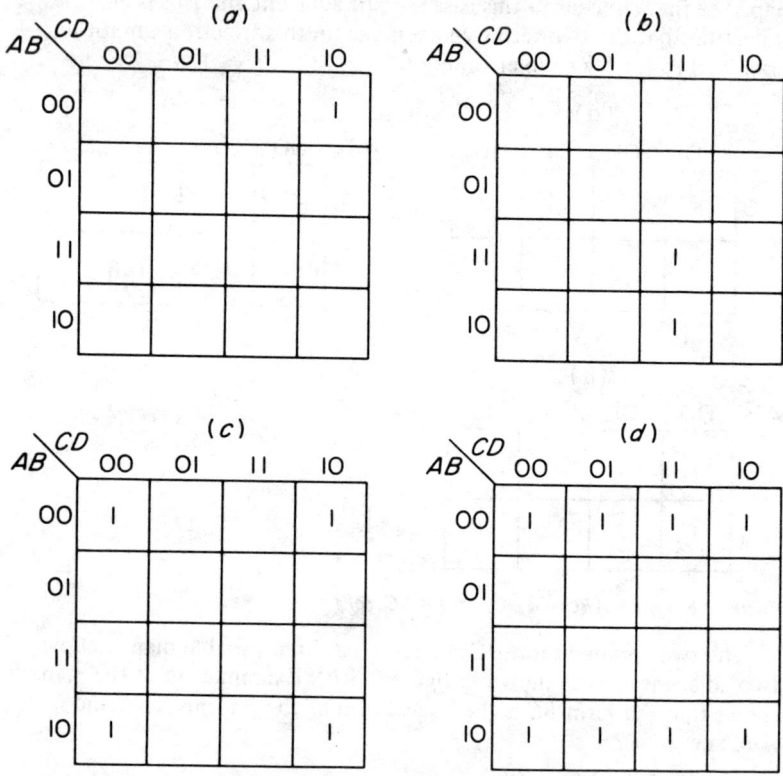

Figure 2.9 (a) $f = \bar{A}\bar{B}C\bar{D}$. (b) $f = ACD$. (c) $f = \bar{B}\bar{D}$. (d) $f = \bar{B}$

eight adjacent cells, respectively, as shown in figures 2.9(*b*, *c*, *d*). Inspection of figure 2.9(*d*) shows that the top and bottom rows of the Karnaugh map may be regarded as adjacent and, as in the case of the three-variable map, the first and last columns of the map may also be regarded as adjacent.

Karnaugh maps and function simplification

2.7 Simplification of Boolean functions

The process of simplifying a Boolean function with the aid of a Karnaugh map is simply a process of finding adjacencies on the function plot. This is best explained with an example. Simplify

$$f = \sum 0, 1, 2, 3, 4, 6, 7, 8, 12, 13.$$

In this example the function has been represented as the logical sum of a number of P-terms, each of which has been represented under the summation sign by its subscript. The plot of the function is shown in

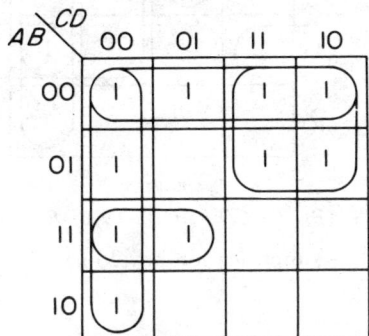

Figure 2.10. $f = \bar{A}\bar{B} + \bar{C}\bar{D} + \bar{A}C + AB\bar{C}$

figure 2.10 on which there are four separate encircled adjacencies of which three are four-cell adjacencies and one is a two-cell adjacency. The four cells on the top row of the map can be represented by the term $\bar{A}\bar{B}$, the four cells in the first column of the map by $\bar{C}\bar{D}$, the top right-hand four cells $\bar{A}C$, and the two-cell adjacency $AB\bar{C}$. Hence the simplified function is given by the equation

$$f = \bar{A}\,\bar{B} + \bar{C}\,\bar{D} + \bar{A}\,C + A\,B\,\bar{C}.$$

The enclosed adjacencies are usually referred to as the *prime implicants* of the simplified function.

It is now clear why the variables C and D and A and B were allocated to the columns and rows in the manner shown. The allocation used ensures that the cells associated with the variables $\bar{C}$, C, $\bar{D}$ and D always lie in two adjacent columns whilst the cells associated with the variables $\bar{A}$, A, $\bar{B}$ and B always lie in two adjacent rows. If the allocation of variables had been made in strict numerical order, i.e. 00, 01, 10, 11, then the cells associated with D, for example, would not have been in adjacent columns and simplification would no longer have been a process of looking for adjacencies.

Simplification of five-variable functions is a little more complicated. As an example, consider the function

$$f = \sum 0, 1, 2, 3, 4, 5, 10, 11, 13, 14, 15, 16, 20, 21, 24, 25, 26, 29, 30, 31$$

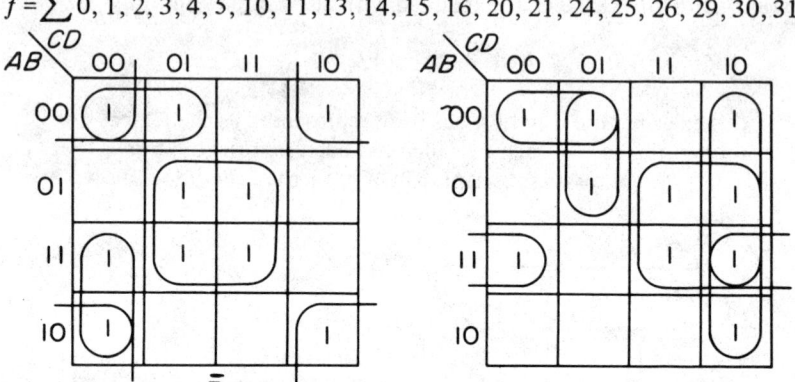

Figure 2.11. $f = \bar{B}\bar{D}\bar{E} + BDE + C\bar{D}E + BCE + \bar{A}\bar{B}\bar{C} + \bar{A}CD + AC\bar{D}E + AB\bar{D}E$

which is shown plotted in figure 2.11; the simplification procedure can be carried out as follows.

(1) First find the simplified functions for each of the two maps $f_{\bar{E}}$ and f_E in the way previously described:

$$f_{\bar{E}} = \bar{B}\bar{D}\bar{E} + BDE + A\bar{C}\bar{D}\bar{E} + \bar{A}\bar{B}\bar{C}\bar{E}$$

and

$$f_E = C\bar{D}E + BCE + AB\bar{D}E + \bar{A}\bar{C}DE + \bar{A}\bar{B}CE$$

(2) The second step is to look for possible combinations between prime implicants on the two maps that will result in an overall simplification of the logical sum of these two functions, $f_{\bar{E}} + f_E$. For example, $\bar{A}\bar{B}\bar{C}\bar{E}$ is a prime implicant of $f_{\bar{E}}$ and $\bar{A}\bar{B}CE$ is a prime implicant of f_E. These two will combine to form one three-variable term $\bar{A}\bar{B}\bar{C}$. It is also possible to add a non-essential prime implicant to the equation for $f_{\bar{E}}$, namely $\bar{A}CD\bar{E}$, which will then combine with the essential prime implicant $\bar{A}CDE$ on the f_E map to form one three-variable term $\bar{A}CD$. Hence the equations for $f_{\bar{E}}$ and f_E may be written as follows:

Karnaugh maps and function simplification

$f_{\bar{E}} = \bar{B}\bar{D}\bar{E} + BDE + A\bar{C}\bar{D}E + \bar{A}\bar{B}\bar{C}\bar{E} + \bar{A}C\bar{D}\bar{E}$ non-essential prime implicant

$f_E = C\bar{D}E + BCE + AB\bar{D}E + \bar{A}\bar{B}\bar{C}E + \bar{A}C\bar{D}E$

                           forms   forms
                           $\bar{A}\bar{B}\bar{C}$ $\bar{A}C\bar{D}$

and the simplified function is

$f = \bar{B}\bar{D}\bar{E} + BDE + C\bar{D}E + BCE + \bar{A}\bar{B}\bar{C} + \bar{A}CD + A\bar{C}\bar{D}\bar{E} + AB\bar{D}E$

2.8 The inverse function

In some cases it is more economical to implement the inverse of the function rather than implement the given function. For example, suppose the given function is

$f(A, B, C, D) = \sum 2, 6, 7, 8, 12, 13$

This is plotted in figure 2.12(a) and the simplified function obtained from this map is

$f = AB\bar{C} + \bar{A}BC + \bar{A}C\bar{D} + A\bar{C}\bar{D}$

To implement this function, four three-input AND gates and one four-input OR gate are required as shown in figure 2.12(b). Besides the gates the circuit requires 17 interconnections.

The inverse function can be represented by 0's in the unmarked cells of figure 2.12(a) but, for clarity, a separate map in figure 2.12(c) shows the plot of the inverse function. From this map the simplified form of the inverse function is obtained and is given by the equation

$\bar{f} = \bar{A}\bar{C} + AC + \bar{B}D$

The implementation of $\bar{f}$ is shown in figure 2.12(d). In order to generate the original function, the inverse function is inverted with the aid of an inverter. The hardware required in this case is three two-input AND gates, one three-input OR gate and an inverter, whilst the number of interconnections required is eleven.

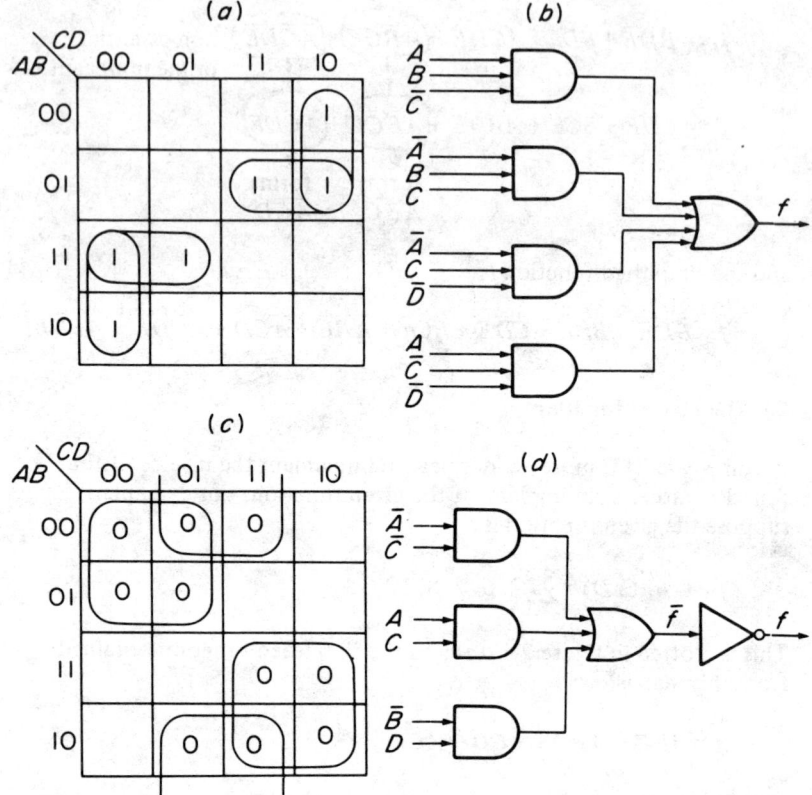

Figure 2.12. (a) $f = AB\bar{C} + \bar{A}BC + \bar{A}C\bar{D} + A\bar{C}\bar{D}$. (b) Implementation of $f = AB\bar{C} + \bar{A}BC + \bar{A}C\bar{D} + A\bar{C}\bar{D}$. (c) $f = \bar{A}\bar{C} + AC + \bar{B}D$. Plot of the inverse function. (d) Implementation of f from the inverted function

2.9 'Don't care' terms

In some logic problems certain combinations of the variables may never occur. For example, the NBCD code shown in figure 2.13(a) is frequently used to represent the decimal digits. This is a four-bit code with 16 possible combinations, only ten of which are used. The remaining six combinations, namely 1010, 1011, 1100, 1101, 1110 and 1111, can't happen in practice and as a consequence can be used for simplification purposes. Such terms are usually referred to as 'don't care' terms or 'can't happen' terms.

Karnaugh maps and function simplification

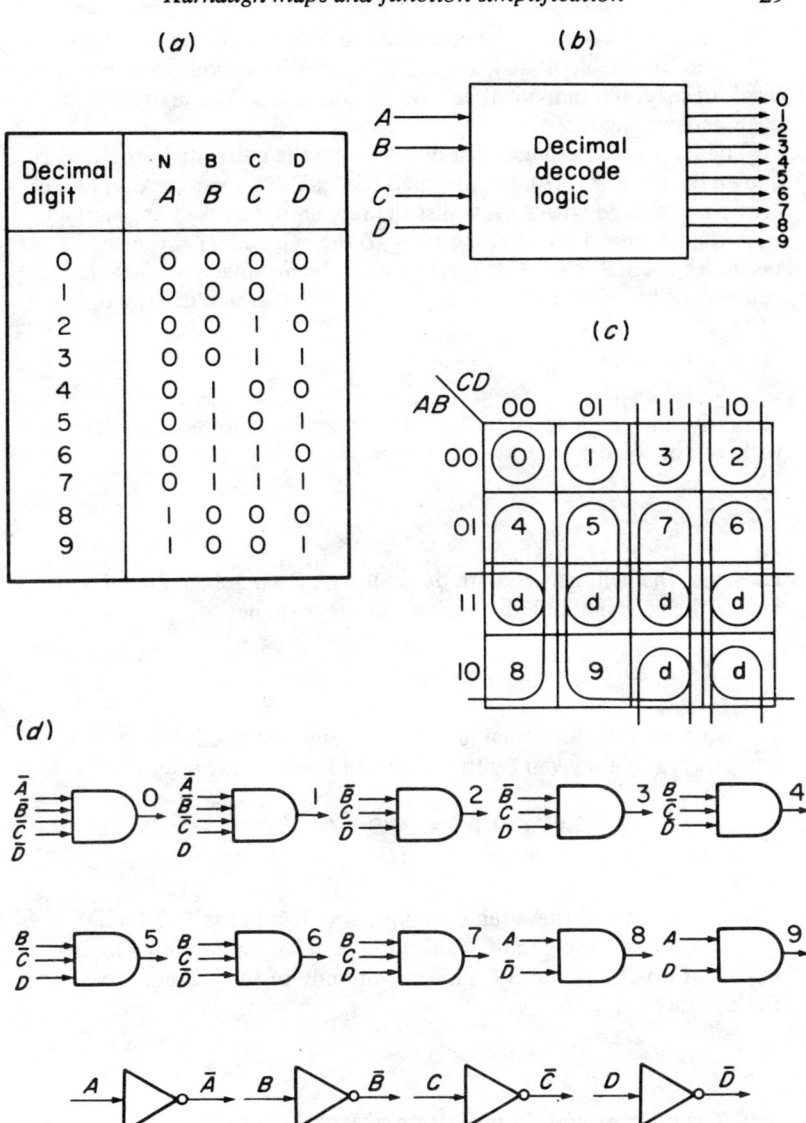

Figure 2.13. (a) The NBCD code. (b) Block schematic for the NBCD to decimal converter. (c) Karnaugh map for determining the decimal decode logic. (d) Implementation of the decimal decode logic circuit

If the NBCD code is to be converted to give a decimal output, as indicated by the block diagram in figure 2.13(b), decode logic has to be used. Clearly, ten individual decode circuits will be required, one for each decimal digit.

The Karnaugh map used for determining the individual functions is shown in figure 2.13(c). On this map the 'can't happen' or 'don't care' terms are marked with a 'd' whilst the remaining ten cells are marked with the decimal digit corresponding to the cell code, i.e. the cell defined by $ABCD = 0000$ is marked with the decimal digit 0. This cell cannot be combined with any of the cells marked 'd', hence

$$0 = \bar{A}\bar{B}\bar{C}\bar{D}$$

Similarly, the cell marked with decimal digit 1 cannot be combined with any of the cells marked 'd', hence

$$1 = \bar{A}\bar{B}\bar{C}D$$

However, the cell marked with decimal digit 2 can be combined with one of the cells marked 'd' as shown in the map, hence

$$2 = \bar{B}C\bar{D}$$

The equations for the remaining decimal digits can be found in the same way, and are given by the following Boolean expressions.

$$3 = \bar{B}CD \quad 5 = B\bar{C}D \quad 7 = BCD \quad 9 = AD$$
$$4 = B\bar{C}\bar{D} \quad 6 = BC\bar{D} \quad 8 = A\bar{D}$$

Implementation of these ten functions is shown in figure 2.13(d).

This decode logic circuit is an example of a combinational logic circuit in which the output is a function only of the present inputs to the circuit.

2.10 The plotting and simplification of P-of-S expressions

It is not possible to plot P-of-S expressions directly onto a Karnaugh map, although if the inverse function is found this will be expressed in S-of-P form and consequently can be plotted directly. The unmarked

Karnaugh maps and function simplification

cells on this map then represent the inverse of the inverse function, i.e. the original function. For example,

$$f = \bar{A} + \bar{B}$$

the dual of this function is

$$f_d = \bar{A} \cdot \bar{B}$$

and the inverse is

$$\bar{f} = AB$$

This can be plotted on a K-map in the normal way except in this case the cell representing AB is marked with a '0' since it is the inverse function. The inverse function is shown plotted in figure 2.14(a), and the unmarked cells on this map represent the complement of the inverse function which is, of course, $f = \bar{A} + \bar{B}$. This is plotted in

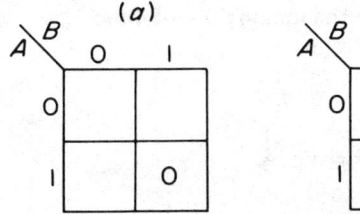

Figure 2.14. (a) $\bar{f} = AB$. (b) $f = \bar{A} + \bar{B}$

figure 2.14(b), each of the cells being marked with a '1', in this case because it is the original function that has been plotted.

For a three-variable P-of-S expression such as

$$f = (A + B + C)(B + \bar{C})$$

the dual function is

$$f_d = ABC + B\bar{C}$$

and the inverse function is

$$\bar{f} = \bar{A}\bar{B}\bar{C} + \bar{B}C$$

32   Karnaugh maps and function simplification

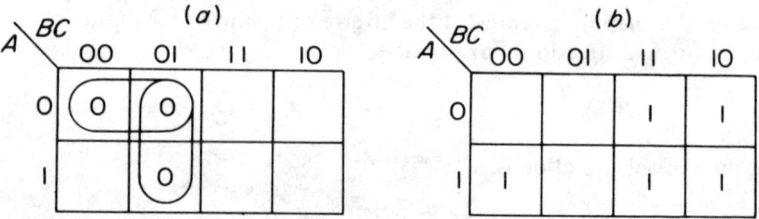

Figure 2.15. (a) $\bar{f} = \bar{A}\bar{B}\bar{C} + \bar{B}C$. (b) $f = (A + B + C)(B + \bar{C})$

This function is plotted in figure 2.15(a) and its inverse (the original function) is plotted in figure 2.15(b).

The inverse function can be simplified using the methods described above. There are two prime implicants on the map of $\bar{f}$ and the minimal form of the inverse function is given by

$$\bar{f} = \bar{A}\bar{B} + \bar{B}C.$$

Reinverting this expression will convert it back into the P-of-S form and the expression thus obtained will be the minimal P-of-S function. For this example the dual of the inverse is given by

$$\bar{f}_d = (\bar{A} + \bar{B})(\bar{B} + C)$$

and the inverse of the inverse function is given by

$$f = (A + B)(B + \bar{C})$$

and this is the minimal P-of-S form of the original function $f = (A + B + C)(B + \bar{C})$.

Problems

2.1 Expand the following Boolean functions into their canonical form:
(a) $f_1 = \bar{A}B + C$
(b) $f_2 = AB + \bar{A}C + A\bar{B}C$
(c) $f_3 = B + CD + AB\bar{D} + \bar{A}\bar{B}C\bar{D}$.

2.2 Simplify the following Boolean functions algebraically:
(a) $f_1 = P_1 + P_2 + P_5 + P_6$

Karnaugh maps and function simplification 33

    (b) $f_2 = P_0 + P_1 + P_2 + P_3 + P_7$
    (c) $f_3 = P_3 + P_5 + P_6 + P_7$

2.3 (a) Express the three-variable function $f = P_0 + P_1$ as a product of S-terms.
    (b) Express the three-variable function $f = S_0 S_1 S_2 S_5 S_6 S_7$ as a sum of P-terms.
    (c) Determine the inverse function of $f = P_3 + P_5 + P_6 + P_7$ and express it as a product of S-terms.

2.4 Find the minimum S-of-P expression for
    (a) $f_1 (A, B, C) = \Sigma\ 0, 1, 3, 4, 6, 7$
    (b) $f_2 (A, B, C, D) = \Sigma\ 0, 1, 2, 3, 7, 8, 9, 11, 12, 15$
    (c) $f_3 (A, B, C, D) = \Pi\ 0, 4, 5, 6, 7, 8, 9, 10$
    (d) $f_4 (A, B, C, D, E) = \Sigma\ 0, 1, 3, 5, 6, 7, 8, 9, 10, 15, 16, 20,$
                                         $21, 22, 23, 24, 25, 28, 29, 30, 31.$

2.5 Minimise the following functions using the 'don't care' terms for simplification wherever possible.
    (a) $f(A, B, C) = \Sigma\ 3, 5 +$ dc terms $(0, 7)$
    (b) $f(A, B, C, D) = \Sigma\ 1, 2, 3, 5, 6, 7, 10, 11 +$ dc terms
        $(9, 12, 15).$
    (c) $f(A, B, C, D) = \Pi\ 0, 4, 7, 11, 14 +$ dc terms $(6, 8, 9, 13)$
    (d) $f(A, B, C, D, E) = \Sigma\ 4, 5, 6, 7, 12, 14, 16, 20, 21, 24, 26,$
                                   $27, 31 +$ dc terms $(0, 11, 19, 22, 30)$

2.6 Find the minimum P-of-S form of the following functions
    (a) $f(A, B, C) = \Sigma\ 0, 1, 2, 5, 7$
    (b) $f(A, B, C, D) = \Sigma\ 0, 1, 9, 10, 11$
    (c) $f(A, B, C, D, E) = \Sigma\ 1, 2, 5, 6, 10, 11, 14, 15, 16, 17, 20, 21$
    (d) $f(A, B, C, D) = \Sigma\ 5, 7, 9, 10, 11 +$ dc terms $(2, 13, 15)$

2.7 Find the minimum P-of-S form of the logical product $F = F_1 F_2$ of the following pairs of functions
    (a) $F_1 (A, B, C, D) = \Sigma\ 1, 3, 5, 7, F_2 (A, B, C, D) = \Sigma\ 2, 3, 6, 7$
    (b) $F_1 (A, B, C, D) = \Sigma\ 1, 3, 5, 6, 8, 10, 11, 12, 13$
       $F_2 (A, B, C, D) = \Sigma\ 0, 3, 5, 8, 9, 11, 13, 15$
    (c) $F_1 (A, B, C) = \Pi\ 0, 3, 6, 7, F_2 (A, B, C) = \Pi\ 1, 3, 7$

2.8 The XS3 code shown in the table below is used to represent the ten decimal digits. Develop the decode logic for converting from XS3 to decimal.

d	A	B	C	D
0	0	0	1	1
1	0	1	0	0
2	0	1	0	1
3	0	1	1	0
4	0	1	1	1
5	1	0	0	0
6	1	0	0	1
7	1	0	1	0
8	1	0	1	1
9	1	1	0	0

3

NAND and NOR logic

3.1 Introduction

The gates dealt with in the two previous chapters have been the AND, OR and inverter gates. In practice, many logic circuits are implemented using *NAND* and *NOR gates* simply because the basic gates in some logic families such as TTL and CMOSL are NAND and NOR gates. It is true that the AND and OR gates exist in these families but there is a very much smaller selection of them and they are usually more expensive. Furthermore, the AND and OR gates in these families usually have longer delay times because they have more stages than the NAND and NOR gates and they may also have a higher power dissipation. For the reasons given above, this chapter will concentrate on the use of NAND and NOR gates for the implementation of Boolean functions since, generally speaking, the logic designer will use these types of gate in his circuit design unless there is a particular reason for not doing so.

3.2 The NAND function

The NAND function is defined by the Boolean equation

$$f = \overline{A}\,\overline{B}$$

and the truth table for the function is shown in figure 3.1(*a*) whilst the conventional diagram for a two-input NAND gate is shown in figure 3.1(*b*). An examination of the truth table shows that the output of a

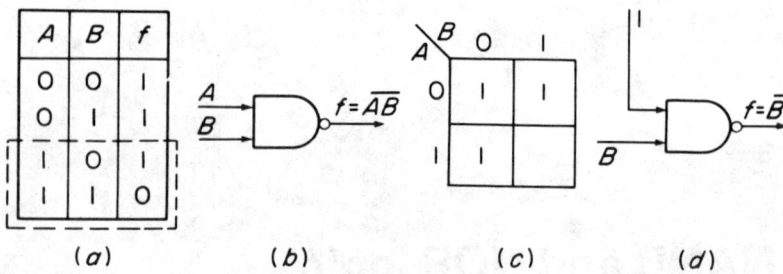

Figure 3.1. (a) Truth table for the NAND function. (b) Conventional diagram for a NAND gate. (c) K-map for the NAND function. (d) The NAND gate used as an inverter

two-input NAND gate is logical '1' if either or both inputs are held at logical '0' and it is a general rule for NAND gates having any number of inputs that if any one of these inputs is held at logical '0' the output of the gate is logical '1'.

A plot of the two-variable NAND function on a two-variable Karnaugh map is shown in figure 3.1(c). Each cell, except the one for the term AB, is marked with a '1' on this map. From the discussion in Chapter 2 it is clear that these three cells represent the complement of the term AB, which is also obviously $\overline{AB}$, the equation of the NAND function.

Inspection of the truth table reveals that the NAND function may also be represented by the equation

$$f = \overline{A}\overline{B} + \overline{A}B + A\overline{B}$$

$$= \overline{A} + A\overline{B}$$

$$= \overline{A} + \overline{B}$$

Hence

$$\overline{AB} = \overline{A} + \overline{B}$$

which is merely a restatement of de Morgan's theorem.

If the A input of the two-input NAND gate is permanently connected to logical '1' then the only valid rows in the truth table for the NAND function are those enclosed by dotted lines in figure 3.1(a). An examination of these rows shows that if $B = 0$ then $f = 1$, and that if $B = 1$ then $f = 0$. In other words the output is the inverse of the input.

NAND and NOR logic

Hence a NAND gate can be used as an inverter if all of the inputs except one are connected to logical '1', as illustrated in figure 3.1(d).

3.3 The implementation of AND and OR functions using NAND gates

Implementation of the AND function using NAND gates is simply performed by connecting two NAND gates in the cascade, the first one performing the NAND operation while the second one is used as an inverter, as shown in figure 3.2.

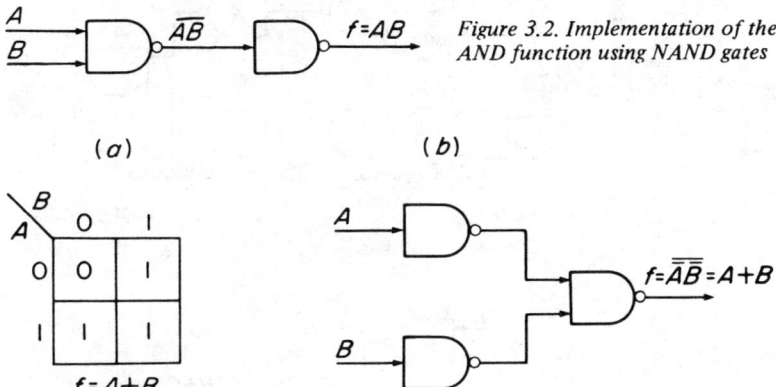

Figure 3.2. Implementation of the AND function using NAND gates

Figure 3.3. (a) K-map plot of the OR function. (b) Implementation of the OR function using NAND gates

The OR function $f = A + B$ is shown plotted on the K-map in figure 3.3(a). Inspection of the map shows that the inverse of the OR function is represented by the cell marked with a '0'. Hence

$$\bar{f} = \bar{A}\,\bar{B}$$

from which it follows that

$$f = \overline{\bar{A}\,\bar{B}}$$

This equation shows that the OR function can be implemented by performing the NAND operation on the inverted variables. The implementation of the OR function using NAND gates is illustrated in figure 3.3(b).

3.4 The implementation of S-of-P expressions using NAND gates

The function $f = AB + CD$ is referred to as a two-level sum-of-products expression. The implementation of this function using AND/OR logic is shown in figure 3.4(a) and the diagram shows that there are two levels of logic in this circuit, the first level consisting of the OR gate, and the second of the two AND gates.

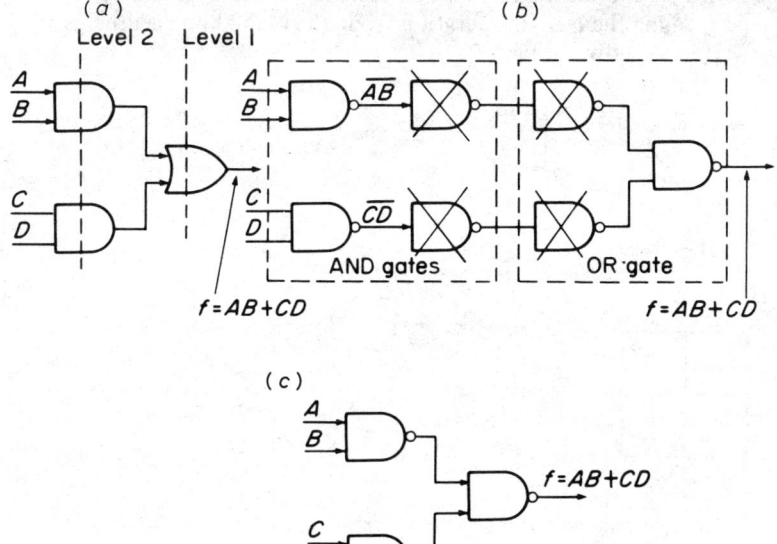

Figure 3.4. (a) $f = AB + CD$ implemented with AND/OR logic. (b) Transformation of the AND/OR circuit to a NAND circuit. (c) The NAND implementation of $f = AB + CD$

This circuit can be translated into a NAND circuit by using the transformations developed in 3.3. The translation is shown in figure 3.4(b), where the first block enclosed by dotted lines represents the two AND gates, whilst the second block constitutes the OR gate. It can be seen that in both branches of the circuit there are two single input NAND gates in cascade and these will simply produce a double inversion of the signals $\overline{AB}$ and $\overline{CD}$. As a consequence the four gates shown crossed through are redundant and the circuit reduces to that shown in figure 3.4(c). This diagram shows that there is a one-to-one translation from the AND/OR configuration to the corresponding NAND configuration.

NAND and NOR logic

Even a complicated-looking function such as

$$f = (A + \bar{B}D)C + (C + \bar{D})(A + C)B$$

can be regarded as a two-level sum-of-products since it can be expressed in the following form:

$$f = PQ + SRT$$

where

$$P = A + \bar{B}D \quad S = C + \bar{D}$$
$$Q = C \quad\quad\quad\ R = A + C$$
$$T = B$$

Hence the implementation must be of the form shown in figure 3.5(a).

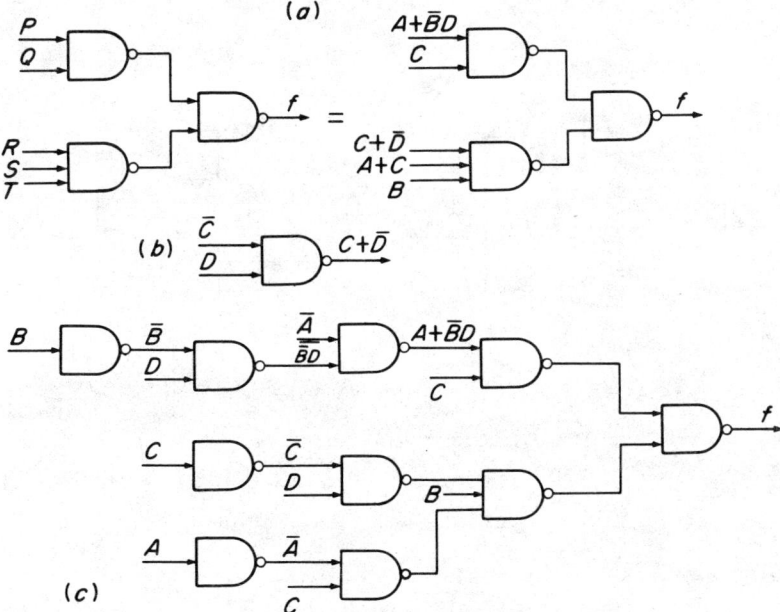

Figure 3.5. (a) Basic circuit for the implementation of $f = (A + \bar{B}D)C + (C + \bar{D})(A + C)B$. (b) The NAND implementation of $\bar{C} + D$. (c) Implementation of $f = (A + \bar{B}D)C + (C + \bar{D})(A + C)B$

In order to generate such terms as $C+\bar{D}$ using NAND logic, a NAND gate is required whose inputs are the inverse of the variables appearing at the outputs as shown in figure 3.5(b). The complete circuit for the implementation of the given function is shown in figure 3.5(c).

The technique described above for the implementation of a Boolean function using NAND gates does not necessarily lead to the minimal NAND implementation. However, sometimes by using a factorisation process it is a simple matter to produce a NAND implementation which leads to a circuit that requires a smaller number of gates. For example, consider the function

$$f = A\bar{C} + A\bar{B} + \bar{C}D$$

Direct implementation of this function as a two-level S-of-P leads to the circuit shown in figure 3.6(a), which requires one three-input NAND gate, three two-input NAND gates, and two single-input NAND gates. (In practice, of course, multi-input NAND gates with their unused inputs connected to logical '1' will be used as inverters).

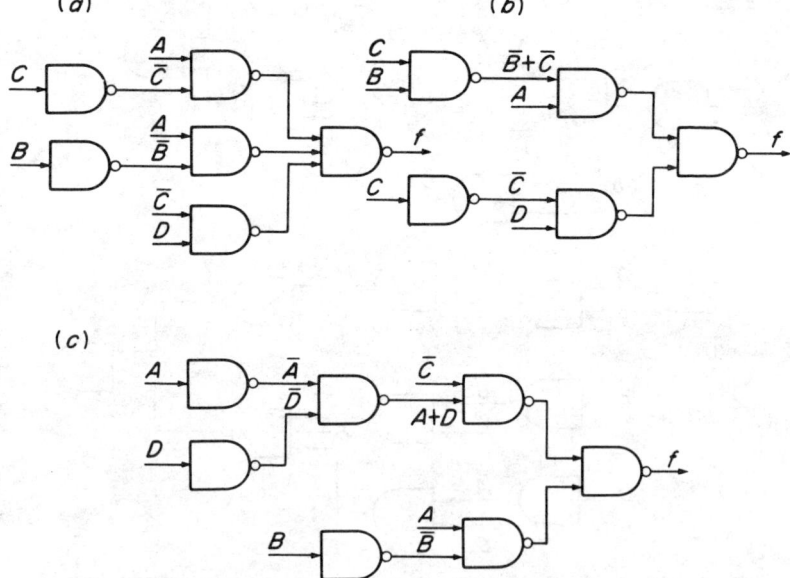

Figure 3.6. (a, b, c) Three ways of implementing $f = A\bar{C} + A\bar{B} + \bar{C}D$

NAND and NOR logic

However, the expression can be written in the form

$$f = A(\bar{B} + \bar{C}) + \bar{C}D$$

and this function can be implemented using five NAND gates as shown in figure 3.6(b). Alternatively, the expression may be written in the form

$$f = \bar{C}(A + D) + A\bar{B}$$

and this can be implemented in the form shown in figure 3.6(c) which requires seven NAND gates.

It is clear in this case that the implementation shown in figure 3.6(b) uses the smallest number of NAND gates. A formal method for determining the minimum NAND gate implementation, given particular gate input (fan-in) restrictions, has been described in Zissos. The method depends upon the merging process in which two brackets such as $(A + \bar{B})(A + \bar{C})$ can be merged into a single term $A + \bar{B}\bar{C}$. In some cases, merging reduces the number of gates required in an implementation, while in other cases it leads to an increase. The Zissos method is a systematic way of determining the effect of merging on the number of gates required for the NAND implementation of a given function.

3.5 The NOR function

The NOR function is defined by the Boolean equation

$$f = \overline{A + B}.$$

The truth table is shown in figure 3.7(a) and the conventional diagram

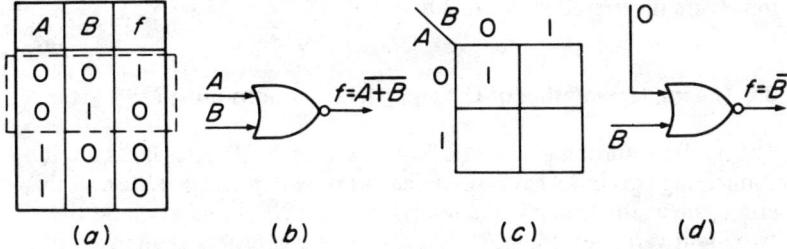

Figure 3.7. (a) Truth table for the NOR function. (b) Conventional diagram for a NOR gate. (c) K-map plot for the NOR function. (d) The NOR gate used as an inverter

for a two-input NOR gate is shown in figure 3.7(b). Examination of the truth table shows that if any one or both of the inputs is logical '1' then the output of the gate is logical '0', and hence there is a general rule for NOR gates having any number of inputs, which states that if any of the inputs are held at logical '1' the output of the gate will be logical '0'.

The NOR function for two variables is shown plotted on the two-variable K-map in figure 3.7(c), and the plot shows that there is an alternative way of expressing the NOR function, namely

$$f = \overline{A}\,\overline{B} = \overline{A+B}$$

If the A input of the two input NOR gate is permanently connected to logical '0' then only the first two rows of the truth table shown in figure 3.7(a) are valid. Inspection of these two rows shows that if $B = 0$ then $f = 1$, and if $B = 1$, then $f = 0$. Hence the NOR gate with a single active input acts as an inverter as shown in figure 3.7(d).

It is good practice never to leave the unused inputs to any gate floating. There are two methods of dealing with unused inputs. Firstly, in the case of NAND gates they can be connected to logical '1' whilst for NOR gates they can be connected to logical '0'. Alternatively,

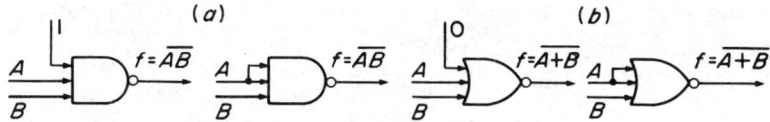

Figure 3.8. *Methods for connecting unused inputs on (a) NAND gates, and (b) NOR gates*

unused inputs in both NAND and NOR gates can be connected to one of the used inputs. The alternative methods of dealing with unused inputs are illustrated in figure 3.8.

3.6 The implementation of OR and AND functions using NOR gates

The implementation of the OR function using NOR gates is achieved by connecting two NOR gates in cascade as shown in figure 3.9(a). In this arrangement the first NOR gate performs the NOR operation on the two input variables A and B while the second gate acts as an inverter.

The circuit for generating the AND function can be developed as follows. Since NOR gates are being used for the implementation of the function the output gate will be a NOR gate whose output is $f = AB$,

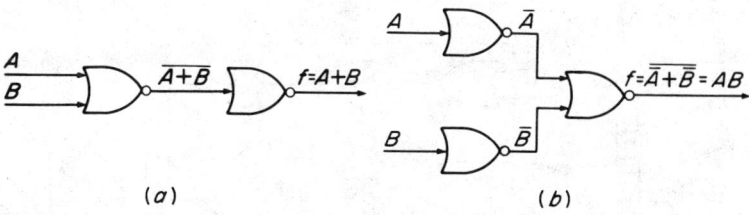

Figure 3.9. Implementation of (a) the OR function, and (b) the AND function, using NOR gates

as shown in figure 3.9(b). In order to obtain this output, the inputs to the gate should be $\bar{A}$ and $\bar{B}$ since $\overline{\bar{A} + \bar{B}} = AB$. To complete the circuit the output NOR gate is preceded by two further NOR gates, both used as inverters, one for each variable.

It is perhaps worth noting at this point that if the output of the NOR gate is the product of n Boolean variables, there will be n input lines to the gate, one for each variable appearing in the output product. The signal attached to each input line will be the complement of the variable appearing in the output product; i.e. if the variable appears as $\bar{C}$ in the output it will appear as C on an input line.

3.7 The implementation of P-of-S expressions using NOR gates

A function such as $f = (A + B)(C + D)$ is called a two-level product-of-sums expression. The implementation of this function using OR/AND logic is shown in figure 3.10(a). This circuit can be converted to a NOR circuit using the transformations developed in the last section as shown in figure 3.10(b). An examination of this circuit shows that four of the NOR gates in this implementation are redundant since they are merely producing a pair of double inversions, hence they have been crossed through in figure 3.10(b).

The simplest form of the circuit using NOR gates is shown in figure 3.10(c) and it can be seen that there is a one-to-one transformation from the OR/AND circuit to the corresponding NOR circuit.

3.8 The implementation of S-of-P expressions using NOR gates

Generally speaking, a Boolean function is expressed in the sum-of-products form and if the function is to be implemented using NOR

NAND and NOR logic

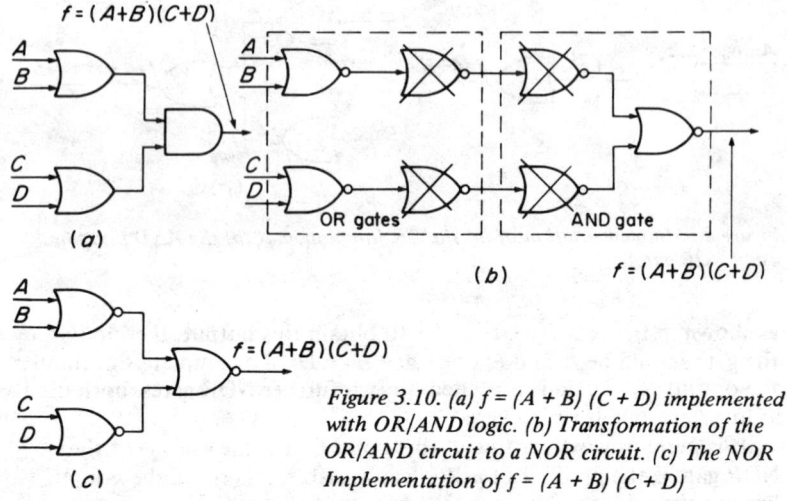

Figure 3.10. (a) $f = (A + B)(C + D)$ implemented with OR/AND logic. (b) Transformation of the OR/AND circuit to a NOR circuit. (c) The NOR Implementation of $f = (A + B)(C + D)$

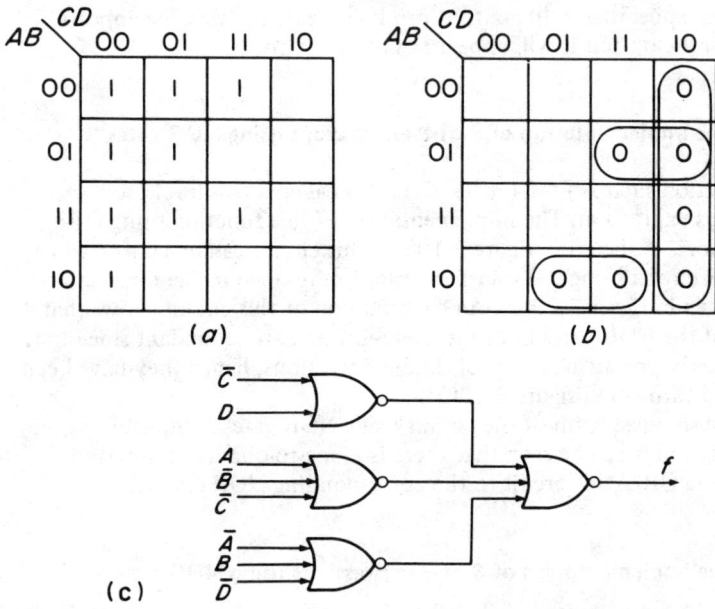

Figure 3.11. (a) Plot of $f = \Sigma\ 0, 1, 3, 4, 5, 8, 12, 13, 15$. (b) Simplication of $\bar{f} = C\bar{D} + \bar{A}BC + A\bar{B}C$. (c) Implementation of $f = (\bar{C} + D)(A + \bar{B} + \bar{C})(\bar{A} + B + \bar{D})$

NAND and NOR logic

gates it has to be converted to the product-of-sums form. For example, if it is required that the function

$$f = \sum 0, 1, 3, 4, 5, 8, 12, 13, 15$$

is to be implemented using NOR gates, the first step is to plot the given function on a K-map as shown in figure 3.11(a), each of the cells designated by the above equation being marked with 1's. The complementary function is then obtained by plotting 0's in the vacant cells of the map shown in figure 3.11(a). For simplicity this function is shown plotted on a separate map in figure 3.11(b).

From the map the minimum form of the complementary function is given by

$$\bar{f} = C\bar{D} + \bar{A}BC + A\bar{B}D.$$

Hence

$$\bar{f}_d = (C + \bar{D})(\bar{A} + B + C)(A + \bar{B} + D)$$

and

$$f = (\bar{C} + D)(A + \bar{B} + \bar{C})(\bar{A} + B + \bar{D})$$

This is the minimal P-of-S form of the given Boolean function and it is shown implemented using NOR gates in figure 3.11(c).

3.9 Gate expansion

The number of inputs to a NAND gate can be increased by two methods: the first involves the use of AND circuits synthesised from NAND gates, as shown in figure 3.12(a), and in the second, AND gates can be used to achieve the required expansion as shown in figure 3.12(b). A similar technique is used to increase the number of inputs to a NOR gate and the method is illustrated in figure 3.12(c).

The ability to expand the number of inputs to a gate depends upon the type and the fan-in of the gates available in a particular logic family. For example, in the type 54/74 TTL logic family, NAND gates with up to eight inputs are available; hence it would be possible to expand so that a NAND gate with 64 inputs could be obtained. On the other hand, AND gates having a maximum of only four inputs are available

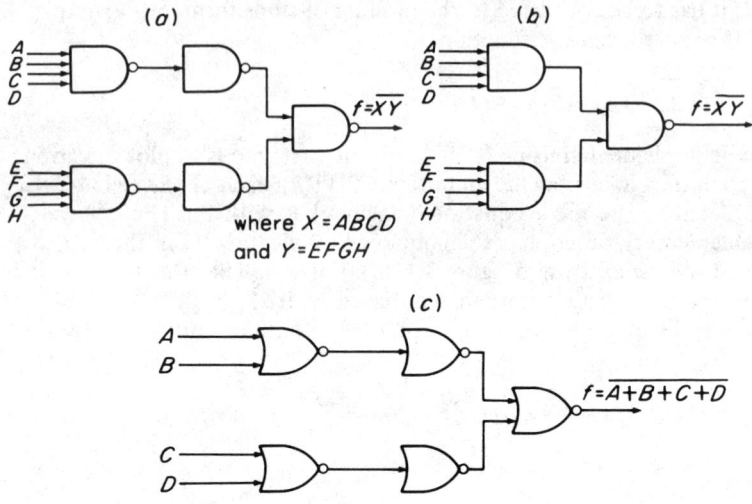

Figure 3.12. (a) Expansion of NAND gate inputs by means of AND circuits synthesised from NAND gates. (b) Expansion of NAND gate inputs using AND gates. (c) Expansion of NOR gate inputs by means of OR circuits synthesised from NOR gates

in the same series, which means that using the second method a NAND gate with a maximum of 32 inputs is possible.

3.10 Miscellaneous gates

A circuit performing the AND-OR-invert operation is illustrated in figure 3.13, and is available in the type 54/74 TTL family. It is a two-level logic circuit, the first level consisting of a NOR gate which

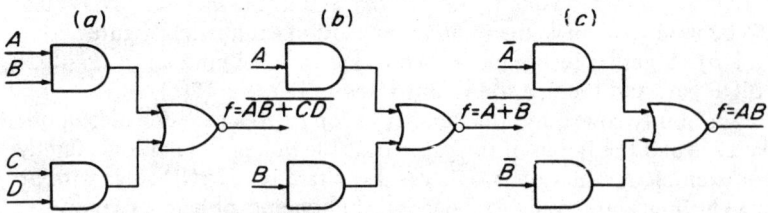

Figure 3.13. The A-O-I module (a), used as a NOR gate (b), and as an AND gate (c).

performs the OR and inversion operation, while the gates in the second level perform the AND operation. If necessary the A-O-I circuit can be used as a NOR gate simply by limiting the input to each AND gate to one as shown in figure 3.13(*b*). Alternatively, if the AND gates are fed with the inverted variables as shown in figure 3.13(*c*) the A-O-I circuit behaves as an AND gate.

In the high-speed section of the type 54/74 TTL family it is possible to obtain AND-OR gates. A diagram of this circuit is shown in figure 3.14(*a*). Again, this is a two-level logic circuit, the first level consisting of an OR gate and the second of two AND gates.

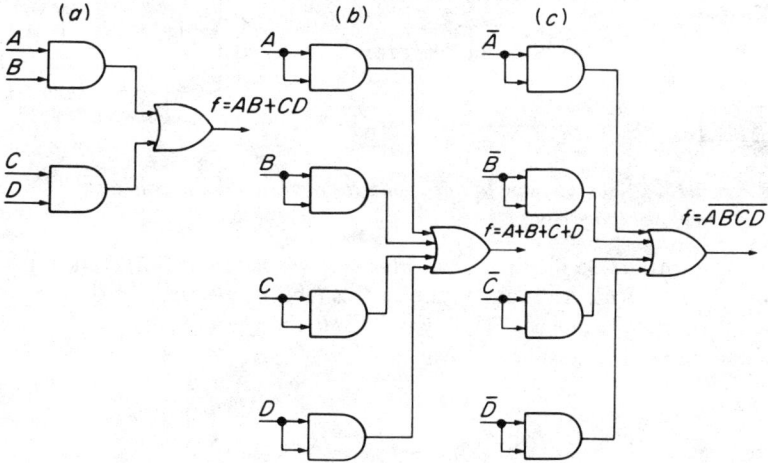

Figure 3.14. (a) The AND-OR gate. (b) The AND-OR circuit used as an OR gate. (c) The AND-OR module used as a NAND gate

If the inputs to the AND gates are limited to one as shown in figure 3.14(*b*) the circuit will act as an OR gate or, alternatively, if the inputs to the AND gates are the inverse of the variables used in figure 3.14(*b*) the AND-OR circuit acts as a NAND gate, as illustrated in figure 3.14(*c*).

Another commonly available type of gate is the open-collector gate; this is used with an external output pull-up resistor which enables several open-collector outputs to be connected in combination with one external pull-up resistor.

In figure 3.15(*a*) the outputs of two two-input open-collector NAND gates are connected to the same external pull-up resistor R_L. This circuit configuration performs the logic operation of collector dotting which is in effect an AND process. The collector dot may be regarded as part of an AND gate which has no physical reality and is as a consequence referred to as a phantom AND gate.

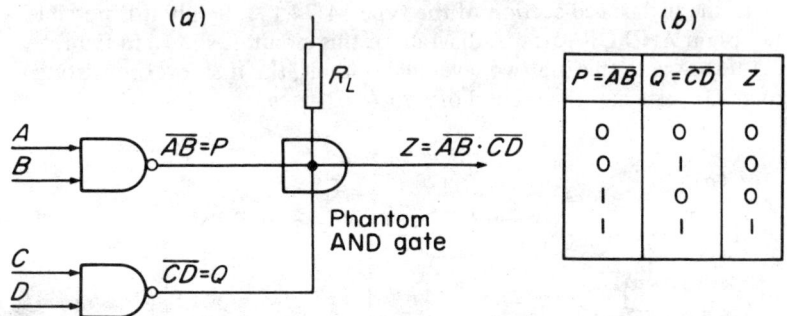

Figure 3.15. (a) Combination of open collector gates. (b) Truth table for the collector dotting operation

The truth table for the circuit illustrated in figure 3.15(a) is shown in figure 3.15(b). An examination of this table shows that the output Z of the circuit is logical '1' only if $P = \overline{AB}$ and $Q = \overline{CD}$ are simultaneously logical '1'. Hence

$$Z = PQ$$
$$= \overline{AB} \cdot \overline{CD}$$

If the output Z is inverted

$$\overline{Z} = AB + CD$$

Alternatively, if the inverted variables had been applied to the inputs of the NAND gates shown in figure 3.15(a), then the output Z would have been

$$Z = \overline{\overline{A}\,\overline{B}} \cdot \overline{\overline{C}\,\overline{D}}$$
$$= (A + B)(C + D)$$

and the inverse of this expression is

$$\overline{Z} = \overline{A}\,\overline{B} + \overline{C}\,\overline{D}$$

3.11 The tri-state gate

A gate commonly used in microcomputer systems is the tri-state gate illustrated in figure 3.16(a). The gate shown is a tri-state inverter and its truth table is shown in figure 3.16(b). This table shows that the gate performs the inversion operation on the input signal providing the enable signal E is a logical '1'. There is a third state which occurs if the enable signal E is logical '0' and this is referred to as the high-impedance state.

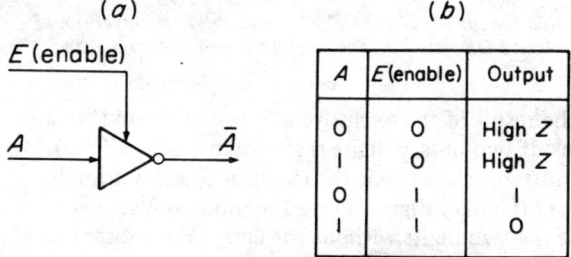

Fgiure 3.16. (a) The tri-state gate. (b) Truth table for the tri-state inverter

The tri-state inverter can be used to transfer data to one of the lines of the data bus used in a microcomputer system, when it is enabled. If the gate is disabled it presents a high impedance to the bus line and hence does not load it, thus allowing the line to be made available to some other device which requires to transfer data to the microcomputer system.

3.12 The exclusive-OR gate

The exclusive-OR function is defined by the Boolean equation

$$f = \overline{A} B + A \overline{B}$$

which is usually written

$$f = A \oplus B,$$

where the symbol $\oplus$ is used to indicate the exclusive-OR operation. The truth table for the exclusive-OR function is shown in figure 3.17(a) and the conventional way of representing the function diagrammatically is shown in figure 3.17(b).

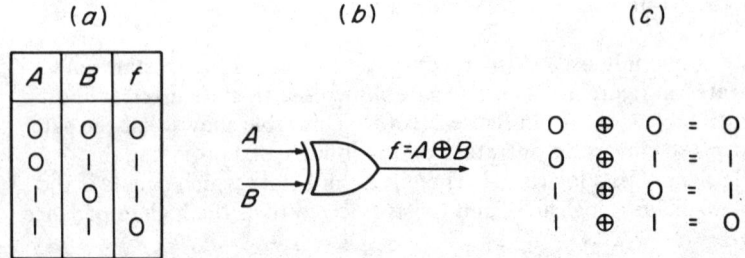

Figure 3.17. (a) Truth table for the exclusive-OR function. (b) The conventional representation of the exclusive-OR gate. (c) The modulo-2 sum of two binary digits

A table showing the result of the exclusive-OR operation on the four possible combinations of two binary digits is shown in figure 3.17(c). An examination of this table shows that the result of the operation is the modulo-2 sum of two binary digits, where the modulo-2 sum is defined as the sum of the two digits without the carry. For example, $1 + 1 = 10$, but the modulo-2 sum is simply 0.

The K-map plot of the function is shown in figure 3.18(a), while the complementary function is plotted on the K-map of figure 3.18(b). An examination of the map in figure 3.18(b) reveals that the complementary function is expressed by the Boolean equation

$$\bar{f} = \bar{A}\,\bar{B} + A\,B$$

which is conventionally written

$$\bar{f} = A \odot B$$

where the symbol ⊙ is used to indicate coincidence of the exclusive-NOR operation. A diagrammatic representation of the exclusive-NOR gate is shown in figure 3.18(c), and the truth table of the function is given in figure 3.18(d). This tabulation indicates that the complementary function $\bar{f}$ has the value of logical '1' when the two digits A and B are numerically equal.

It is immediately apparent from the foregoing that the exclusive-OR function has a direct application to the process of binary addition, whilst the exclusive-NOR function is used when comparing the magnitudes of two binary numbers.

There are a number of ways of implementing the exclusive-OR function. For example, it can be fabricated from AND, OR and inverter

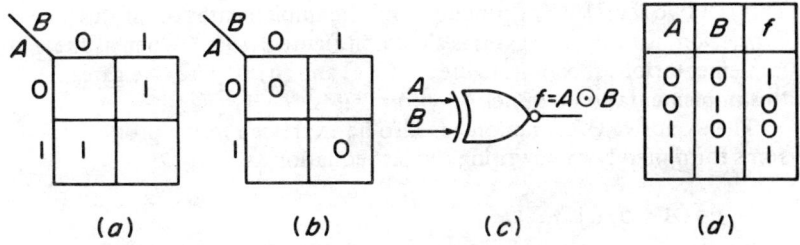

Figure 3.18. K-map plots of (a) the exclusive-OR function, and (b) the exclusive-NOR function. (c) Diagrammatic representation of the exclusive-NOR function. (d) Truth table for the coincidence function

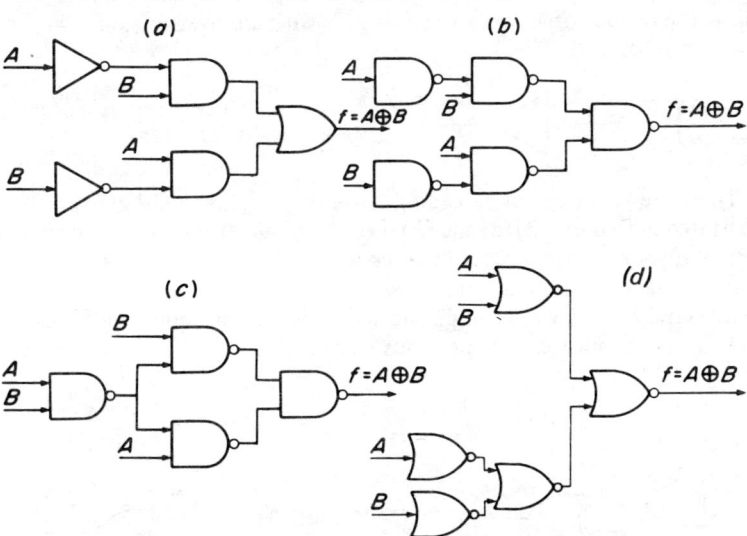

Figure 3.19. (a) AND/OR, (b) NAND, (c) minimal NAND, and (d) NOR implementation of the exclusive-OR function

gates as illustrated in figure 3.19(a), or, alternatively, an exclusive-OR circuit can be made up from NAND gates as shown in figure 3.19(b). However, the exclusive-OR function $f = \bar{A} B + A \bar{B}$ can be manipulated algebraically into another form which leads to a simpler implementation.

Adding $A \bar{A}$ and $B \bar{B}$ to the right-hand side of the above equation gives

$$f = \bar{A} B + B \bar{B} + A \bar{B} + A \bar{A}$$
$$= B (\bar{A} + \bar{B}) + A(\bar{A} + \bar{B})$$

This is a two-level sum-of-products and the implementation of this expression using NAND gates is shown in figure 3.19(c). Whereas the implementations shown in figures 3.19(a) and (b) require five gates, this implementation requires only four gates.

The exclusive-OR function can also be expressed in the product-of-sums form merely by rewriting the last equation as

$$f = (A + B)(\bar{A} + \bar{B})$$

and this expression is in a suitable form for implementing with NOR gates as shown in figure 3.19(d).

The simplest way of generating the exclusive-NOR function is to follow the output of an exclusive-OR gate with an inverter as shown in figure 3.20.

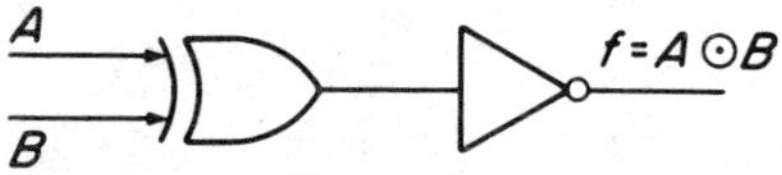

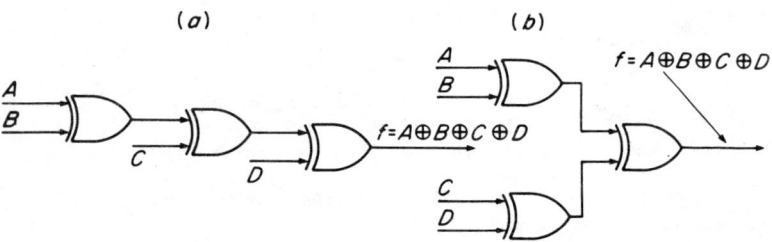

Figure 3.20. Implementation of the exclusive-NOR function

There are two ways of cascading two-input exclusive-OR gates, as illustrated in figures 3.21(a) and (b). For the first method, the exclusive-ORing of four variables requires three levels of logic. On the other hand, the second method, using exactly the same number of gates, requires only two levels of logic and will therefore produce less time delay at its output than the previous circuit.

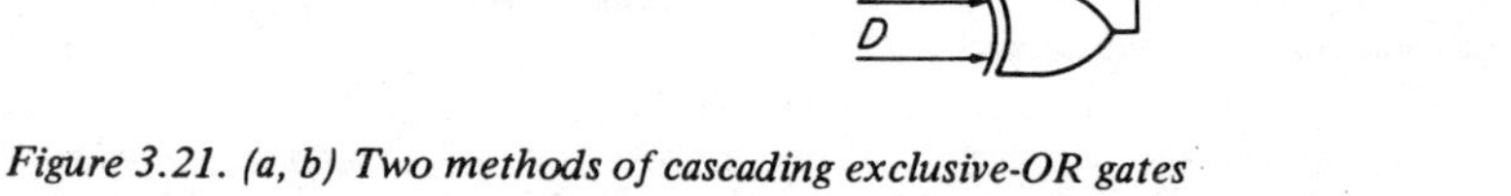

Figure 3.21. (a, b) Two methods of cascading exclusive-OR gates

The exclusive-OR gate can also be used as a controlled inverter as shown in figure 3.22(a) and (b), where binary data is fed to the two lines labelled D_0 and D_1 and the other inputs to the gates are supplied with a control signal M which may be either 0 or 1.

If $M = 0$ the outputs of the two gates are 1 and 0 respectively, as shown in figure 3.22(a). On the other hand, if the control signal $M = 1$

NAND and NOR logic

then the outputs of the two gates are 0 and 1 respectively. Hence, for $M = 0$, the information on the two data lines D_0 and D_1 appears at the outputs of the two gates, while for $M = 1$ the outputs of the two gates are the inverse of the inputs. This particular circuit arrangement is often used in conjunction with an adder circuit and it allows the adder circuit to perform complement arithmetic.

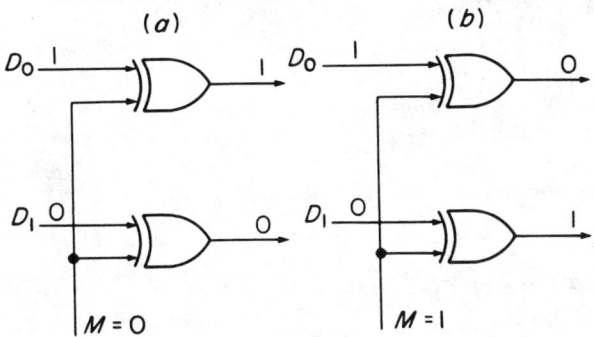

Figure 3.22. *The exclusive-OR gate as a controlled inverter. (a) Transmission, and (b) inversion*

When plotted on a K-map, the exclusive-OR function has a characteristic pattern. For three variables A, B and C, the function can be manipulated algebraically, as follows:

$$f = A \oplus B \oplus C$$
$$= (\bar{A} B + A \bar{B}) \oplus C$$
$$= \overline{(\bar{A} B + A \bar{B})} C + (\bar{A} B + A \bar{B}) \bar{C}$$
$$= \bar{A} \bar{B} C + \bar{A} B \bar{C} + A \bar{B} \bar{C} + A B C$$

These terms are shown plotted on a K-map in figure 3.23 and the plot exhibits a chequer-board pattern.

The cells occupied by 1's on this map are those for which the binary values of the variables when added have a modulo-2 sum of 1. For

A\BC	00	01	11	10
0		1		1
1	1		1	

Figure 3.23. *Characteristic K-map plot of the exclusive-OR function*

example, $\bar{A}\,\bar{B}\,C = 001$ and the modulo-2 sum of these three binary digits is one.

There are a number of rules associated with the exclusive-OR operation and they can be developed using algebraic techniques as shown below

$$A \oplus A = \bar{A}A + A\bar{A} = 0$$
$$A \oplus \bar{A} = \bar{A}\bar{A} + AA = 1$$
$$A \oplus 0 = \bar{A}\cdot 0 + A\cdot 1 = A$$
$$A \oplus 1 = \bar{A}\cdot 1 + A\cdot 0 = \bar{A}$$

and, similarly, it can be shown that

$$A \oplus B \oplus AB = A + B$$
$$A \oplus B \oplus 1 = \overline{A \oplus B} = A \odot B$$

Problems

3.1 Implement the following functions using NAND gates:
 (a) $f_1 = A\bar{B} + (\bar{B} + \bar{C})\bar{A}$
 (b) $f_2 = (AB + C)(B + \bar{D}) + A(\bar{B} + C)(D + \bar{E})$

3.2 Minimise the following functions and implement the minimised function using NAND gates:
 (a) $f(A, B, C) = \Sigma\ 0, 1, 2, 3, 4, 5, 6$
 (b) $f(A, B, C, D) = \Sigma\ 0, 2, 8, 9, 10, 12, 13, 14$
 (c) $f(A, B, C, D, E) = \Sigma\ 8, 9, 10, 11, 15, 16, 17, 18, 19, 20, 21,$
    $22, 23, 24, 25, 26, 27, 31$

3.3 Implement the following Boolean functions using NOR gates:
 (a) $f_1 = A(\bar{A} + \bar{B})(B + \bar{C}D)$
 (b) $f_2 = \bar{A}(B + C + DE)(\bar{B} + CD + \bar{A}E)$

3.4 Find the minimum product-of-sums form for each of the following functions and implement using NOR gates:
 (a) $f(A, B, C) = \Sigma\ 0, 2, 4, 6, 7$
 (b) $f(A, B, C, D) = \Sigma\ 0, 1, 2, 3, 4, 9, 10, 13, 14$
 (c) $f(A, B, C, D, E) = \Sigma\ 0, 1, 2, 3, 4, 6, 10, 11, 12, 13, 14, 15,$
    $16, 29, 31.$

3.5 Using a simple factoring technique, implement each of the following functions in as many ways as possible using NAND gates:
(a) $f_1 = B\bar{C}D + \bar{B}CD + A$
(b) $f_2 = \bar{A}C + BC + \bar{A}\bar{D}$
(c) $f_3 = A\bar{B}D + A\bar{B}\bar{C} + CD$

3.6 Implement the following functions using NAND gates having a maximum fan-in of three:
(a) $f_1 = AB\bar{C} + AD + B\bar{C}\bar{D} + AC\bar{D}$
(b) $f_2 = AB + \bar{A}D + BD + \bar{C}D + AC$
(c) $f_3 = ABCD + \bar{A}B\bar{C}D + AB\bar{D} + CD$

3.7 Implement the following functions using NOR gates having a maximum fan-in of three:
(a) $f_1 = (\bar{A} + B)(C + D)(B + \bar{C})(A + D)(\bar{A} + C)$
(b) $f_2 = (A\bar{C} + BC)(\bar{A} + \bar{C})$
(c) $f_3 = \bar{A}B + B\bar{C}D + A\bar{B}D$

3.8 Express the following equations in their minimal sum-of-products form:
(a) $f_1 = A(A \oplus B \oplus C)$
(b) $f_2 = A(A \odot B \odot C)$
(c) $f_3 = A + (A \oplus B \oplus C)$
(d) $f_4 = A + (A \odot B \odot C)$

3.9 Prove the following identities:
(a) $\overline{A \oplus B \oplus C} = A \oplus B \odot C$
(b) $\overline{A \odot B \odot C} = A \odot B \oplus C$
(c) $(A \oplus B \oplus AB)(A \oplus C \oplus AC) = A + BC$

4

Combinational logic design

4.1 Introduction

A combinational logic circuit can be described by the block schematic shown in figure 4.1. It consists of a rectangular block containing logic gates only, and to which the input signals $I_1, I_2, \ldots, I_n$ and from

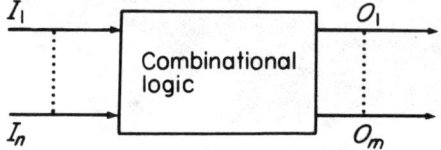

Figure 4.1. Block schematic of a combinational circuit

which the output signals $O_1, O_2, \ldots, O_m$ are either logical '0' or '1'. Each output is a function of some or all of the input variables. Hence

$$O_1 = f(I_1, I_2, \ldots, I_n)$$
$$O_2 = f(I_1, I_2, \ldots, I_n)$$

and

$$O_m = f(I_1, I_2, \ldots, I_n).$$

If it is assumed that all the logic gates used in this combinational logic circuit are without time delay then the outputs $(O_1, O_2, \ldots, O_m)$ will appear on their respective output terminals at the same instant as the input signals are applied at the input terminals. Furthermore, the output signals will be maintained until such time as the input signals

Combinational logic design

are removed. In practice, of course, a finite delay is associated with every gate in the combinational logic circuit and hence the outputs will appear on the individual output terminals a short time after the application of the input signals. The delay for each output will depend upon the individual gate delays and the number of gate levels associated with it.

4.2 The half-adder

One of the simplest combinational logic circuits is the half-adder, which is used for adding together the two least significant digits in a binary sum as indicated in figure 4.2(a). There are four possible combinations of two binary digits A and B that can be added together, and these are shown in figure 4.2(b). The sum of the two digits has been found in each case and it will be noticed that for the last case

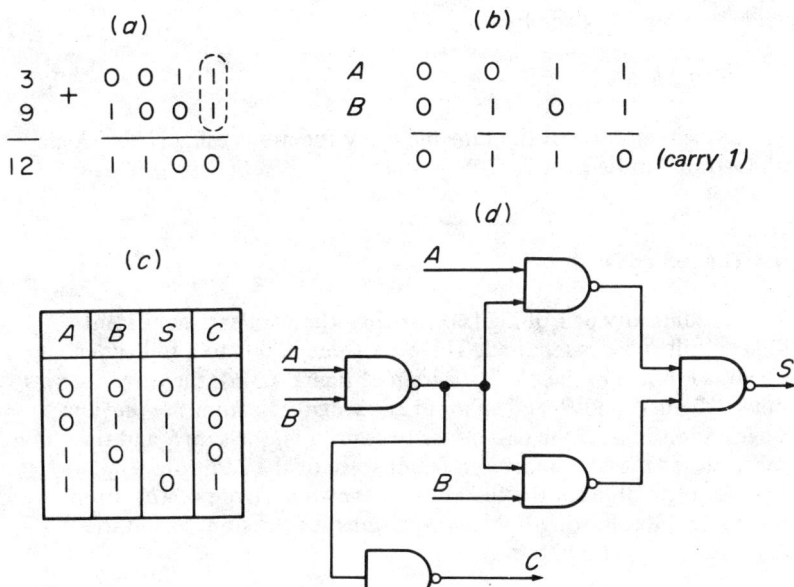

Figure 4.2. (a) Binary addition. The half-adder is used for adding together the two least significant bits (dotted). (b) The addition of the four possible combinations of two binary digits A and B (with a carry of 1 to the next most significant stage of addition). (c) Truth table for the half-adder. (d) NAND implementation of the half-adder

when $A = 1$ and $B = 1$ a carry is generated which must be propagated to the next most significant stage of the addition where it is added into the sum for that stage.

The additions shown in figure 4.2(b) are tabulated in the truth table of figure 4.2(c) where the columns headed A and B give every possible combination of the two binary digits to be added, while the third and fourth columns of this table give the sum (S) and carry (C), respectively. An examination of this table shows that the Boolean equation for the sum is

$$S = \bar{A}B + A\bar{B}$$

these being the two combinations of the variables which give a value of $S = 1$. This function is simply the two-input exclusive-OR function

$$S = A \oplus B$$

whilst the carry is given by

$$C = AB$$

The implementation of the sum and carry functions using NAND logic is illustrated in figure 4.2(d).

4.3 The full adder

When adding any pair of digits other than the two least significant digits a full adder is required. There are three inputs to a full adder circuit as shown in the block diagram of figure 4.3(a); these are the two binary digits A and B, and an input carry digit C_{in} from the previous stage. Additionally, the circuit has two outputs, the sum S and the carry-out to the next most significant stage of the addition, C_{out}.

The truth table for the full adder is shown in figure 4.3(b). From this table the following Boolean equations for the sum, S, and the carry-out, C_{out}, are obtained.

$$S = \bar{A}\bar{B}C_{in} + \bar{A}B\bar{C}_{in} + A\bar{B}\bar{C}_{in} + ABC_{in}$$

and

$$C_{out} = \bar{A}BC_{in} + A\bar{B}C_{in} + AB\bar{C}_{in} + ABC_{in}$$

Combinational logic design 59

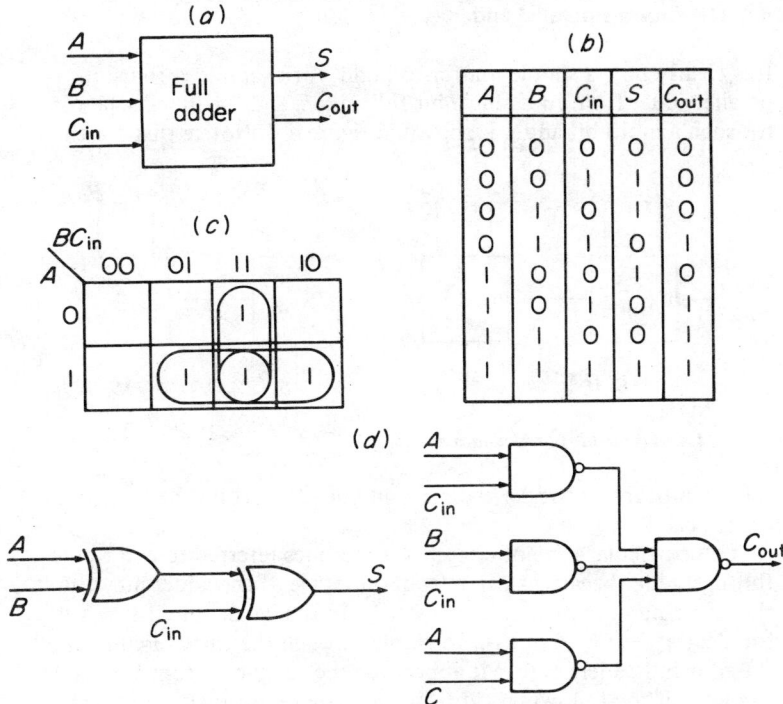

Figure 4.3. The full adder. (a) Block diagram. (b) Truth table. (c) K map plot for C_{out}. (d) Implementation

The equation for the sum may be rewritten as

$$S = \overline{A}(\overline{B}C_{in} + B\overline{C}_{in}) + A(\overline{B}\,\overline{C}_{in} + BC_{in})$$

$$= \overline{A}(B \oplus C_{in}) + A(\overline{B \oplus C_{in}})$$

$$= A \oplus B \oplus C_{in}$$

The carry equation is plotted on the Karnaugh map shown in figure 4.3(c) and has been simplified in the normal way. From the K-map the simplified carry-out equation may be written as

$$C_{out} = AC_{in} + BC_{in} + AB$$

An implementation of the full adder is shown in figure 4.3(d).

4.4 The four-bit parallel adder

It is clearly now a simple matter to build, for example, a four-bit parallel adder from four single-bit full adders and the block schematic for such a multi-bit adder is shown in figure 4.4. Notice that for the

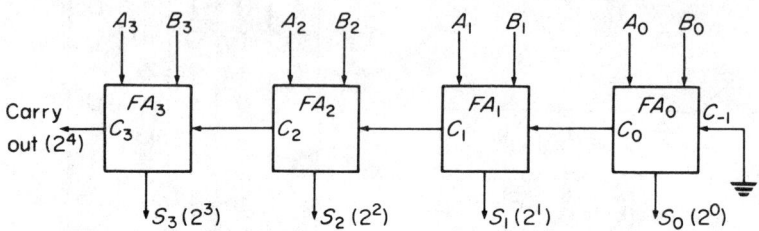

Figure 4.4. A four-bit parallel adder

least significant full adder the carry-in line C_{-1} is connected to ground, i.e. logical '0'.

This particular type of circuit is sometimes referred to as a ripple-through adder because a carry from one stage of the adder may ripple through a number of succeeding stages. In the worst case it is possible for a carry generated in FA_0 to ripple through the carry circuits of all the four full adders before it appears as the carry-out from FA_3. For example, if the following addition has to be performed

$$\begin{array}{rr} A & 1111 \quad\quad 15 \\ B + 1001 \quad\quad 9 \\ \hline 11000 \quad\quad 24 \end{array}$$

a carry is generated in the least significant stage of the addition and it ripples through each successive stage of the addition until it appears as the carry output from the most significant stage where it becomes the sum digit for what is in effect the fifth stage of addition. Under these circumstances C_3 ripples through four two-level logic circuits and hence the sum cannot be completed until eight gate delays have occurred. For this kind of adder the maximum delay is directly proportional to the number of stages n.

4.5 The carry look-ahead adder

Since the delay time of a ripple-through adder increases with the number of bits to be added, several techniques have been developed for

Combinational logic design

reducing addition time. One of these, the carry look-ahead technique, since it is a combinational logic problem, will be described here.

The carry output equation for a full adder is

$$C_{out} = \bar{A}BC_{in} + A\bar{B}C_{in} + AB\bar{C}_{in} + ABC_{in}$$

which can be expressed as follows

$$C_{out} = (A \oplus B)C_{in} + AB$$

or as

$$C_{out} = PC_{in} + G$$

where $P = A \oplus B$ and is referred to as the propagation term and $G = AB$ is referred to as the generation term. If $G = 1$, then $A = 1$ and $B = 1$ and a carry is generated in the stage defined by the C_{out} equation. Alternatively, if the carry into the stage $C_{in} = 1$ and either A or B is 1 then the input carry will be propagated to the next stage. For a four-bit adder the generate and propagate terms for each stage are

$$G_0 = A_0 B_0 \quad P_0 = A_0 \oplus B_0$$
$$G_1 = A_1 B_1 \quad P_1 = A_1 \oplus B_1$$
$$G_2 = A_2 B_2 \quad P_2 = A_2 \oplus B_2$$
$$G_3 = A_3 B_3 \quad P_3 = A_3 \oplus B_3$$

while the carries for the various stages are

$$C_0 = P_0 C_{-1} + G_0$$
$$C_1 = P_1 C_0 + G_1$$
$$C_2 = P_2 C_1 + G_2$$

and

$$C_3 = P_3 C_2 + G_3$$

Substituting for C_0 in the C_1 equation, etc, leads to the following equations:

$$C_1 = P_1 P_0 C_{-1} + P_1 G_0 + G_1$$

$$C_2 = P_2P_1P_0C_{-1} + P_2P_1G_0 + P_2G_1 + G_2$$
$$C_3 = P_3P_2P_1P_0C_{-1} + P_3P_2P_1G_0 + P_3P_2G_1 + P_3G_2 + G_3.$$

The sum for the least significant stage of the adder is given by

$$S_0 = A_0 \oplus B_0 \oplus C_{-1}$$
$$= P_0 \oplus C_{-1}$$

and for the succeeding stages by

$$S_1 = P_1 \oplus C_0$$
$$S_2 = P_2 \oplus C_1$$
$$S_3 = P_3 \oplus C_2.$$

A block diagram can now be drawn for the four-bit carry look-ahead adder which consists of three sections, one for generating the propagate and generate terms, a second for generating the various carry terms, and a third for generating the sum terms. The block schematic diagram is shown in figure 4.5(a) and this has been translated into hardware in figure 4.5(b).

If the circuit had been implemented using NAND logic then each exclusive-OR gate in figure 4.5(b) would require four NAND gates for implementation. The propagation and generation section consists in effect of four half-adders, each of which requires five NAND gates as shown in figure 4.2(d). Hence this section requires 20 NAND gates for implementation. For the carry section the AND and OR gates can be replaced by NAND gates on a one-for-one basis as shown in Chapter 3, and an additional NAND gate will be required for inverting the single G terms in the carry equations; hence this section requires 18 NAND gates for its implementation. Finally, the sum section consists of four exclusive-OR gates each of which requires four NAND gates and the total number of NAND gates required for the implementation of this section is 16. In all, the implementation of the adder requires 54 NAND gates.

One stage of a four-bit ripple-through adder consists of two exclusive-OR gates and four NAND gates as shown in figure 4.3(d), hence each stage of this kind of adder requires twelve NAND gates in all, and overall the four-bit ripple-through adder requires 48 NAND gates for implementation.

Combinational logic design

(a)

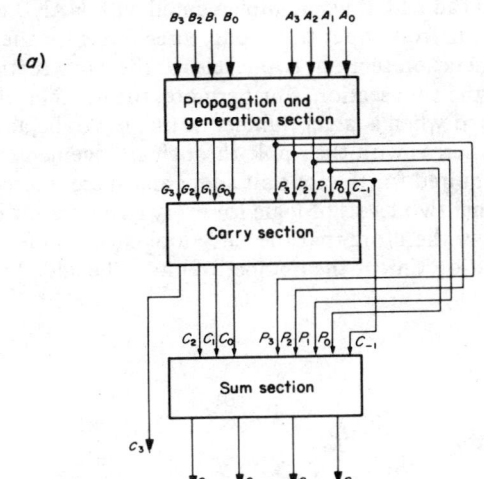

(b)

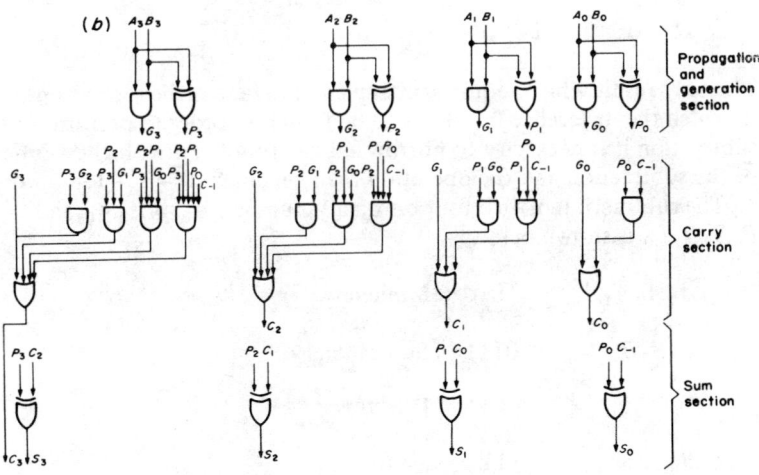

Figure 4.5. (a) Block schematic, and (b) hardware implementation for a four-bit carry look-ahead adder

The carry look-ahead adder, when implemented with NAND gates, has eight levels of logic from input to output; three levels for the propagation and generation section; two levels for the carry section, and three levels for the sum section. Furthermore, the number of levels of logic required when a larger number of bits has to be added does not increase, whereas with the ripple-through arrangement, eight levels of logic are required for the four-bit adder and there is a requirement for an additional two levels of logic for every extra pair of bits to be summed. It follows therefore that the carry look-ahead adder will provide a faster addition time if the number of bits to be added is greater than four.

4.6 The full subtractor

The binary subtraction of the four possible combinations of two binary digits A and B is shown below.

A	0	0	1	1
B	0	1	0	1
D	0	1	1	0

The only result which requires an explanation here is the second one, in which the difference $0 - 1$ has to be found. In order to perform this subtraction it is necessary to borrow a digit from the next highest order of the subtraction and the operation then becomes $10 - 1 = 1$.

The subtraction of two four-bit binary numbers $A_3 A_2 A_1 A_0$ and $B_3 B_2 B_1 B_0$ is shown below:

$A_3 A_2 A_1 A_0$ 1010   Minuend

$B_3 B_2 B_1 B_0$ 0111   Subtrahend

                     0011   Difference
                     ↙↙↙
B_{in} 111

$A_3 A_2 A_1 A_0$ is called the *minuend* and $B_3 B_2 B_1 B_0$ is called the *subtrahend*. In the least significant place of the subtraction a difference

Combinational logic design

can only be obtained by borrowing a 1 from the next most significant stage. The difference is now 1 and this is entered in the least significant place of the difference. However, the borrow has to be repaid and this process is illustrated on the B_{in} row of the above subtraction.

In the next most significant stage of the subtraction the difference that has to be obtained is $A_1 - (B_1 + B_{in})$, which in this case is $1 - 10$ and this leads to a difference of 1 with a borrow transferred to the next most significant stage of the subtraction.

It is now possible, using the above rules, to construct the truth table for a full subtractor as shown in figure 4.6(a). From the truth table the following equation is obtained for the difference D:

$$D = \overline{A}\overline{B}B_{in} + \overline{A}B\overline{B}_{in} + A\overline{B}\overline{B}_{in} + ABB_{in}$$

and simplifying,

$$D = \overline{A}\,(\overline{B}B_{in} + B\overline{B}_{in}) + A(\overline{B}\overline{B}_{in} + BB_{in})$$

$$= \overline{A}(B \oplus B_{in}) + A(\overline{B \oplus B_{in}})$$

$$= A \oplus B \oplus B_{in}.$$

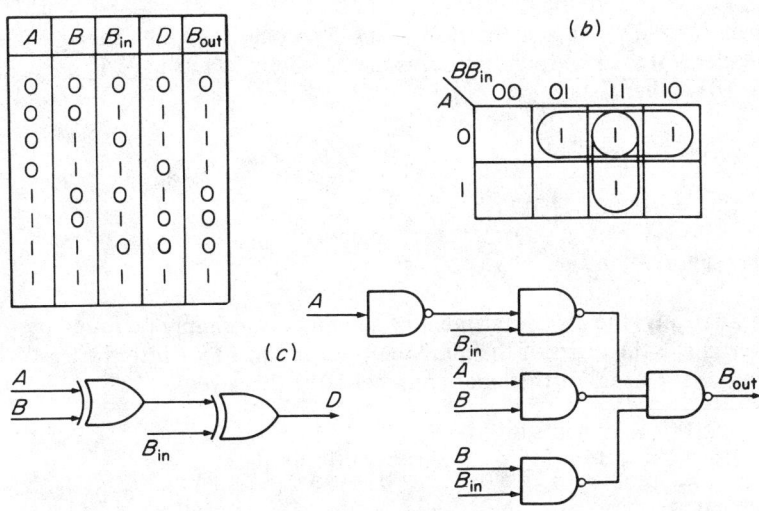

Figure 4.6. (a) Truth table for a full subtractor. (b) K-map plot of B_{out}. (c) Logic implementation of the full subtractor

The B_{out} function is shown plotted directly on the K-map in figure 4.6(b). Simplifying in the manner described above leads to the following equation for the borrow-out

$$B_{out} = \overline{A}B_{in} + \overline{A}B + BB_{in}$$

Both the difference and the borrow-out functions are shown implemented in figure 4.6(c).

It would be unusual to find a full subtractor, such as the one just designed, in a digital machine. Obviously, if possible, it would be more convenient if the same hardware could be used for both addition and subtraction. If complement arithmetic is used then an adder circuit can be used for both addition and subtraction, and because this leads to an economy of hardware it is normal practice to use complement arithmetic when performing arithmetical processes in digital machines.

4.7 The 2's complement

For an n-bit number, the *2's complement* is defined as

$$N^* = 2^n - N$$

where N^* is 2's complement of N, N is the number for which the complement is to be found, and n is the number of digits in N. If $N = 1011$, then

$$N^* = 2^4 - 1011$$

$$= 10000 - 1011$$

$$N^* = 00101$$

Alternatively, the 2's complement of a number can simply be found by inverting all the digits of the number N and adding in a 1 into the least significant place. For the binary number 1011 this gives

```
0100    All digits inverted
   1    1 added into the least significant place
————
0101
```

One other method of finding the 2's complement is also available.

Using this method sense the lowest order 1 in the number to be converted, then invert all the higher-order digits leaving the 1 and the lower-order digits unchanged.

The 2's complement is sometimes referred to as the *radix complement* since 2 is the radix or base of the binary number system.

4.8 The 1's complement

The *1's complement* or the *diminished radix complement* is defined as

$$N^{**} = (2^n - 1) - N$$

where N^{**} is the 1's complement of N. For the binary number 1011

$$N^{**} = (10000 - 1) - 1011$$
$$= 0100$$

The simplest method of determining the 1's complement is to invert all the binary digits in the given number.

4.9 Binary representations for arithmetic units

In a digital machine a binary number can be represented in any one of three ways:

    (*a*) sign and magnitude notation,
    (*b*) 2's complement notation,
    (*c*) 1's complement notation.

The sign of the number in each of these three methods is represented by a binary signal, 0 for a positive number and 1 for a negative number. This sign bit is positioned at the left-hand end of the number and is separated from the numerical part of the number by a comma.

For the sign and magnitude representation, the first bit is the sign and the remaining bits represent the numerical value in binary form. For example,

    $+13 = 0,1101$

and

    $-13 = 1,1101$

68 Combinational logic design

The design of logic networks to perform arithmetic operations with sign and magnitude binary numbers is not easy, and as a consequence it is more usual to use either the 2's or 1's complement notations.

In the 2's complement system positive numbers are represented by their magnitude with the appropriate sign bit attached to the left-hand end of the number, whilst a negative number is represented by its 2's complement form, again with the appropriate sign bit attached. Hence

$+13 = 0,1101$

and

$-13 = 1,0011$

Similarly, in the 1's complement system a positive number is represented by its magnitude and the appropriate sign bit whilst a negative number is represented by its 1's complement form again with sign bit attached. Thus

$+13 = 0,1101$

and

$-13 = 1,0010$

4.10 Addition and subtraction using 2's complement arithmetic

Addition is carried out in all cases irrespective of whether the numbers are positive or negative. The sign bits are included in this addition and any carry-out from the sign bit position is ignored. If the resulting answer is positive the sign bit is 0 and the numerical part of the number is expressed in magnitude form. If, however, the resulting answer is negative, the sign bit is 1 and the numerical part of the number is expressed in 2's complement form. These results are illustrated by the examples shown in figure 4.7(a) for the four possible combinations of the positive and negative values of the decimal numbers 7 and 5.

An adder/subtractor using 2's complement arithmetic is illustrated in figure 4.7(b). The number $A_3A_2A_1A_0$ will be the augend in the addition mode, and the minuend for the subtraction mode, whilst the number $B_3B_2B_1B_0$ will be the addend in the addition mode and the subtrahend in the subtraction mode. The circuit requires five full

Combinational logic design

```
+7    0,0111
+5    0,0101
───   ──────
+12   0,1100
```

```
+7    0,0111         -7    1,1001
-5    1,1011         +5    0,0101
───   ──────         ──    ──────
+2    0,0010         -2    1,1110
```

```
                     -7    1 1001
                     -5    1 1011
                     ───   ──────
                     -12   1,0100
```

(a)

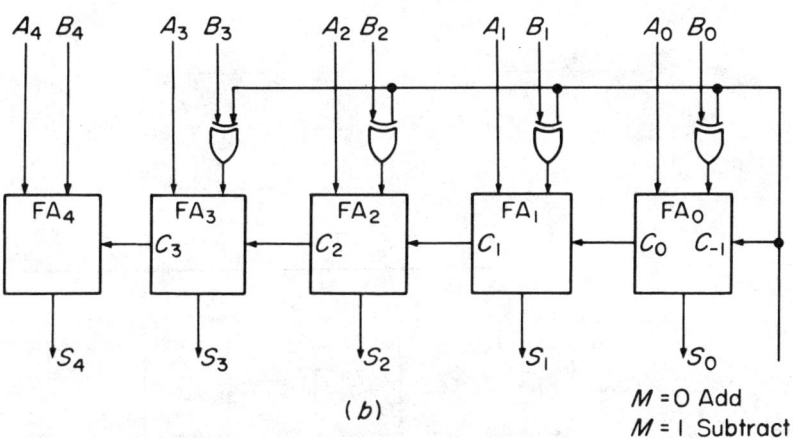

(b)

$M = 0$ Add
$M = 1$ Subtract

Figure 4.7. (a) Addition and subtraction using 2's complement arithmetic. (b) An adder/subtractor using 2's complement arithmetic

adders, FA_4 being used for the addition of the sign digits A_4 and B_4. A controlled exclusive-OR inverter is used for inverting the digits B_3, B_2, B_1 and B_0 in the subtraction mode, i.e. when $M = 1$. The signal M is also applied to the carry input terminal of the least significant full adder. When $M = 0$, $C_{-1} = 0$ and FA_0 behaves as a half-adder, however when $M = 0$, $C_{-1} = 1$, and a 1 is added to the least significant digit of the 1's complement of the subtrahend to form its 2's complement.

4.11 Binary multiplication

One of the simplest and fastest methods of multiplying employs a combinational logic circuit which is composed of AND gates, half-adders and full adders. The method depends upon the fact that the rules of Boolean multiplication are identical to those of binary multiplication and as a consequence a series of AND gates can be used for forming the products that occur in the binary multiplication process.

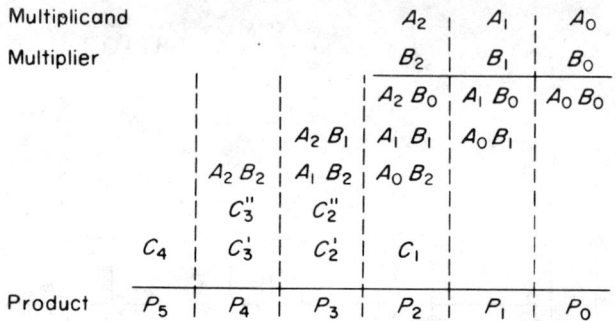

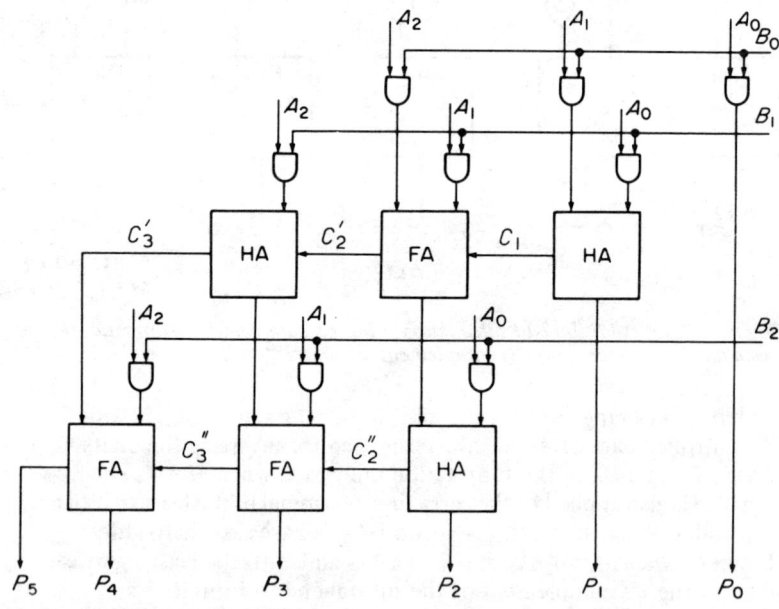

Figure 4.8. (a) Binary multiplication. (b) A parallel binary multiplier

Combinational logic design 71

If, for example, two three-bit binary numbers $A_2A_1A_0$ and $B_2B_1B_0$ are to be multiplied together, the pencil and paper method that would normally be employed is illustrated in figure 4.8(a). For the first row of the multiplication the least significant bit of the multiplier multiplies each of the multiplicand bits in turn, forming three product terms A_2B_0, A_1B_0 and A_0B_0, and the next row of products is obtained by first shifting left and then multiplying each of the multiplicand bits by the next most significant bit of the multiplier, and so on, until all the multiplier bits have been used. The vertical columns are then added in turn to produce the various product digits P_5 to P_0. For each of these additions, allowance must be made for the possibility of a carry from the preceding stage. In the second column from the right, for example, A_1B_0 and A_1B_1 are added together to form P_1, but in doing so there is a possibility of a carry C_1 being produced which has to be added into the next most significant vertical column. For the third column from the right two additions have to be performed since there are four digits to be added. This raises the possibility of producing two carries, C_2' and C_2'', which are carried forward to the next vertical column for adding in, and so on.

The implementation of the multiplier is shown in figure 4.8(b). For this kind of parallel multiplier the amount of combinational logic required increases rapidly with the number of bits in the multiplier and the multiplicand. Before the advent of LSI chips the amount of combinational logic required was a deterrent to using this technique, but now that LSI circuits are readily available fast multiplier chips using this method are obtainable.

4.12 Code conversion

A fairly common requirement of a digital system is the facility to convert from one code to another. Decimal information is frequently expressed in binary-coded form, a very widely used code being NBCD (naturally binary-coded decimal), as shown tabulated in figure 4.9(a).

Four binary digits are required to represent one decimal digit in this code. There are 2^4 possible combinations of four binary digits of which only ten are used in the NBCD code. Hence there are six forbidden code combinations and the code is said to contain some redundancy.

Another code employed in digital systems is the 2-out-of-5 code, in which each combination contains two 1's and as it happens in a five-bit code there are only ten combinations available that contain two 1's, conveniently one for each of the decimal digits. This code has been tabulated by the side of the NBCD code in figure 4.9(a).

72 Combinational logic design

Because of the 2-out-of-5 property this code affords protection against transmission errors which may occur in a digital communication channel. For this reason the NBCD code would be converted to 2-out-of-5 before transmission and at the receiving end each code combination would be checked to verify the 2-out-of-5 property. The circuit used for code conversion would be purely combinational and a block diagram for it is shown in figure 4.9(b).

(a)

Decimal digit	NBCD				2-out of -5				
	A	B	C	D	P	Q	R	S	T
0	0	0	0	0	1	1	0	0	0
1	0	0	0	1	0	0	0	1	1
2	0	0	1	0	0	0	1	0	1
3	0	0	1	1	0	0	1	1	0
4	0	1	0	0	0	1	0	0	1
5	0	1	0	1	0	1	0	1	0
6	0	1	1	0	0	1	1	0	0
7	0	1	1	1	1	0	0	0	1
8	1	0	0	0	1	0	0	1	0
9	1	0	0	1	1	0	1	0	0

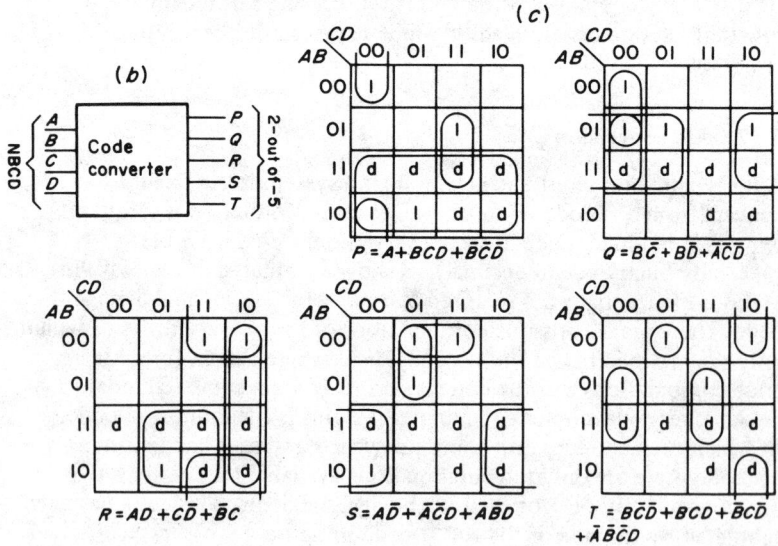

(d)

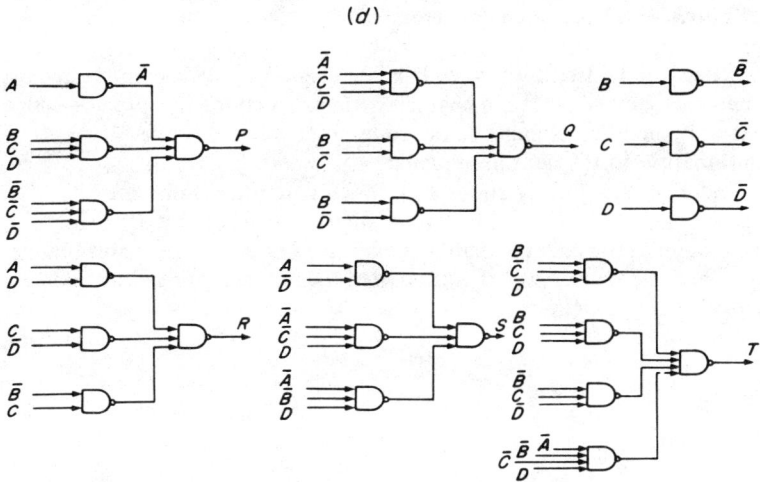

Figure 4.9. (a) The NBCD and 2-out-of-5 codes. (b) Block schematic of NBCD to 2-out-of-5 code converter. (c) K-maps for the NBCD to 2-out-of-5 converter. (d) Implementation of the NBCD to 2-out-of-5 converter

The tabulations in figure 4.9(a) may be regarded as the truth table for the converter. From this table the K-maps for the output signals P, Q, R, S and T can be plotted directly, as shown in figure 4.9(c). Each cell marked with a d on this map represents those combinations of four binary digits that do not exist in the NBCD code, and hence should never happen. These cells are treated as 'don't care' terms and are used in the simplification process.

From the maps in figure 4.9(c) the equations of the output signals are

$$P = A + BCD + \bar{B}\bar{C}\bar{D}$$

$$Q = B\bar{C} + B\bar{D} + \bar{A}\bar{C}D$$

$$R = AD + C\bar{D} + \bar{B}C$$

$$S = A\bar{D} + \bar{A}CD + \bar{A}BD$$

$$T = B\bar{C}\bar{D} + BCD + \bar{B}C\bar{D} + \bar{A}\bar{B}CD$$

These functions are shown implemented with NAND logic in figure 4.9(d).

4.13 Binary to Gray code converter

Another widely used type of code is the single-step code commonly called the *Gray code*. It is a characteristic property of this type of code that only one digit changes as the code progresses from one combination to the next in sequence.

There are many Gray codes, but one in particular is obtained by taking the modulo-2 sum of adjacent digits of four-bit binary code. For example, the decimal digit 3 is expressed as 0011 in four-bit binary code, and the conversion of this code to its corresponding Gray code is demonstrated below.

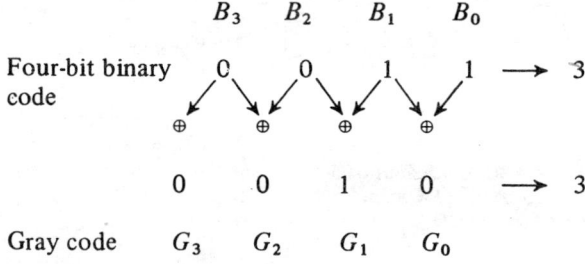

It is always assumed when making this type of conversion that the bit that does not exist and which has been used to form G_3 is a 0.

An examination of the above conversion reveals that the Gray code digits G_3, G_2, G_1 and G_0 are related to the binary digits B_3, B_2, B_1, and B_0 by the following logic equations

$$G_0 = B_0 \oplus B_1 \qquad G_2 = B_2 \oplus B_3$$

$$G_1 = B_1 \oplus B_2 \qquad G_3 = B_3$$

The logic required for implementing these equations is shown in figure 4.10(*a*) and the corresponding codes are tabulated in figure 4.10(*b*). One chip only is required for this conversion, namely the quad two-input exclusive-OR gate which is available in the type 74 TTL logic family.

This particular form of Gray code is called *reflected binary*. If a horizontal line is drawn immediately below the Gray code combination 0100 in the tabulation of figure 4.10(*b*), it can be seen that the last three digits of the code are reflected about this horizontal line.

The logic equations for the reverse conversion are obtained by plotting the K-map for each of the binary digits B_3, B_2, B_1 and B_0

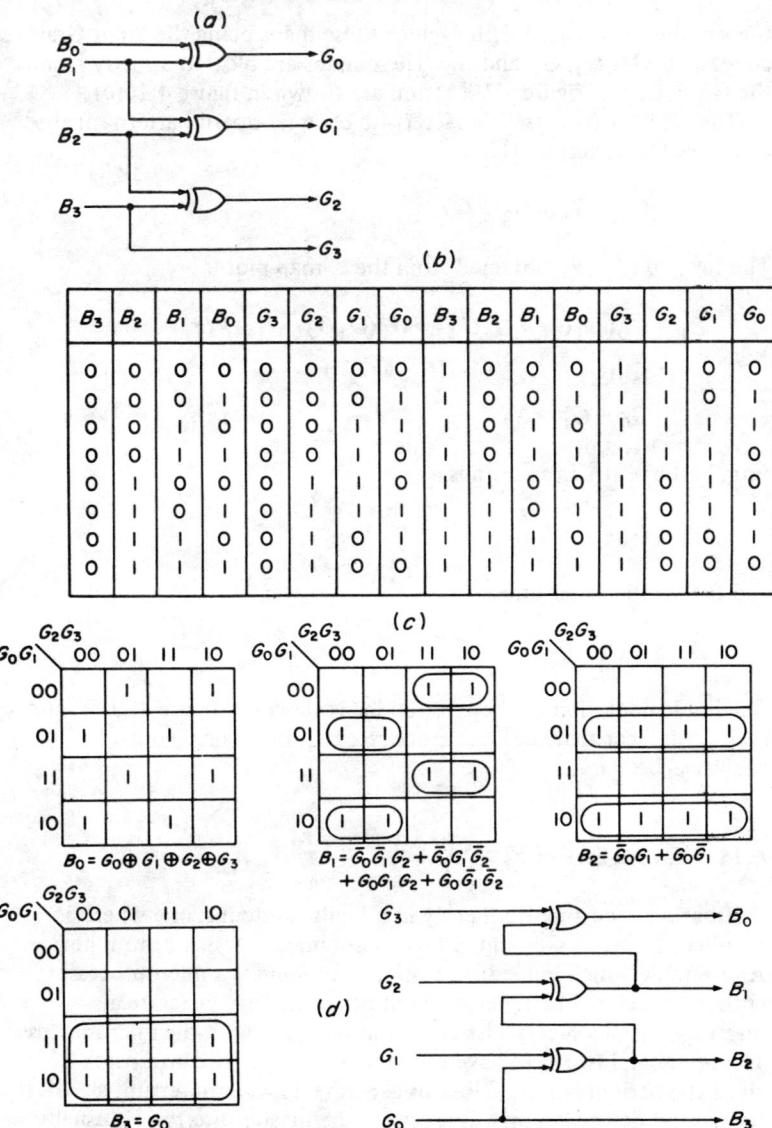

Figure 4.10. (a) Code conversion circuit of a four-bit binary to a reflected binary (Gray) code. (b) The four-bit binary and reflected binary (Gray) codes. (c) K-maps for the reflected binary (Gray) code to four-bit binary conversion. (d) Code conversion circuit of the reflected binary (Gray) to four-bit binary code

the variables associated with each of these maps being the input Gray code digits G_3, G_2, G_1 and G_0. These maps are plotted directly from the tabulation of figure 4.10(b) and are shown in figure 4.10(c).

The plot for B_0 is the characteristic chequer-board pattern for the exclusive-OR function. Hence

$$B_0 = G_0 \oplus G_1 \oplus G_2 \oplus G_3$$

The function for B_1 obtained from the K-map plot is

$$B_1 = \bar{G}_0(\bar{G}_1 G_2 + G_1 \bar{G}_2) + G_0(G_1 G_2 + \bar{G}_1 \bar{G}_2)$$
$$= \bar{G}_0(G_1 \oplus G_2) + G_0(\overline{G_1 \oplus G_2})$$
$$= G_0 \oplus G_1 \oplus G_2$$

For B_2 the plotted function is

$$B_2 = \bar{G}_0 G_1 + G_0 \bar{G}_1 = G_0 \oplus G_1$$

and for B_3 the function is

$$B_3 = G_0$$

The implementation of these functions is shown in figure 4.10(d) and again, it is clear that the code converter requires a single quad two-input exclusive-OR gate.

4.14 Interrupt sorters

A situation occurring frequently in a digital system is one where a number of devices are required to communicate with a central device. An example which immediately springs to mind is a microprocessor with its associated peripherals. Each peripheral can generate an interrupt signal when it wishes to communicate with the microprocessor. The peripheral is said to have raised an interrupt. The interrupts from all of the peripherals are ORed to generate a master interrupt signal. If the central device is a microprocessor, the master interrupt is usually called the interrupt request signal. This signal requests the microprocessor to interrupt its current activity and jump to the peripheral's service routine.

The interrupt request informs the microprocessor that one of the

Combinational logic design

peripherals in the system wishes to communicate with it. It is then the function of the microprocessor to identify which of the peripherals wishes to communicate with it. One method is to use an interrupt sorter circuit, as shown in figure 4.11.

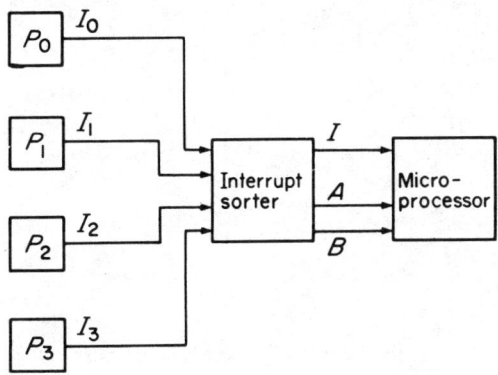

Figure 4.11. *Block schematic of an interrupt sorting system*

In this diagram, the individual peripherals each have their own interrupts I_0, I_1, I_2 and I_3 respectively. These signals are fed to the interrupt sorter which generates the master interrupt signal.

$$I = I_0 + I_1 + I_2 + I_3$$

and identifies the peripheral which has raised an interrupt. In this case there are four peripherals and hence two address lines A and B are sufficient to distinguish between their individual interrupts.

Clearly there are practical situations where a given interrupt signal has to be given priority over the others. In the combinational interrupt sorter to be designed here, it will be assumed the higher the suffix of the interrupt, the higher its priority.

The block diagram of a simple four interrupt sorter is shown in figure 4.12(*a*). The circuit is required to generate a master interrupt signal *I*, to indicate the presence of one or more interrupts, and also the address signals A and B which are able to identify which of the four interrupts are raised.

The truth table for the interrupt sorter is shown in figure 4.12(*b*). A 'd' in the truth table is used to denote a 0 or a 1 and its use in the second row of this table indicates that $I = 1, A = 1$ and $B = 1$ if $I_3 = 1$ irrespective of whether interrupts I_0, I_1 and I_2 are raised or not.

From the truth table the following equations can be written

$$I = I_0 + I_1 + I_2 + I_3$$

and

$$A = I_3 + \overline{I}_3 \overline{I}_2 I_1$$

$$= I_3 + \overline{I}_2 I_1$$

and

$$B = I_3 + \overline{I}_3 I_2$$

$$= I_3 + I_2$$

The implementation of the interrupt sorter using NAND logic is shown in figure 4.12(c).

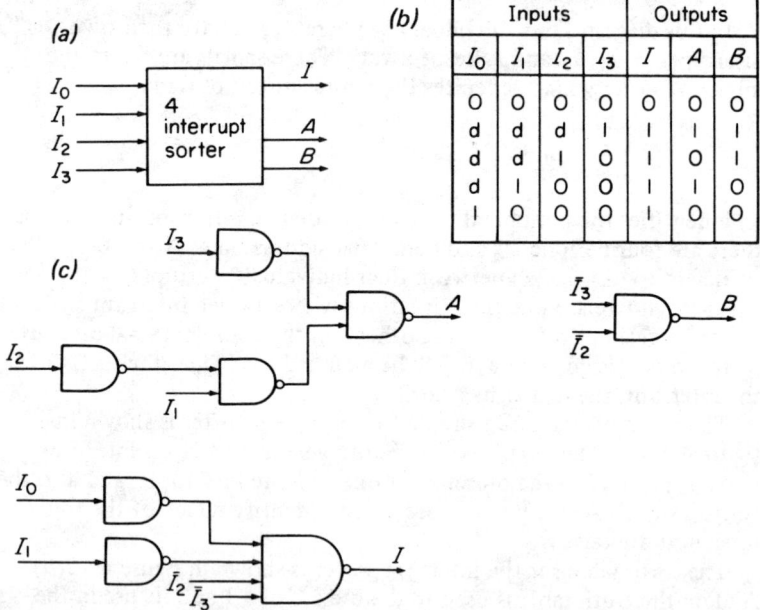

Figure 4.12. (a) Block diagram of a 4-interrupt sorter. (b) Truth table for a 4-interrupt sorter. (c) The implementation of a 4-interrupt sorter

4.15 Daisy-chaining

A common hardware method of identifying the peripheral which wishes to communicate with the central device is the *daisy-chain technique*. In essence this is a hardware polling technique. A block schematic illustrating the method is shown in figure 4.13(a). The central

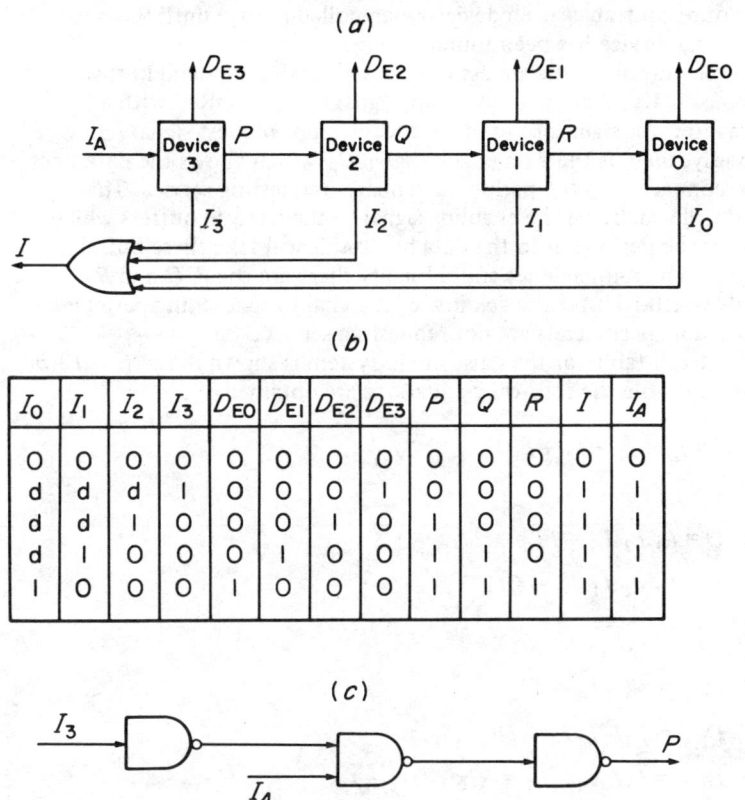

Figure 4.13. The daisy-chaining technique. (a) Block diagram. (b) Truth table. (c) Daisy chain logic for device 3

device, when it receives an interrupt request, acknowledges it with an interrupt acknowledge signal I_A which is in the first instance passed to the peripheral with the highest priority, in this case device 3. If device 3 is requesting service it will identify itself and the interrupt acknowledge signal will not be passed to device 2. If, on the other hand, it is not device 3 which has raised its interrupt, it will pass the acknowledge signal to the peripheral with the next highest priority, in this case device 2. This procedure continues, each device being polled in turn until the one requesting service has been found.

Three output signals are associated with each peripheral in this system. Firstly, there is an interrupt signal which is ORed with all other interrupt signals to produce the interrupt request signal I. Secondly, there is the data enable signal D_E which allows the data lines to be connected to the peripheral which is requesting service. This signal will usually be the enabling signal to the tri-state buffers which connect the peripheral to the data bus and it will take these buffers out of the high-impedance state. Finally there are the P, Q and R signals which transfer the acknowledge signal to succeeding peripherals if preceding peripherals are not requesting service.

The truth table for the daisy-chain system is shown in figure 4.13(b). From this table the following equations are obtained:

$$I = I_0 + I_1 + I_2 + I_3$$
$$P = \bar{I}_3 I_A$$
$$Q = \bar{I}_3 \bar{I}_2 I_A = P\bar{I}_2$$
$$R = \bar{I}_3 \bar{I}_2 \bar{I}_1 I_A = Q\bar{I}_1$$

$$D_{E3} = I_3 I_A$$
$$D_{E2} = \bar{I}_3 I_2 I_A = P I_2$$
$$D_{E1} = \bar{I}_3 \bar{I}_2 I_1 I_A = Q I_1$$
$$D_{E0} = \bar{I}_3 \bar{I}_2 \bar{I}_1 I_0 I_A = R I_0$$

The implementation of these equations for one peripheral, in this case device 3, is shown in figure 4.13(c).

Problems

4.1 Design combinational logic circuits whose inputs are NBCD code and whose outputs will detect:

(1) input digits divisible by 3;
(2) numbers that are greater than or equal to 7;
(3) numbers that are less than 4.
Implement each of the circuits designed with NAND gates.

4.2 Develop combinational logic circuits which can be used to compare the magnitude of two binary digits A and B. The circuits designed should be able to indicate whether (i) $A > B$; (ii) $A = B$ and (iii) $A < B$.
Extend the design so that two four-digit numbers $A_3A_2A_1A_0$ and $B_3B_2B_1B_0$ can be compared in the same way.

4.3 Develop a combinational logic circuit which will convert four-bit binary numbers into their corresponding 2's complement form.

4.4 Design an NBCD to seven-segment decoder which is able to accept decimal information expressed in NBCD and generates outputs which select segments in the seven-segment indicator for displaying the appropriate decimal digit. The disposition of

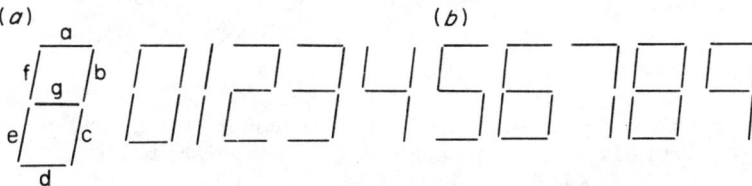

Figure P4.4. (a) Disposition of segments. (b) Segmental representation of decimal digits

the seven segments is shown in figure P4.4(a) and the segmental representation of each decimal digit is shown in figure P4.4(b).

4.5 Design a combinational logic circuit that will generate the square of all the combinations of a three-bit binary number $A_2A_1A_0$.

4.6 Design a combinational logic circuit that will generate the 9's complement of each of the decimal digits, where the decimal digits are expressed in the NBCD code.

4.7 A block of flats has four floors, and it is to be arranged that the lights for the stair well can be switched on or off at any floor level when the switch on that level is operated. Design a logic circuit to satisfy the above specification.

Combinational logic design

4.8 A three-bit binary code is to be transmitted over a land line to a receiver, and to protect the code against errors an extra bit called a parity bit is added to each code combination so that there are always an odd number of 1's in the transmitted combination.

Develop a logic circuit which will generate the parity bit, when required, at the transmitting end of the line, and additionally, develop a second logic circuit which when situated at the receiving end of the line can be used for checking the parity of each code combination.

4.9 A majority logic circuit is to be developed which will indicate when the majority of digits in a five-bit number are 1. Develop the circuit and implement your design using NAND gates.

4.10 Develop the Gray codes corresponding to
(a) the 2–4–2–1 code;
(b) the XS3 code.
N.b. the XS3 code is obtained from the NBCD code by adding 0011 to each NBCD code combination.

4.11 Develop a logic circuit for converting the XS3 code into its corresponding Gray code.

4.12 Obtain the decimal equivalents of the numbers (a) 0,1011, and (b) 1,0110101, if it is assumed they are expressed in
(i) sign and magnitude form;
(ii) 2's complement form;
(iii) 1's complement form.

4.13 Design a binary multiplier that multiplies a four-bit number $B = B_3B_2B_1B_0$ by a three-bit number $A = A_2A_1A_0$. The circuit is to be implemented using AND gates and full adders.

4.14 Positive and negative numbers are represented by eight-bit words in a digital machine which performs its arithmetic operations using the 2's complement system. How would the numbers +21 and −29 appear in the machine registers?

5

Single-bit memory elements

5.1 Introduction

A digital logic circuit is usually made up of combinational elements such as NAND and NOR gates and memory elements which might be single bit memory elements such as discrete flip-flops or, alternatively, an array of flip-flops such as might be found in a shift register.

With the introduction of memory elements as components in digital circuits, an additional variable, time, has been introduced and must be taken into account when dealing with digital circuit problems. In effect, logic operations can now be performed sequentially, information being stored in a memory element and being released at some particular instant so that it can take part in a controlled combinational operation. Circuits operating in this way are called *sequential circuits*. Some sequential circuits are controlled by a repetitive clock signal, in which case the circuit is called a *synchronous* or, alternatively, a *clock-driven circuit*. Other sequential circuits are controlled by random events in which case they are called *asynchronous* or *event-driven circuits*.

The basic characteristic of any flip-flop is that it has two stable states which can be represented by logical '0' or logical '1' respectively. There are a number of flip-flops in common usage in digital circuits. They are called T, SR, JK and D type flip-flop. This chapter is concerned with the logical behaviour of these various types of flip-flop.

5.2 The T flip-flop

This flip-flop is conventionally represented by the diagram shown in figure 5.1(a). The device has one input T and complementary outputs Q and $\bar{Q}$.

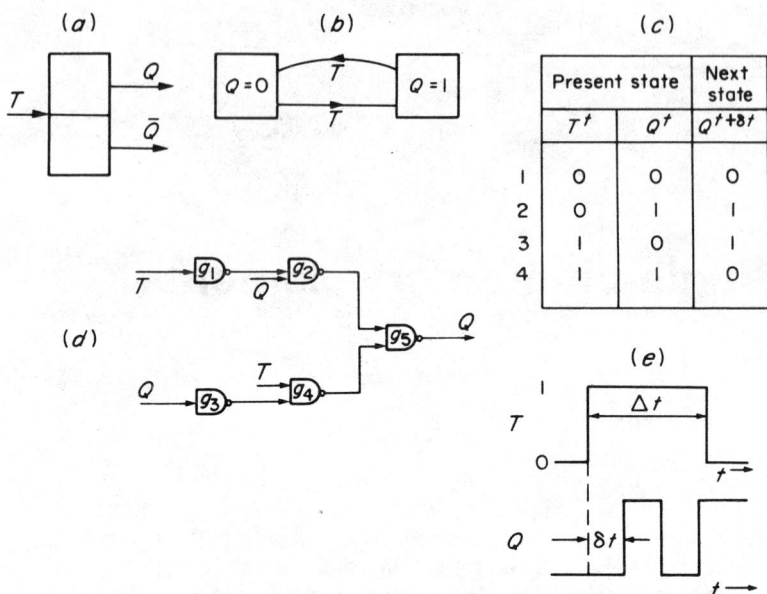

Figure 5.1. The T flip-flop. (a) Conventional diagram. (b) External state diagram. (c) State table. (d) Implementation. (e) The race condition

One way of representing the behaviour of a T flip-flop is by means of an external state diagram such as the one shown in figure 5.1(b). The external states of the flip-flop are represented by squares in this diagram. There are two possible states of the flip-flop as far as the outside world is concerned, $Q = 0$ and $Q = 1$, and these external states are allocated to the two squares indicated in figure 5.1(b). Transitions between states are described by full lines and an arrowhead on the line indicates the direction of the transition. The signal causing the transition is placed by the arrowhead. Examination of figure 5.1(b) shows that if the flip-flop is in the state $Q = 0$ and it receives a signal T then it will make a transition to the alternative state of $Q = 1$. Alternatively, if in the state $Q = 1$ and the flip-flop receives a T signal, it will make a transition to $Q = 0$. The output is said to 'toggle', i.e. it changes state every time it receives a signal at the T input.

The concept of the state diagram is of fundamental importance to the digital designer. If a digital problem can be expressed in terms of a state diagram, or a series of state diagrams, then there is usually no difficulty in transforming such diagrams into a hardware implementation. There are, however, two types of state diagram. The one shown in figure 5.1(b) is an example of an external state diagram. In this case the behaviour of the flip-flop has been described in terms of the two states it presents to the outside world and there has been no reference to the internal states which the circuit might take up as a transition is made from $Q = 0$ to $Q = 1$ and vice versa. It is, in fact, possible to describe the behaviour of a T flip-flop in terms of its internal states, such a description is referred to as an internal state diagram. The internal state diagram is particularly useful to the digital designer and the development and use of internal state diagrams will be dealt with more fully in a later chapter.

A second important method of describing the behaviour of a T flip-flop is by means of a *state table*. The state table for a T flip-flop is shown in figure 5.1(c), where the table is divided into two halves, the left-hand half which is concerned with the present state of the input T^t and the present state of the flip-flop Q^t, and the right-hand half which is concerned with the next state of the flip-flop $Q^{t+\delta t}$. The next state of the flip-flop depends upon the present state of the flip-flop and the present state of the input. Furthermore, a change of state at the output will occur at some time δt after the change of the input signal T at time t.

The state table has been constructed in the same way as a truth table. The first two columns consist of all the possible combinations of the present state of T and the present state of Q and the first row of the table should be interpreted as follows:

If the value of T has not changed at time t and remains as $T = 0$, then there will be no change in the state of the flip-flop and Q at $t + \delta t = 0$.

Similarly, the interpretation of the entries in row 3 would be as follows:

If the value of T has changed from 0 to 1 at time t then the state of the flip-flop which was $Q = 0$ at time t will change to $Q = 1$ at time $t + \delta t$.

An equation describing the behaviour of the flip-flop and sometimes referred to as the characteristic equation of the flip-flop, may be extracted from the state table. The equation is obtained by writing down the present state combinations which will give a value of $Q^{t+\delta t} = 1$. Hence

$$Q^{t+\delta t} = (\overline{T}Q + T\overline{Q})^t$$
$$= (T \oplus Q)^t$$

This equation is a two-level sum-of-products and can be implemented with NAND gates as shown in figure 5.1(d).

There are two important points that should be observed at this juncture. First, the equation taken from the state table is a Boolean equation but with a difference from the combinational equations that have been seen hitherto. Time has been introduced into the equation and the value of Q on the right-hand side of the equation may well be different from the value of Q on the left-hand side of the equation simply because these two values of Q are being observed at different times. The second important point is that the circuit shown in figure 5.1(d) appears to be a conventional combinational circuit. However, inspection of the circuit shows that its output is fed back to the input and hence a feedback loop exists.

Unfortunately, the circuit shown in figure 5.1(d) will not operate satisfactorily for the following reason. If

$$Q^t = 0 \text{ and } T^t = 1, \quad \text{then } Q^{t+\delta t} = 0 \cdot 0 + 1 \cdot 1 = 1$$

Hence there is an $0 \to 1$ transition of Q at time $t + \delta t$. If it is assumed that T has not yet changed back to 0 at $t + \delta t$, then $Q^{t+\delta t} = 1$ and $T^{t+\delta t} = 1$. Hence

$$Q^{t+2\delta t} = 0 \cdot 1 + 1 \cdot 0 = 0$$

and there is a $1 \to 0$ transition of Q at time $t + 2\delta t$.

The output Q will continue toggling between 0 and 1 in this manner until such time as the T signal terminates and becomes 0 once again as shown in figure 5.1(e). Furthermore, the final value of Q at the termination of the T signal will depend upon the time duration of T and also the circuit delay δt, and hence the output state Q is said to be indeterminate.

When a $1 \to 0$ transition takes place in T then there is no change in the output Q of the flip-flop. That this is so can be seen with the aid of the characteristic equation. For if $Q^t = 0$ and T makes a $1 \to 0$ transition at time t, then

$$Q^{t+\delta t} = 1 \cdot 0 + 0 \cdot 0 = 0$$

or, alternatively, if $Q^t = 1$ and T makes a transition from $1 \to 0$ at time t, then

$$Q^{t+\delta t} = 1 \cdot 1 + 0 \cdot 0 = 1$$

The cause of the instability in the T flip-flop shown in figure 5.1(d) is due to a race condition. If the new value of Q at $t + \delta t$ is fed back to the inputs of gates g_2 and g_3 before the termination of the T signal then a further change in Q will take place. The situation is described by the time diagram shown in figure 5.1(e). Inspection of the diagram shows that if $\Delta t \geqslant \delta t$ then instability will occur. A more satisfactory way of synthesising a T flip-flop will be described later in this chapter.

5.3 The SR flip-flop

The SR flip-flop is shown symbolically in figure 5.2(a), the set and reset inputs being labelled S and R respectively, and the complementary outputs are labelled Q and $\overline{Q}$.

The state table for the flip-flop is shown in figure 5.2(b). In the first three columns of this table all combinations of the present states of S, R and Q are shown, i.e. their states at time t. The fourth column contains a tabulation of the next state of the flip-flop, i.e. its state at time $t + \delta t$.

Examination of this table shows that a change of flip-flop state occurs in rows 4 and 5 only. In row 4 the flip-flop is being reset or turned off, i.e. its state is changing from 1 to 0 as a consequence of the application of a reset input $R = 1$. In row 5 the flip-flop is being set or turned on, i.e. its state is changing from 0 to 1 as a result of the application of a set input $S = 1$. For rows 1 and 2, $S = 0$ and $R = 0$, and consequently there is no change in the state of the flip-flop and the entries in the last column are 0 and 1 respectively. On row 3, $R = 1$ and this signal in normal circumstances would turn the flip-flop off, however the flip-flop is already turned off since $Q^t = 0$ and consequently the signal $R = 1$ leaves the flip-flop state unchanged. Similarly, on row 6, $S = 1$, and this signal would normally turn the flip-flop on, but $Q^t = 1$, i.e. the flip-flop is already turned on and consequently there will be no change in the state of the flip-flop. Finally, with this type of flip-flop it is forbidden for S and R to be logical '1' simultaneously, this restriction being expressed algebraically by $SR = 0$.

From the table the turn-on condition is given by

$$\text{turn-on} = S\,\overline{R}\,\overline{Q}$$

and the turn-off condition is given by

$$\text{turn-off} = \overline{S}\,R\,Q$$

With the aid of these two conditions the external state diagram can be constructed as shown in figure 5.2(c). The diagram shows that a

Single-bit memory elements

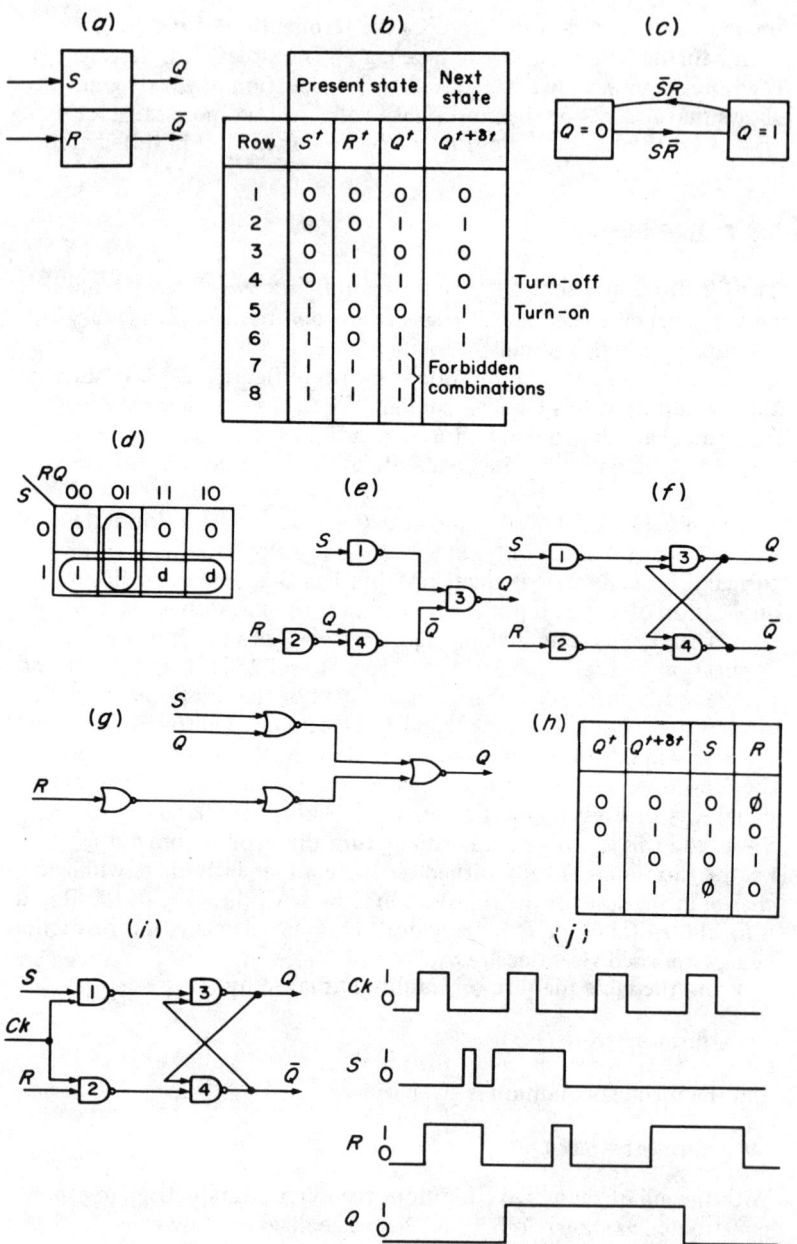

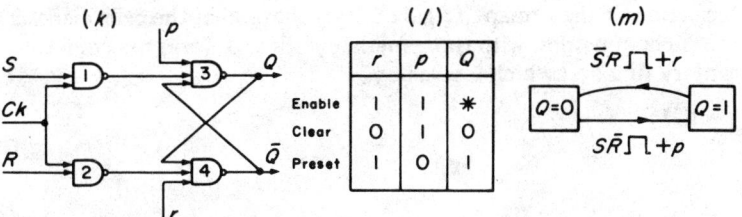

Figure 5.2. The SR flip-flop. (a) Symbolic representation. (b) State table. (c) External state diagram. (d) K-map plot. (e) Implementation of the NAND sequential equation. (f) Conventional representation of the SR flip-flop. (g) Implementation of the NOR sequential equation. (h) Steering table for the SR flip-flop. (i) The clocked SR flip-flop. (j) Timing diagram. (k) The preset and clear controls. (l) Table describing the behaviour of the preset and clear controls. (m) External state diagram including preset and clear signals.

transition is made from $Q = 0$ to $Q = 1$ if $S\bar{R} = 1$, and the reverse transition is made if $\bar{S}R = 1$.

The characteristic equation of the flip-flop, or as it is sometimes called, the NAND sequential equation, is obtained by taking the logical sum of the combinations in the truth table for which $Q^{t+\delta t} = 1$, in conjunction with the term $SR = 0$. This does not affect the value of $Q^{t+\delta t}$ but leads to a simpler equation for it. Hence

$$Q^{t+\delta t} = (\bar{S}\bar{R}Q + S\bar{R}\bar{Q} + S\bar{R}Q + SR)^t$$

and the plot of this function is shown in figure 5.2(d) in which the term SR has been plotted as a 'can't happen' condition, this being indicated by a 'd' in the appropriate cells. Simplifying the expression for $Q^{t+\delta t}$ leads to the NAND sequential equation

$$Q^{t+\delta t} = (S + \bar{R}Q)^t,$$

where S is termed the turn-on set of Q, and R is termed the turn-off set of Q.

The implementation of this equation using NAND gates is shown in figure 5.2(e) and it appears in its conventional form in figure 5.2(f). The equation of the complementary output is

$$\bar{Q}^{t+\delta t} = \overline{(S + \bar{R}Q)^t} = \{\bar{S}(R + \bar{Q})\}^t$$

Now the output of gate 4

$$f^{t+\delta t} = \bar{Q}^{t+\delta t} + R^t$$

and substituting for $\bar{Q}^{t+\delta t}$ gives

$$f^{t+\delta t} = (\bar{S}\bar{Q} + \bar{S}R + R)^t = (\bar{S}\bar{Q} + R)^t$$

An inspection of the K-map in figure 5.2(*d*) shows that the cells marked with a 0 in conjunction with those marked with a 'd' form the complementary function which is given by

$$\overline{Q}^{t+\delta t} = (S\overline{Q} + R)^t.$$

Hence

$$f^{t+\delta t} = \overline{Q}^{t+\delta t}$$

and the output of gate 4 is the complementary output.

A second form of the sequential equation is obtained by excluding the product SR from the equation for $Q^{t+\delta t}$ so that

$$Q^{t+\delta t} = (\overline{S}\overline{R}Q + S\overline{R}\overline{Q} + S\overline{R}Q)^t.$$

These three terms are represented by the three 1's plotted in the K-map shown in figure 5.2(*d*) and, when simplified, the above equation reduces to

$$Q^{t+\delta t} = (S\overline{R} + Q\overline{R})^t$$
$$= \{(S+Q)\overline{R}\}^t$$

This is a two-level product of sums and is implemented using NOR gates as shown in figure 5.2(*g*).

The behaviour of the *SR* flip-flop can be described in a slightly different way by means of the steering table shown in figure 5.2(*h*). This table shows every possible output transition which can occur in the first two columns, whilst the last two columns give the values of *S* and *R* which will produce these transitions. For example, in the first row the 0 → 0 transition will occur providing $S = 0$ and $R = 0$ or 1. Since *R* can be either 0 or 1 this is indicated in the *R* column by the symbol ∅. For the second row the 0 → 1 transition is generated if $S = 1$. Since *S* and *R* cannot be simultaneously 1, it follows that the entry for *R* will be 0. The entries for the other two rows can be determined in a similar fashion.

By means of the simple modification shown in figure 5.2(*i*) the *SR* flip-flop can be clocked. An examination of this diagram shows that if $Ck = 0$ the outputs of g_1 and g_2 will always be logical '1', irrespective of the present values of *S* and *R* or of any changes which may occur in these two inputs. The flip-flop can only change its output during a

Single-bit memory elements

clock transition and assuming zero gate delay, the output Q will change state on the leading edge of a clock pulse when Ck is changing from 0 to 1 as illustrated in figure 5.2(j).

Besides the S, R and Ck inputs an SR flip-flop may have one or two additional controls which allow it to assume one of its two states irrespective of whether $Ck = 0$ or $Ck = 1$. These controls are usually called 'clear' and 'preset'. Most commercially available flip-flops are provided with a clear control whereas the preset control is not nearly as common. The operation of these controls is described by the table shown in figure 5.2(l) and it should be observed that in the circuit of figure 5.2(k) these signals are active when low.

With both controls at logical '1' the flip-flop is enabled and operates in the normal way. If $r = 0$ and $p = 1$ the output $\bar{Q}$ of g_4 in figure 5.2(k) becomes $\bar{Q} = 1$. Hence $Q = 0$ and the flip-flop is unconditionally reset. If $r = 1$ and $p = 0$ the output Q of g_3 becomes $Q = 1$ and the flip-flop is now preset. Logical '0' signals on the p and r lines will automatically override signals on the S and R lines. The inclusion of these two controls will lead to a modified external state diagram as illustrated in figure 5.2(m).

It should be observed by the reader that if a preset facility is required when a p input is not provided by the manufacturer it is possible to interchange the Q and $\bar{Q}$ output terminals and also the S and R input terminals. The clear terminal can then be used as a preset control.

5.4 The JK flip-flop

The symbolic representation of the JK flip-flop is shown in figure 5.3(a) and the state table describing its logical operation in figure 5.3(b). The operation of this flip-flop differs in one respect from that of the SR flip-flop in that it is allowable for J and K to be simultaneously equal to logical '1'. For example, if $J = K = 1$, the flip-flop 'toggles', that is, in row 7 the flip-flop changes state from 0 to 1 whilst in row 8 the converse action takes place. In rows 4 and 5 normal reset and set operations take place as described for the SR flip-flop in § 5.3.

An examination of the state table shows that the flip-flop is turned on in rows 5 and 7, whilst it is turned off in rows 4 and 8.

The turn-on set of Q: $S = J\bar{K}\bar{Q} + JK\bar{Q}$

$$= J\bar{Q}$$

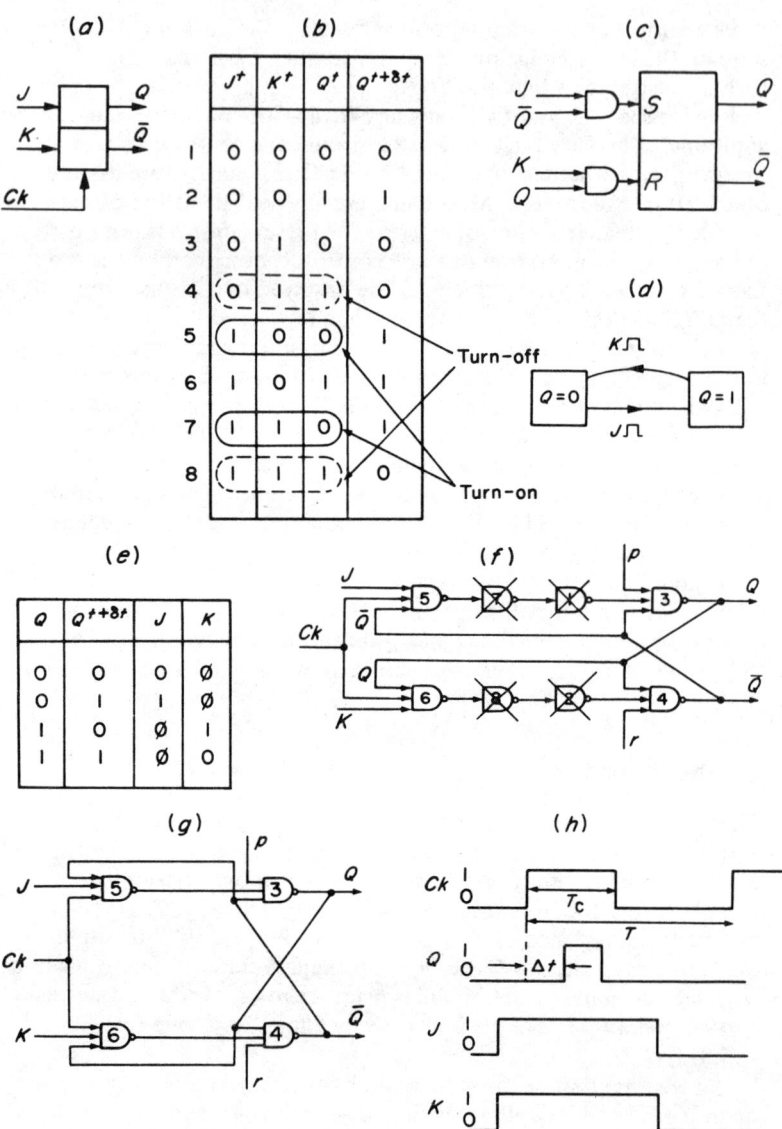

Figure 5.3. The JK flip-flop (a) Symbolic representation. (b) State table. (c) Representation of a JK flip-flop by an SR flip-flop preceded by two AND gates. (d) External state diagram. (e) Steering table. (f) A NAND implementation of the JK flip-flop. (g) Reduced form of the JK flip-flop. (h) The race in the JK flip-flop.

Single-bit memory elements

and the turn-off set of Q: $R = \overline{J}KQ + JKQ$

$\qquad\qquad\qquad\qquad\quad = KQ$

These two equations indicate that a *JK* flip-flop may be regarded as an *SR* flip-flop preceded by two AND gates which implement the turn-on and turn-off functions respectively, as illustrated in figure 5.3(*c*).

The state diagram describing the terminal behaviour of the flip-flop is shown in figure 5.3(*d*). Assuming a clocked flip-flop in the state Q = 0 with J = 1 and *Ck* changing from 0 to 1, it makes a transition to the state Q = 1. Similarly, if in the state Q = 1, with K = 1 and *Ck* changing from 0 to 1, it makes a transition to Q = 0.

A steering table for the *JK* flip-flop is shown in figure 5.3(*e*). Comparing the steering tables of the *SR* and *JK* flip-flops in figures 5.2(*h*) and 5.3(*e*) respectively, it will be observed that the *JK* flip-flop has more $\emptyset$ or 'don't care' input conditions and consequently this type of flip-flop leads to simpler logic when used in the design of clock-driven circuits.

A *JK* flip-flop can be implemented by connecting the output of the two AND gates in figure 5.3(*c*) to the S and R inputs of the *SR* flip-flop shown in figure 5.2(*i*). The Q and $\overline{Q}$ outputs of this flip-flop and its clock connections are fed to the inputs of the two AND gates in conjunction with the J and K lines as shown in figure 5.3(*f*). Notice that the AND gates are formed from two pairs of NAND gates in cascade, namely g_5 and g_7, and g_6 and g_8. Clearly, gates g_7 and g_1 and gates g_8 and g_2 give a double inversion. These gates are therefore redundant and can be omitted from the circuit, thus reducing the *JK* flip-flop to an assembly of four gates only, as shown in figure 5.3(*g*).

Comparison of figures 5.1(*d*) and 5.3(*g*) shows that the *JK* flip-flop exhibits the same instability as the *T* flip-flop since in both cases the outputs Q and $\overline{Q}$ are fed back to the input. The instability occurs in the *JK* flip-flop when J and K are simultaneously set to logical '1' and the flip-flop is clocked. Examination of the state table shown in figure 5.3 shows that the *JK* flip-flop behaves like a *T* flip-flop when this condition exists.

If the Q output changes before the termination of the clock pulse then the input conditions to gates g_5 and g_6 also change and this will lead to a further change in the output of Q. As a consequence, Q is indeterminate at the termination of the clock pulse. A race exists in the circuit between the initial change in Q and the termination of the clock pulse, as illustrated in figure 5.3(*h*). The condition for the race is that $\Delta t < T_c$. Since in practice this condition usually exists, it is clear that the circuit in figure 5.3(g) will not operate satisfactorily and for this reason the *master/slave connection* has been developed.

A block diagram of a master/slave *JK* flip-flop is shown in figure 5.4. It consists of two separate flip-flops, the master and the slave. The master is clocked in the normal way whilst the clock signal is inverted before being applied to the slave. Changes in the output of the master will take place on the rising edge of the clock pulse and these changes are transmitted to the input of the slave. However, no change can occur at the output of the slave until the rising edge of the inverted clock pulse which is, of course, the trailing edge of the clock pulse. Consequently, changes in Q and $\bar{Q}$ which are fed back to the

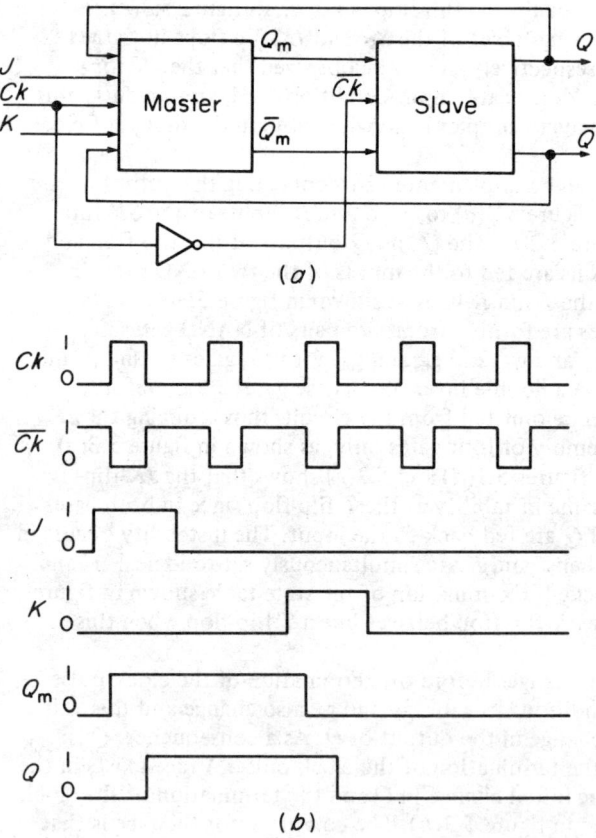

Figure 5.4. (a) The master/slave JK flip-flop. (b) Timing diagram.

input do not take place until the trailing edge of the clock pulse and the race condition previously described has now been averted. The time diagram shown in figure 5.4(b) describes the action of the flip-flop.

It is now a relatively simple matter to develop a T-type flip-flop from a master/slave JK flip-flop. All that is required is that the J and K inputs should be permanently connected to logic '1' as illustrated in figure 5.5(a), and that the flip-flop should have the T signal connected to the clock input.

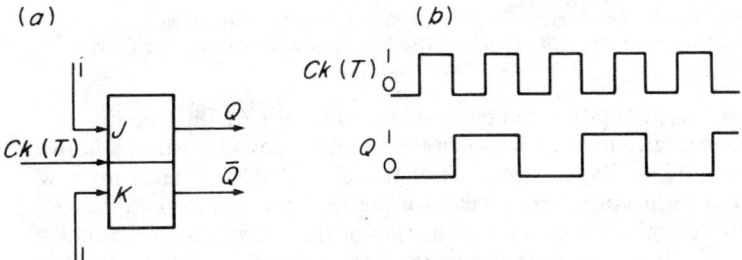

Figure 5.5. (a) The T-type flip-flop. (b) Timing diagram

On the trailing edge of every T pulse the flip-flop will change state as illustrated in figure 5.5(b). The flip-flop is behaving now in a toggling mode in the sense that the Q output is alternately taking up the 0 and 1 states.

This circuit is the basis of all counting circuits; it is in fact, a scale-of-two counter and an examination of the Q waveform shows that its frequency is half that of the T signal and hence it is sometimes called a divide-by-two circuit.

5.5 The D flip-flop

The conventional schematic diagram for this kind of flip-flop is shown in figure 5.6(a). It has one input, namely on the D line, and the output Q at time $t + \delta t$ always takes up the value of the input D at time t. This behaviour is described by the state table shown in figure 5.6(b).

The characteristic equation of the flip-flop is obtained from the state table by taking the logical sum of those combinations of D^t and Q^t which give a value of Q at $t + \delta t$ equal to logical '1'. Hence

$$Q^{t+\delta t} = (D\bar{Q} + DQ)^t$$
$$= D^t$$

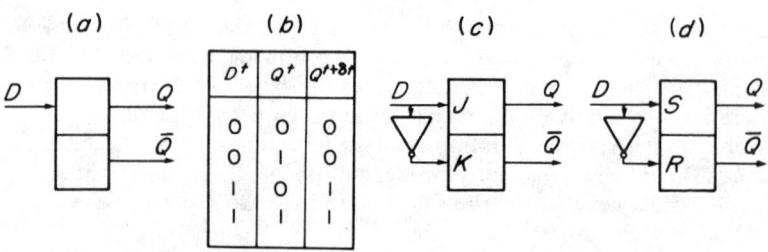

Figure 5.6. (a) The D-type flip-flop. (b) State table. (c) The JK flip-flop connected as a D-type flip-flop. (d) The SR flip-flop connected as a D-type flip-flop

A D-type flip-flop can be constructed from a JK flip-flop by connecting an inverter between the J and K lines as illustrated in figure 5.6(c). By making this connection it is clear that $J = \bar{K}$ and only rows 3–6 inclusive of the state table for the JK flip-flop shown in figure 5.3(b) are now valid. An examination of these rows shows that if $J = D = \bar{K}$ then they reduce to give the state table of the D-type flip-flop shown in figure 5.6(b).

In a similar manner the state table for the SR flip-flop can be reduced to that of a D-type flip-flop by making $S = D = \bar{R}$. Hence a D-type flip-flop can also be constructed from an SR flip-flop by connecting an inverter between the S and R input lines as illustrated in figure 5.6(d).

5.6 The latching action of a flip-flop

The SR flip-flop shown in figure 5.2(i) is sometimes called a latch because as long as the clock is high, the output Q will follow any changes in the input conditions that may take place on the S and R lines provided that the condition $SR = 0$ is always satisfied. When the clock goes low the state of the flip-flop is latched and cannot change until such time as the clock goes high again. This action is illustrated in the timing diagram of figure 5.7(a).

The SR latch can be converted into a D latch by combining the S and R inputs into a single input as shown in figure 5.7(b). When the clock is high in this circuit the output follows the D input, but when the clock goes low the logical value of D is latched at the Q output. This action is described by the timing diagram shown in figure 5.7(c).

A combination of two D-type latches can be used to produce a flip-flop which triggers on the leading edge of the clock pulse. The arrangement shown in figure 5.8(a) consists of two D-type latches, a master

Single-bit memory elements

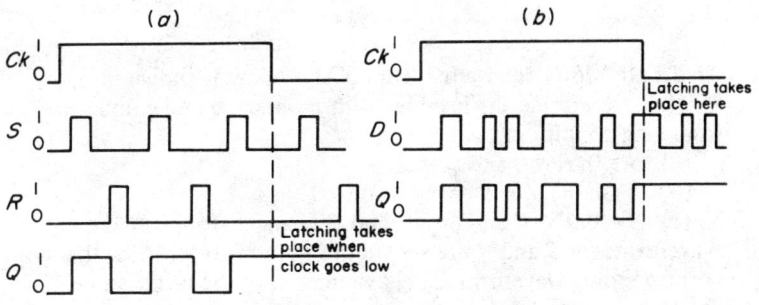

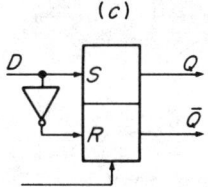

Figure 5.7. (a) Timing diagram illustrating the latching action of an SR flip-flop. (b) The D-type latch. (c) Timing diagrams for the D-type latch

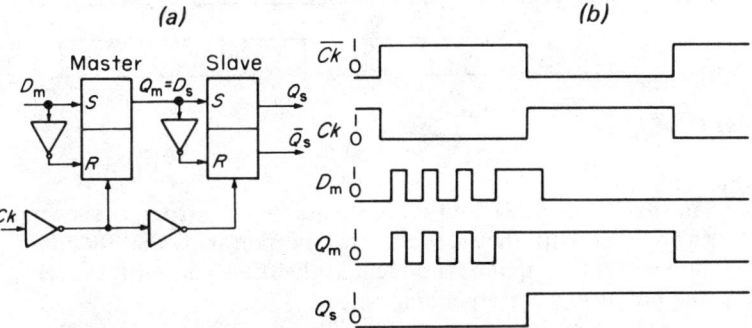

Figure 5.8. (a) A D-type flip-flop triggering on the leading edge of the clock pulse. (b) Timing diagram for the edge-triggered D-type flip-flop

followed by a slave, the output of the master Q_m being connected to the input of the slave D_s. When the clock signal is low the output of the master Q_m follows its input D_m. However, on the rising edge of the clock signal the input to the master is latched at its output and is then transferred to the output of the slave.

Problems

5.1 An *SR* flip-flop constructed from NAND gates is shown in figure P5.1(a). Determine the logic levels at points a, b and c under the following conditions:
    (a) $S = 0, R = 0$ and $Q = 0$.
    (b) As in (a), but S changes from $0 \rightarrow 1$.
    (c) $S = 0, R = 0$ and $Q = 1$, and R changes from $0 \rightarrow 1$.
Waveforms for S and R are shown in figure P5.1(b). Draw the corresponding waveform for Q assuming that the initial value of $Q = 0$.

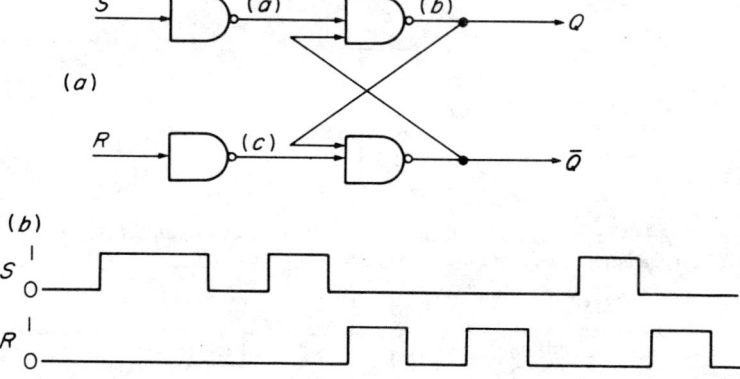

Figure P5.1

5.2 An edge-triggered *D* flip-flop combines two *D* latches, as shown in figure P5.2. With the aid of a timing diagram show that the flip-flop senses the input data present at the rising edge of the clock and produces a corresponding output.

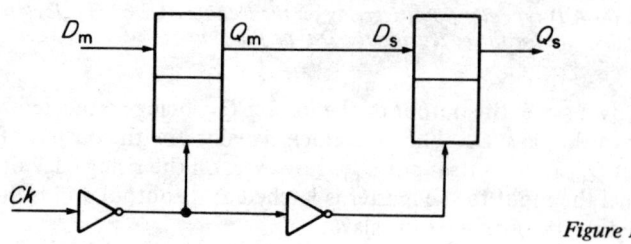

Figure P5.2

Single-bit memory elements

5.3 A master/slave JK flip-flop is shown in figure P5.3. Assuming that the initial condition of the flip-flop is $J = K = Q_m = Q_s = 0$, trace the logic levels through the diagram for the following changes. (N.B. changes in J and K take place in the time intervals between clock pulses.)

(i) $J, 0 \to 1, K, 0 \to 0$ Clock pulse 1 applied
(ii) $J, 1 \to 1, K, 0 \to 1$ Clock pulse 2 applied
(iii) $J, 1 \to 0, K, 1 \to 0$ Clock pulse 3 applied
(iv) $J, 0 \to 1, K, 0 \to 0$ Clock pulse 4 applied

Draw a timing diagram displaying the J, K, Q_m and Q_s waveforms for the period of four clock pulses.

Assuming the same initial conditions, determine the final value of Q_s as the inputs are changed in the following order:

(v) $Ck, 0 \to 1, J, 0 \to 1, Ck, 1 \to 0$
(vi) $J, 0 \to 1, Ck, 0 \to 1, K, 0 \to 1, J, 1 \to 0, Ck, 1 \to 0$

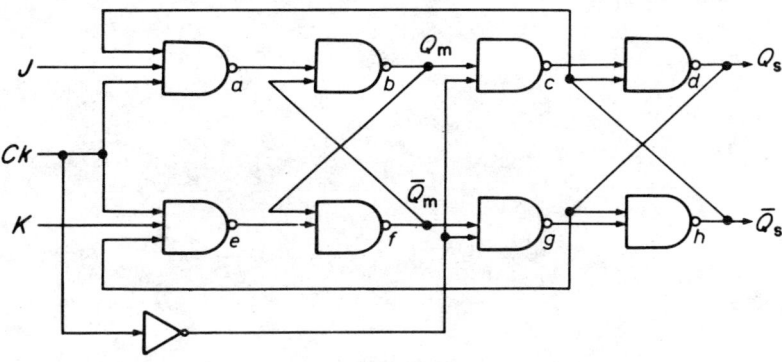

Figure P5.3

5.4 With the aid of external logic, show that a D-type flip-flop can be converted to a JK flip-flop. Construct a timing diagram for the JK flip-flop and show that the circuit produces an output which depends only on the input data present on the rising edge of the clock pulse.

5.5 A JK flip-flop is modified, as shown in figure P5.5 to form a $J'K$ flip-flop. Draw up the state table for this flip-flop and derive its characteristic equation.

Single-bit memory elements

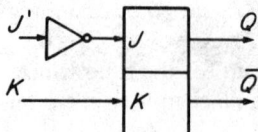

Figure P5.5

The J' and K inputs of the $J'K$ flip-flop are joined together to form another flip-flop. Develop the state table for this flip-flop and derive its characteristic equation.

5.6 Draw the external state diagram for the flip-flop whose characteristic equations are
(a) $Q^{t+\delta t} = (X \oplus Y \oplus Q)^t$
(b) $Q^{t+\delta t} = (X \odot Y \odot Q)^t$

6
Counters

6.1 Introduction

A counting circuit is the simplest form of sequential circuit obtainable. In practice, counting circuits, like all other sequential circuits, are composed of memory elements such as flip-flops and combinational elements such as electronic gates.

All sequential circuits are of two types: (i) synchronous (clock-driven) circuits, and (ii) asynchronous (event-driven) circuits. In synchronous circuits, changes in circuit state are synchronised to the clock pulses, whereas in event-driven circuits, changes in circuit state are governed by events, such as the occurrence of a system fault.

Counting circuits can be in either of the two categories described above. When they are clock-driven they are called synchronous counters, and when they are event-driven they are called either asynchronous or ripple-through counters. In the case of a synchronous counter, the circuit counts clock pulses and stores the number counted in the memory elements, while a ripple-through counter can be used to count irregularly occurring events such as customers entering a bank, and again the number counted is stored in the memory elements.

Counters are fundamental components of digital systems, and can be used in timing, control or sequencing applications. They appear in computer and communications circuits; they can be used for frequency division; in some cases they count in pure binary, in other cases there may be a non-binary count, for example a Gray code counter, or a BCD decade counter.

6.2 Scale-of-two up-counter

The simplest possible counter is the scale-of-two counter which has only two states, namely 0 and 1. Since the ouptut of a flip-flop can

102 Counters

only exist in one of two states it is clear that this counter can be implemented with a single flip-flop.

One design technique is to draw up a state table in which the first column represents the present state of the counter whilst the second column gives the next state of the counter as shown in figure 6.1(a). Examining the table row by row reveals the flip-flop transitions that have to be made as the counter goes from its present state to its next state. Assuming that the designer chooses a JK flip-flop to implement the counter then the J and K inputs required to produce the observed

(a)

Present state A^t	Next state $A^{t+\delta t}$	J_A	K_A
0	1	1	∅
1	0	∅	1

(b)

Q^t	$Q^{t+\delta t}$	J	K
0	0	0	∅
0	1	1	∅
1	0	∅	1
1	1	∅	0

Figure 6.1. Scale-of-two counter. (a) State table. (b) Steering table for a JK flip-flop. (c) Implementation. (d) State diagram

transitions can be obtained from the JK flip-flop steering table shown in figure 6.1(b). Since the entries in the J and K columns of the state table are all either ∅ or 1 it follows that $J_A = K_A = 1$.

The counter is shown implemented in figure 6.1(c) and the state diagram for the counter is illustrated in figure 6.1(d). In this case the state diagram is both the internal and the external state diagram of the counter since $A = 0$ and $A = 1$ represent the internal states of the counter as well as being the externally displayed count.

6.3 Scale-of-four up-counter

A scale-of-four counter has four states and requires two flip-flops for its implementation. The state table for the counter is shown in figure 6.2(a) and the entries in the J and K columns for both flip-flops are obtained by observing the transitions that are being made, row by row, of the state table and then finding the J and K inputs that give these transitions by reference to the steering table for the JK flip-flop.

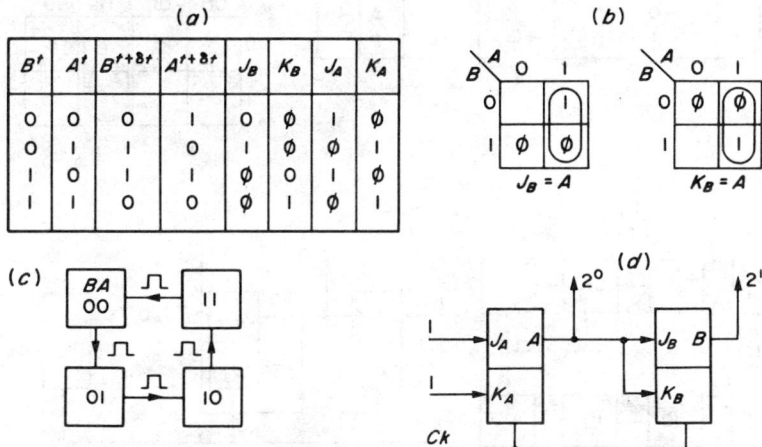

Figure 6.2. Scale-of-four counter. (a) State table. (b) K-map. (c) State diagram. (d) Implementation

Examination of the J_A and K_A columns of the state table show that all the entries are either 0 or 1, hence $J_A = K_A = 1$. To determine J_B and K_B the K-maps are plotted in figure 6.2(b) and after simplification it is found that $J_B = K_B = A$. The implementation of the counter is shown in figure 6.2(c) and the state diagram is illustrated in figure 6.2(d).

6.4 Scale-of-eight up-counter

The design of the A and B stages of this counter are identical to the design of the A and B stages of the scale-of-four up-counter. Adding a further stage to the counter in no way alters the design of the earlier stages. Hence $J_A = K_A = 1$ and $J_B = K_B = A$.

The state table for the counter is illustrated in figure 6.3(a) and the

(a)

C^t	B^t	A^t	$C^{t+\delta t}$	$B^{t+\delta t}$	$A^{t+\delta t}$	J_C	K_C
0	0	0	0	0	1	0	ϕ
0	0	1	0	1	0	0	ϕ
0	1	0	0	1	1	0	ϕ
0	1	1	1	0	0	1	ϕ
1	0	0	1	0	1	ϕ	0
1	0	1	1	1	0	ϕ	0
1	1	0	1	1	1	ϕ	0
1	1	1	0	0	0	ϕ	1

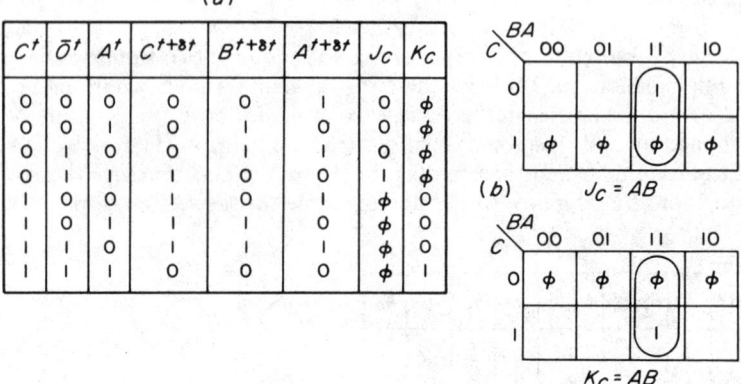

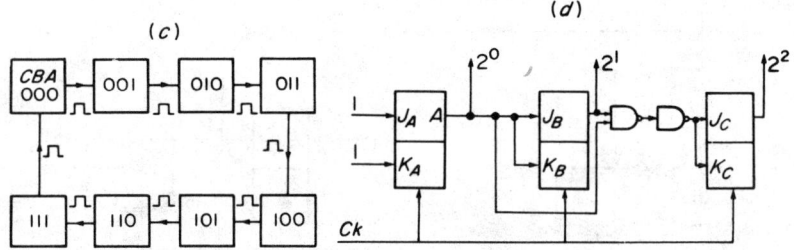

Figure 6.3. Scale-of-eight counter. (a) State table. (b) K-maps. (c) Implementation. (d) State diagram

K-maps for the J and K inputs of the C flip-flop are shown in figure 6.3(b). After simplification it is found that $J_C = K_C = AB$. Implementation of the counter is shown in figure 6.3(c) and the state diagram is given in figure 6.3(d).

6.5 Scale-of-2^N up-counter

The results for the scale-of-two, scale-of-four and scale-of-eight up-counters are tabulated below.

$$J_A = 1 \quad J_B = A \quad J_C = AB = J_B B$$

$$K_A = 1 \quad K_B = A \quad K_C = AB = K_B B$$

and by observation of these equations, it follows that the equations for stages $D, E, \ldots N$ of a scale 2^N up-counter are

$$J_D = ABC = J_C C \quad J_E = ABCD = J_D D \quad J_N = ABC\ldots(N-1) = J_{N-1}(N-1)$$
$$K_D = ABC = K_C C \quad K_E = ABCD = K_D D \quad K_N = ABC\ldots(N-1) = K_{N-1}(N-1)$$

6.6 Series and parallel connection of counters

There are two ways of connecting the inputs to successive flip-flops and these are illustrated in figures 6.4(a) and (b). In the first method the gates providing the J and K inputs to successive flip-flops in the counter are all fed in parallel. As the number of stages in the counter

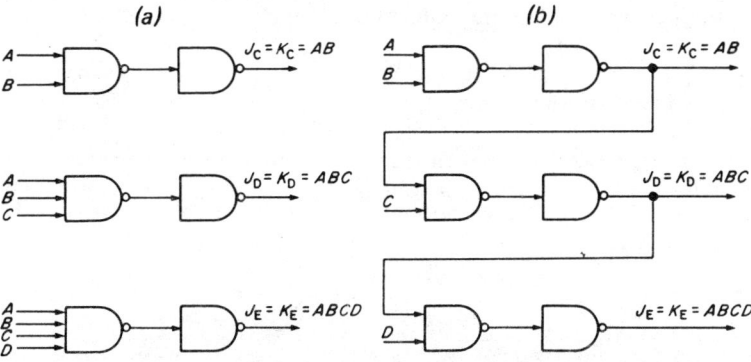

Figure 6.4. Flip-flop input gates. (a) Parallel connection. (b) Serial connection

increases the fan-in to the first NAND gate in the cascade of two increases. However, the gate delay at the input to every flip-flop is identical and equal to $2t_g$ where t_g is the time delay of a single NAND gate.

In the second method the fan-in for the first NAND gate in the cascade is always two, but the gate delay at the inputs to the flip-flops increases as the number of stages in the counter increases. Examination of figure 6.4(b) shows that the gate delay at the J_C input is $2t_g$ at the J_D input is $4t_g$, and so on. Since longer gate delays are experienced if this method of connection is used, it is clear that the upper frequency limit of a counter using this method is lower than that of a counter using the parallel connection.

If the switching times of individual flip-flops is t_f then for the parallel connection the upper frequency limit of the counter is given by

$$f_u = \frac{1}{2t_g + t_f}$$

whilst for the series connection

$$f_u = \frac{1}{2(N-2)t_g + t_f}$$

where N is the number of stages in the counter.

It should also be noted that in the case of the parallel connection the first flip-flop in the counter is required to drive $(N-1)$ gates, the next flip-flop $(N-2)$ gates, and so on, whereas for the serial connection all flip-flops except the last are required to drive one gate only.

6.7 Synchronous down-counters

These can be designed in the same manner as up-counters and the following flip-flop equations are obtained

$$J_A = K_A = 1$$

$$J_B = K_B = \overline{A}$$

$$J_C = K_C = \overline{A}\,\overline{B} = J_B\overline{B}$$

$$J_D = K_D = \overline{A}\,\overline{B}\,\overline{C} = J_C\overline{C}$$

. . .
. . .
. . .

$$J_N = K_N = \overline{A}\,\overline{B}\,\overline{C}\ldots\overline{(N-1)} = J_{N-1}\overline{(N-1)}$$

It is also possible in the case of binary counters to use an up-counter to count down by utilising the complementary flip-flop outputs. This is illustrated for a scale-of-eight counter in figure 6.5.

6.8 Scale-of-five up-counter

A scale-of-five counter has five states. In order to generate five states, three flip-flops are required and this will leave three unused states, as

Counters

d	C	B	A	$\bar{C}$	$\bar{B}$	$\bar{A}$
0	0	0	0	1	1	1
1	0	0	1	1	1	0
2	0	1	0	1	0	1
3	0	1	1	1	0	0
4	1	0	0	0	1	1
5	1	0	1	0	1	0
6	1	1	0	0	0	1
7	1	1	1	0	0	0

Figure 6.5. Using the complementary outputs of a chain of flip-flops to count down

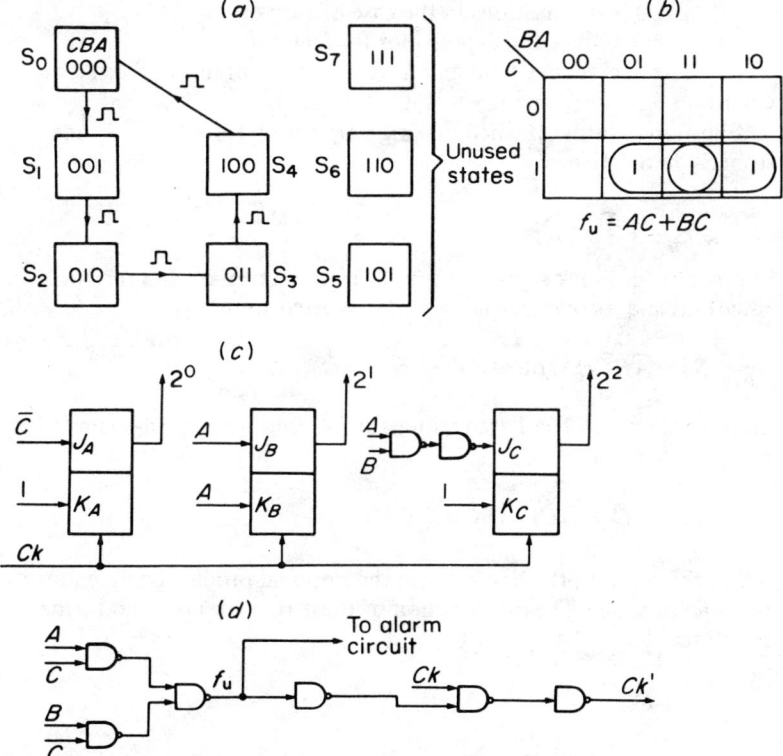

Figure 6.6. (a) State diagram for a scale-of-five counter. (b) K-map for simplification of unused states. (c) Implementation of a scale-of-five counter. (d) Alarm-raising and clock-inhibiting circuit

shown in the state diagram of figure 6.6(a). A K-map for the unused states is shown in figure 6.6(b), and the Boolean equation that represents these unused states is given by

$$f_u = AC + BC$$

It is sometimes more convenient to determine the equations of the inputs to the flip-flops using an algebraic rather than tabular method. A convenient design procedure is as follows.
(1) Determine the S and R expressions for each flip-flop.
(2) Use optional products defined by:
    (a) the unused states;
    (b) $1 \rightarrow 1$ transitions in the case of turn-on;
    (c) $0 \rightarrow 0$ transitions in the case of turn-off
       to reduce the expressions for S and R.
(3) The expressions for J and K can now be obtained by using the equations $S_Q = J_Q \bar{Q}$ and $R_Q = K_Q Q$.

Flip-flop A turns on when making a transition from S_0 to S_1 and from S_2 to S_3. Hence

$$S_A = S_0 \,\sqcap\!\sqcup\, + \, S_2 \,\sqcap\!\sqcup$$

In practice the clock signal is omitted from the turn-on and turn-off equations and its presence is normally inferred so that

$$S_A = \bar{A}\bar{B}\bar{C} + \bar{A}B\bar{C} = \bar{A}\bar{C}$$

Since there are no $1 \rightarrow 1$ transitions for flip-flop A the turn-on equation can be written as

$$S_A = \bar{A}\bar{C} + (AC) + (BC)$$

where the terms in the brackets are the optional products defined by the unused states. The introduction of these terms leads to no further simplification so that

$$J_A = \bar{C}$$

The turn-off equation for flip-flop A is written

$$R_A = S_1 + S_3 = A\bar{B}\bar{C} + AB\bar{C} = A\bar{C}$$

and including the optional products

$$R_A = A\overline{C} + (AC) + (BC) + (\overline{A}\overline{B}C)$$

the last term being an optional product for an $0 \rightarrow 0$ transition. Combining the first two terms gives

$$R_A = A + (BC) + (\overline{A}\overline{B}C)$$

No further simplification is now possible and

$$K_A = 1$$

For the B flip-flop,

$$S_B = S_1 + (S_2) + (AC) + (BC)$$
$$= A\overline{B}\overline{C} + (\overline{A}B\overline{C}) + (AC) + (BC).$$

The first and third terms can be used to form a consensus term $A\overline{B}$ which eliminates the original product $A\overline{B}\overline{C}$. Hence

$$S_B = A\overline{B} + (\overline{A}B\overline{C}) + (AC) + (BC)$$

and

$$J_B = A$$

$$R_B = S_3 + (S_4) + (S_0)(AC) + (BC)$$
$$= AB\overline{C} + (\overline{A}BC) + (\overline{A}\overline{B}\overline{C})(AC) + (BC)$$

The first and fourth terms can be used to form an additional optional product AB which eliminates the parent product $AB\overline{C}$. Hence

$$R_B = AB + (\overline{A}\overline{B}C) + (AC) + (BC)$$

and

$$K_B = A$$

For the C flip-flop

$$S_C = S_3 + (AC) + (BC)$$
$$= AB\bar{C} + (AC) + (BC)$$

and

$$J_C = AB$$

$$R_C = S_4 + (S_0) + (S_1) + (S_2) + (AC) + (BC)$$
$$= \bar{A}\bar{B}C + (\bar{A}\bar{B}\bar{C}) + (A\bar{B}\bar{C}) + (\bar{A}B\bar{C}) + (AC)(+(BC)$$

and after simplification

$$R_C = C + (\bar{A}\bar{B}\bar{C}) + (A\bar{B}\bar{C}) + (\bar{A}B\bar{C}) + (AC) + (BC)$$

from which

$$K_C = 1$$

The implementation of the counter is shown in figure 6.6(c).

If the counter should through faulty operation enter one of the unused states it can be stopped by inhibiting the clock and at the same time an alarm can be raised. This can be done by detecting the entry of the counter into an unused state by implementing the unused state function $f_u = AC + BC$ and using this signal for both stopping the clock and raising an alarm as shown in figure 6.6(d).

The equation for the modified clock signal is

$$Ck' = \bar{f}_u \cdot Ck$$

and when $f_u = 1$ it follows that $Ck' = 0$ and will remain so until such time as the counter has been removed from the unused state.

6.9 Decade binary up-counter

Figure 6.7(a) shows the state diagram for a decade up-counter and the state table for the counter is shown in figure 6.7(b). At the right-hand side of this table the flip-flop input signals have been tabulated. These

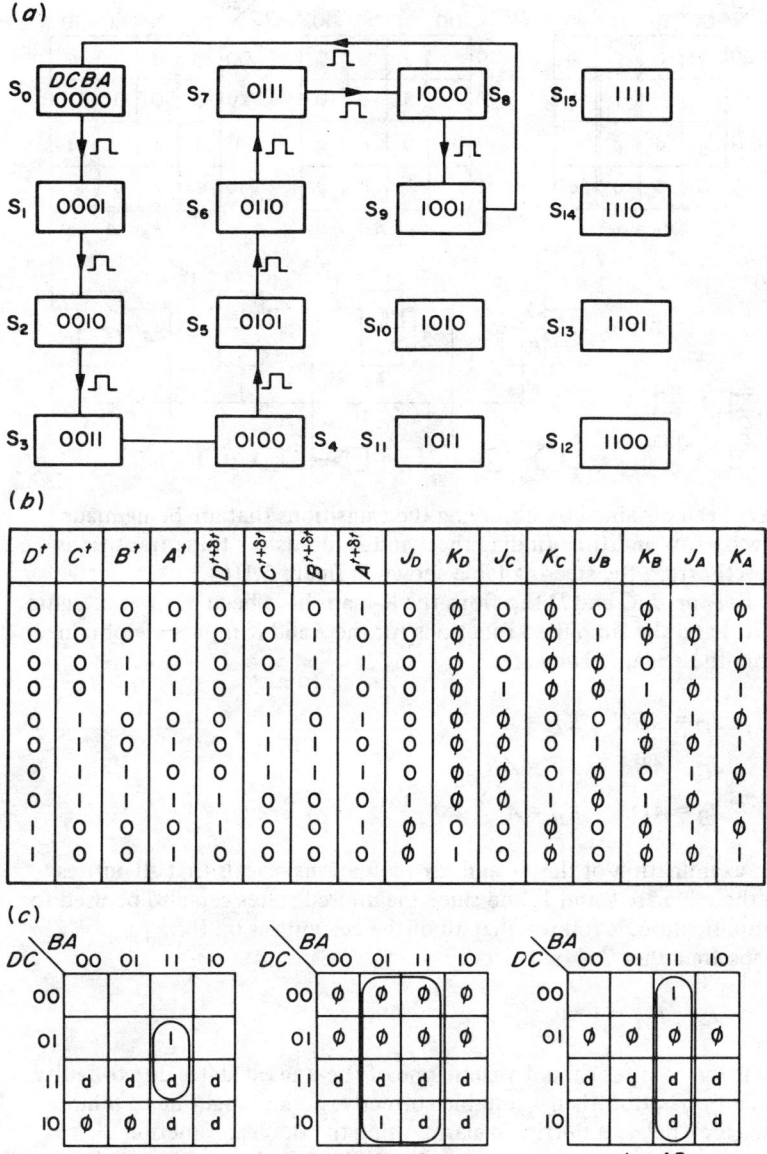

Figure 6.7. Decade binary up-counter. (a) State diagram. (b) State table. (c) K-maps. (d) Circuit implementation

K-maps showing $K_C = AB$, $J_B = A\bar{D}$, $K_B = A$ with BA across top (00, 01, 11, 10) and DC down side (00, 01, 11, 10).

(d) Circuit diagram showing flip-flops J_A/K_A, J_B/K_B, J_C/K_C, J_D/K_D with outputs $2^0, 2^1, 2^2, 2^3$.

have been obtained by observing the transitions that are being made row by row and then finding the J and K inputs for these transitions directly from the steering table shown in figure 6.1(b).

For the B, C and D flip-flops the K-maps have been plotted in figure 6.7(c) and the simplified functions for the J and K inputs are obtained from these maps. They are:

$$J_D = ABC \qquad K_D = A$$
$$J_C = AB \qquad K_C = AB$$
$$J_B = A\bar{D} \qquad K_B = A$$

An examination of the J_A and K_A tabulations reveals that all entries in the table are $\emptyset$ and 1, and since the unused states can also be used for simplification, it follows that all of the cell entries on the J_A and K_A maps are either $\emptyset$, d or 1. Hence,

$$J_A = K_A = 1$$

If the counter should assume one of the unused states due to faulty circuit operation then a suitable corrective action might be to inhibit the clock pulses and trip an alarm, using the Boolean function representing these states, namely $f_u = BD + CD$. A suitable circuit for suppressing the clock pulses is incorporated with the counter implementation shown in figure 6.7(d).

Counters

The equation of the modified clock signal Ck' is given by

$$Ck' = \overline{\overline{Ck} + \overline{BD} + \overline{CD}}$$

and if either BD or $CD = 1$, then $Ck' = 0$ and the counter clock is suppressed.

6.10 Decade binary down-counter

A decade binary down-counter can be designed using the technique described in § 6.9. The flip-flop input signals obtained by using this method are.

$J_D = \overline{A}\,\overline{B}\,\overline{C}$ $K_D = \overline{A}\,\overline{B}\,\overline{C}$

$J_C = \overline{A}D$ $K_C = \overline{A}\,\overline{B}$

$J_B = \overline{A}C + \overline{A}D$ $K_B = \overline{A}$

$J_A = 1$ $K_A = 1$

Implementation of the counter is shown in figure 6.8.

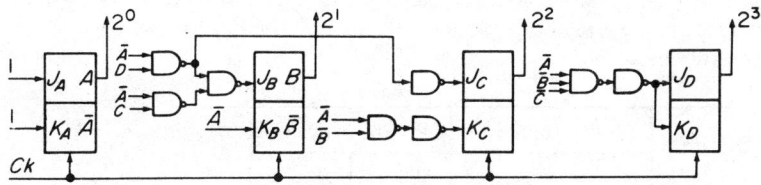

Figure 6.8. Implementation of a decade binary down-counter

6.11 Decade Gray code 'up' counter

Consider the transition from state 0001 to state 0010 in the decade binary 'up' counter and assume that flip-flop B changes faster than flip-flop A. The sequence of changes that take place is:

DCBA

0001

0011 (transient state)

0010

If a 4-to-10 line decoder such as the one shown in block schematic form in figure 6.9 is being used to convert the binary output of the counter to a decimal representation, a spike will occur on the output line marked 3, and this is clearly incorrect circuit operation. This can occur at any point in the binary counting sequence where more than one flip-flop is required to change state during a transition. The difficulty can be eliminated by using a Gray code counter, in which only one flip-flop changes state at each transition.

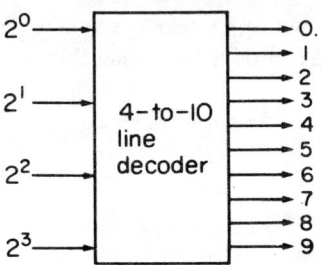

Figure 6.9. *A 4-to-10 line decoder for converting the binary outputs of a counter to decimal form*

A Gray code is one in which only one digit changes as a transition is made from one code combination to the next. As an example of the design of a Gray code counter, the 2-4-2-1 code will be converted to its corresponding Gray code, which will then be used as the basis for the counter design.

Decimal digit	2-4-2-1 code				2-4-2-1 Gray code			
	W	X	Y	Z	D	C	B	A
0	0	0	0	0	0	0	0	0
1	0	0	0	1	0	0	0	1
2	0	0	1	0	0	0	1	1
3	0	0	1	1	0	0	1	0
4	0	1	0	0	0	1	1	0
5	1	0	1	1	1	1	1	0
6	1	1	0	0	1	0	1	0
7	1	1	0	1	1	0	1	1
8	1	1	1	0	1	0	0	1
9	1	1	1	1	1	0	0	0

Figure 6.10. *Table showing the 2-4-2-1 and its corresponding Gray code*

Counters

The 2-4-2-1 code is shown tabulated in figure 6.10 in the columns headed W, X, Y and Z. To convert this code to its corresponding Gray code, the equations developed for a binary to Gray code converter developed in Chapter 4 are used. These are

$$A = Y \oplus Z$$
$$B = X \oplus Y$$
$$C = W \oplus X$$

and

$$D = W$$

The complete 2-4-2-1 Gray code obtained using these equations is tabulated in those columns headed D, C, B and A in figure 6.10.

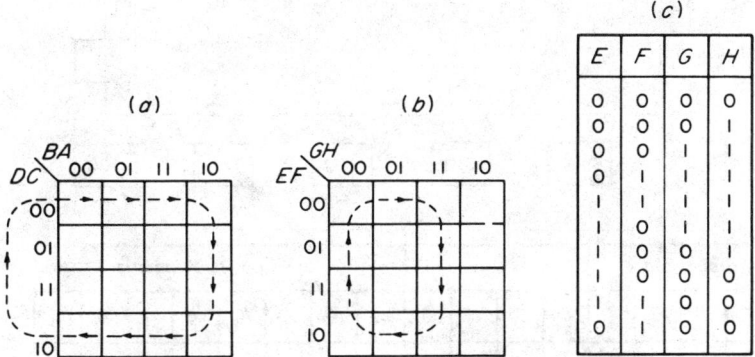

Figure 6.11. (a) The path of the 2-4-2-1 Gray code through the K-map. (b) The path of an alternative Gray code through the K-map. (c) The Gray code tabulation for the K-map plot in (b)

There are many different Gray codes of four variables and the path of the 2-4-2-1- Gray code through the K-map is shown plotted in figure 6.11(a). It will be observed that adjacent code combinations occupy adjacent cells on the K-map. Hence a Gray code having ten combinations, can be developed with the aid of a K-map by simply plotting a closed path that consists of ten adjacent cells, as shown in figure 6.11(b). The Gray code corresponding to this path can be read from the map and is tabulated in figure 6.11(c).

The state diagram for the counter is shown in figure 6.12(a) and the state table which shows the present and next states of the counter is illustrated in figure 6.12(b). This table gives the transitions for each flip-flop as the counter goes from one state to the next, and with the aid of the steering table shown in figure 6.1(b) the flip-flop input signals J and K can be obtained for each transition. A tabulation of these signals is shown on the right-hand side of the state table in figure 6.12(b).

Eight K-maps, one for each of the flip-flop input signals are shown

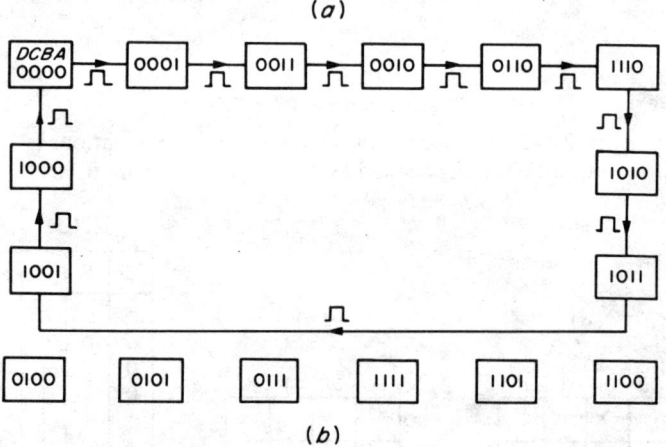

Present state				Next state				Flip-flop input signals							
D	C	B	A	D	C	B	A	J_D	K_D	J_C	K_C	J_B	K_B	J_A	K_A
0	0	0	0	0	0	0	1	0	ø	0	ø	0	ø	1	ø
0	0	0	1	0	0	1	1	0	ø	0	ø	1	ø	ø	0
0	0	1	1	0	0	1	0	0	ø	0	ø	ø	0	ø	1
0	0	1	0	0	1	1	0	0	ø	1	ø	ø	0	0	ø
0	1	1	0	1	1	1	0	1	ø	ø	0	ø	0	0	ø
1	1	1	0	1	0	1	0	ø	0	ø	1	ø	0	0	ø
1	0	1	0	1	0	1	1	ø	0	0	ø	ø	0	1	ø
1	0	1	1	1	0	0	1	ø	0	0	ø	ø	1	ø	0
1	0	0	1	1	0	0	0	ø	0	0	ø	0	ø	ø	1
1	0	0	0	0	0	0	0	ø	1	0	ø	0	ø	0	ø

Figure 6.12. The 2-4-2-1 Gray code counter. (a) State diagram. (b) State table. (c) K-maps. (d) Implementation

Counters

plotted in figure 6.12(c). Cells representing the unused states are marked with a d and the rest of the cell entries are obtained from the tabulation in figure 6.12(b). Simplification is performed using the techniques previously described and the following flip-flop input equations are obtained

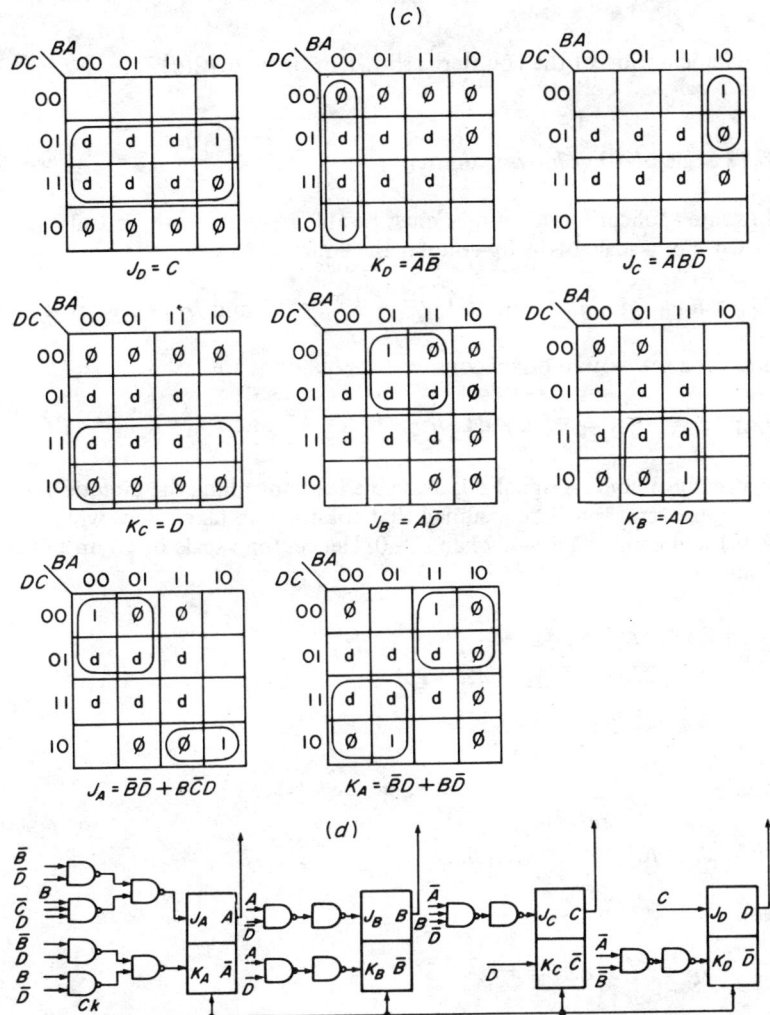

Figure 6.12. Cont.

$$J_A = \bar{B}\bar{D} + B\bar{C}D \qquad K_A = \bar{B}D + B\bar{D}$$
$$J_B = A\bar{D} \qquad K_B = AD$$
$$J_C = \bar{A}B\bar{D} \qquad K_C = D$$
$$J_D = C \qquad K_B = \bar{A}\bar{B}$$

Implementation of the counter is shown in figure 6.12(*d*).

6.12 Scale-of-16 up/down counter

In some applications a counter must be able to count both up and down. For a scale-of-16 up-counter the equations are

$$J_{Au} = K_{Au} = 1; J_{Bu} = K_{Bu} = A; J_{Cu} = K_{Cu} = AB \text{ and } J_{Du} = K_{Du} = ABC$$

and for a scale-of-16 down-counter the equations are

$$J_{Ad} = K_{Ad} = 1; J_{Bd} = K_{Bd} = \bar{A}; J_{Cd} = K_{Cd} = \bar{A}\bar{B} \text{ and } J_{Dd} = K_{Dd} = \bar{A}\bar{B}\bar{C}.$$

Normally, a control signal Z is available for controlling the direction of the count, and it will be assumed that counting up takes place when $Z = 1$ and counting down when $Z = 0$. Hence, for a scale-of-16 up/down counter

$$J_A = K_A = 1$$
$$J_B = ZJ_{Bu} + \bar{Z}J_{Bd} = ZA + \bar{Z}\bar{A}$$
$$K_B = ZK_{Bu} + \bar{Z}K_{Bd} = ZA + \bar{Z}\bar{A}.$$

Similarly,

$$J_C = K_C = ZAB + \bar{Z}\bar{A}\bar{B}$$

and

$$J_D = K_D = ZABC + \bar{Z}\bar{A}\bar{B}\bar{C}$$

The implementation of a scale of 16 up/down counter is shown in figure 6.13.

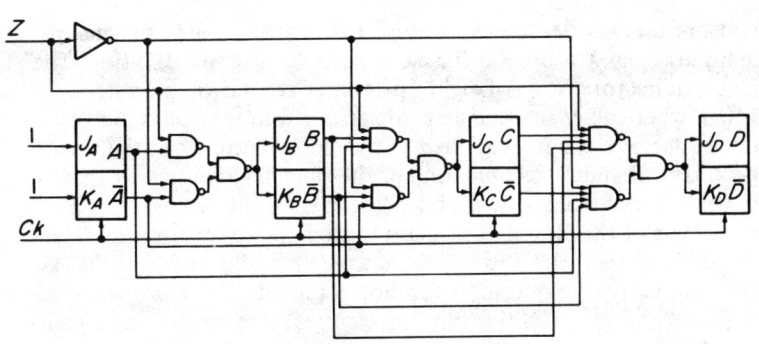

Figure 6.13. *Implementation of a scale-of-16 up/down counter*

6.13 Asynchronous binary counters

One of the simplest types of counter to design is the 'ripple-through' or asynchronous binary counter. For counts of powers of 2 the basic arrangement consists of T flip-flops (or alternatively JK flip-flops with J and K permanently connected to logical '1') connected in cascade as shown in figure 6.14(a). As can be seen from the diagram, the output of

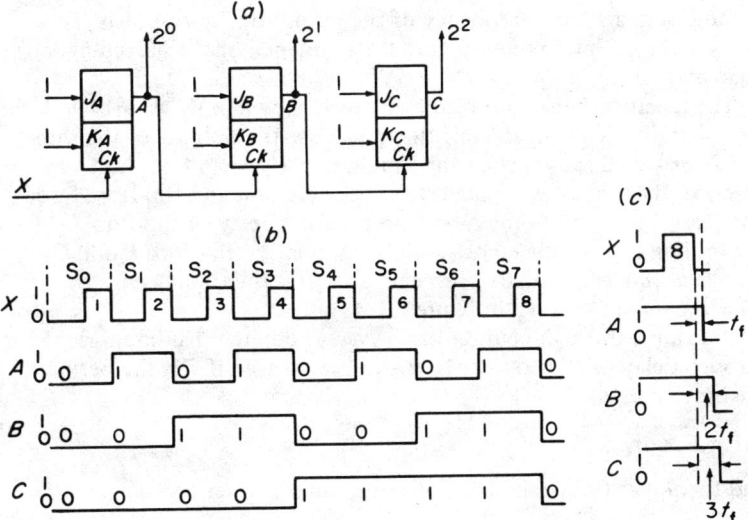

Figure 6.14. *(a) Implementation of, and (b) timing diagrams for a three-stage ripple-through counter. (c) The ripple-through effect in the counter*

each flip-flop provides the clock signal for the next one in the chain. The input signal X is used as the clock pulse for the first flip-flop. The time-diagrams for a scale-of-eight up-counter are shown in figure 6.14(b), where all changes of state are assumed to take place on the trailing edge of the signals applied to the clock terminals of the three flip-flops A, B and C. Examination of the time diagrams shows that flip-flop A changes state on each trailing edge of the input pulses X. The output of flip-flop A is used as the clock pulse for flip-flop B, and a change in state of this flip-flop occurs on the trailing edge of the A pulses. Similarly, the output of flip-flop B provides the clock pulse input for flip-flop C and this flip-flop changes state on the trailing edge of the B pulses.

The various states of the counter and the binary digits associated with each state are marked on the time diagrams for the outputs A, B and C.

A scale-of-eight counter is alternatively called a divide-by-eight circuit. An inspection of the time diagrams shows that the output of flip-flop C produces one pulse for every eight input pulses to flip-flop A. Hence it follows that if the frequency at the input is f then the frequency at the ouptut of flip-flop C is $f/8$, i.e. the input frequency divided by eight. Since flip-flop B output produces two pulses for every eight input pulses then the frequency of its output is $f/4$ and, similarly, the frequency of the output of flip-flop A is $f/2$. Every stage in this counter divides the frequency of the preceding stage by two.

The idealised behaviour of the circuit is shown in figure 6.14(b). On the trailing edge of the eighth input pulse the outputs of the three flip-flops are all shown changing simultaneously from 1 to 0. In practice, these changes ripple through the counter and flip-flop A does not change to 0 until time t_f, the propagation delay of flip-flop A, after the trailing edge of the eighth X pulse. Similarly, flip-flops B and C change at times $2t_f$ and $3t_f$, respectively, after the trailing edge of the eighth X pulse as shown in figure 6.14(c).

If a ripple-through counter has n stages, then the maximum ripple-through delay of the counter is nt_f. Consequently, if T is the period of the input pulses,

$$T \geqslant nt_f$$

and the upper frequency limit of the counter is given by

$$f_u = \frac{1}{T} \leqslant \frac{1}{nt_f}.$$

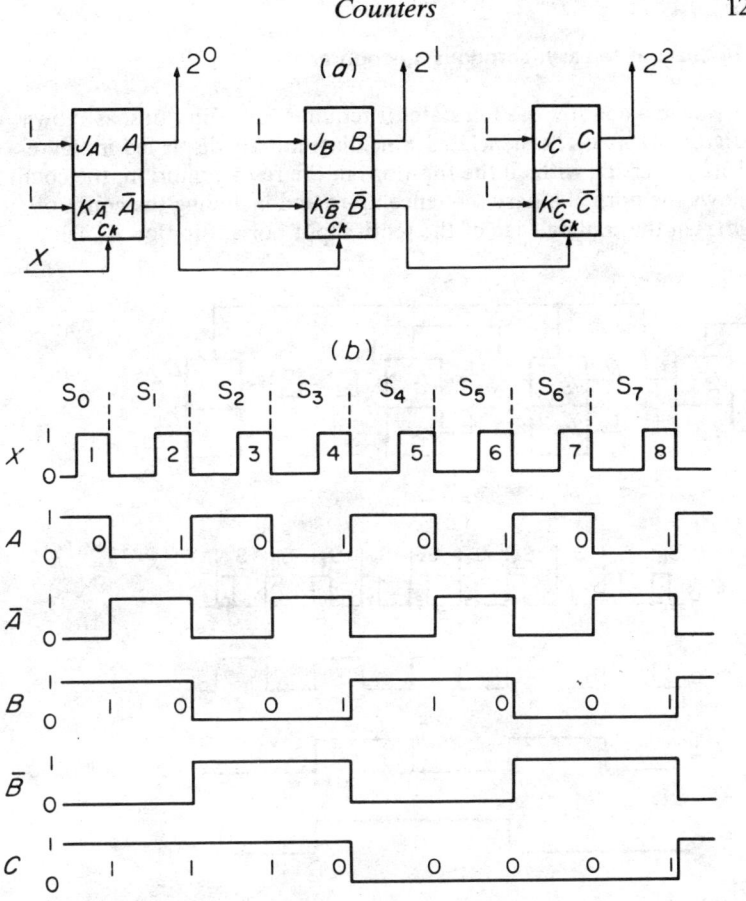

Figure 6.15. *Scale-of-eight ripple-through down-counter.* (a) *Implementation.* (b) *Timing diagrams*

If the circuit of figure 6.14(a) is modified so that signals $\bar{A}$ and $\bar{B}$ are used as the clock signals for flip-flops B and C, respectively, the circuit will operate as a down-counter. The implementation of the down-counter is shown in figure 6.15(a) and the timing diagrams are shown in figure 6.15(b). Binary digits have been inserted on the A, B and C waveforms to illustrate the down-count. In this case the clock signal for flip-flop B is $\bar{A}$ and the flip-flop changes its state on the trailing edge of this signal. Similarly, the clock signal for flip-flop C is $\bar{B}$ and its state changes on the trailing edge of the $\bar{B}$ signal.

6.14 Scale-of-ten asynchronous up-counter

Since such a counter has ten states it requires four flip-flops, as shown in figure 6.16(a). The associated time diagrams are displayed in figure 6.16(b). Starting with all the flip-flops in the reset condition, the count follows the normal binary sequence up to and including the count of eight. On the trailing edge of the tenth input pulse, flip-flop A makes a

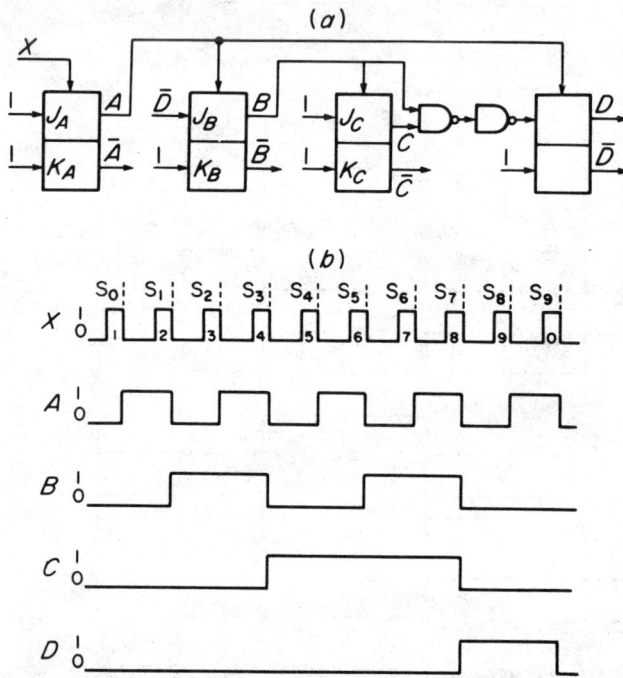

Figure 6.16. (a) Implementation of a scale-of-ten asynchronous up-counter. (b) Timing diagrams for a scale-of-ten counter

transition from 1 to 0, which would normally induce a transition in flip-flop B, changing its state from 0 to 1. However, at this instant $J_B = \bar{D} = 0$, and consequently flip-flop B remains in the reset condition. At the same instant it is also necessary to reset flip-flop D. Now $J_D = BC = 0$ and $K_D = 1$, hence when A makes a 1 to 0 transition at the end of S_9 the flip-flop D is reset. All the flip-flops are now in the reset condition and are ready for the arrival of the first pulse of the next counting cycle.

The counter may also be described as a divide-by-ten circuit, in the sense that ouptut D produces one pulse for every ten input pulses at the clock terminal of flip-flop A.

6.15 Asynchronous resettable counters

An alternative method of counting is to use a resettable counter. A scale-of-N counter of this type is allowed to count up to the number N and a logic signal representing this number is used to clear all the flip-flops in the counter. The state diagram for a resettable scale-of-tive counter is shown in figure 6.17(a). The counter remains in each of the

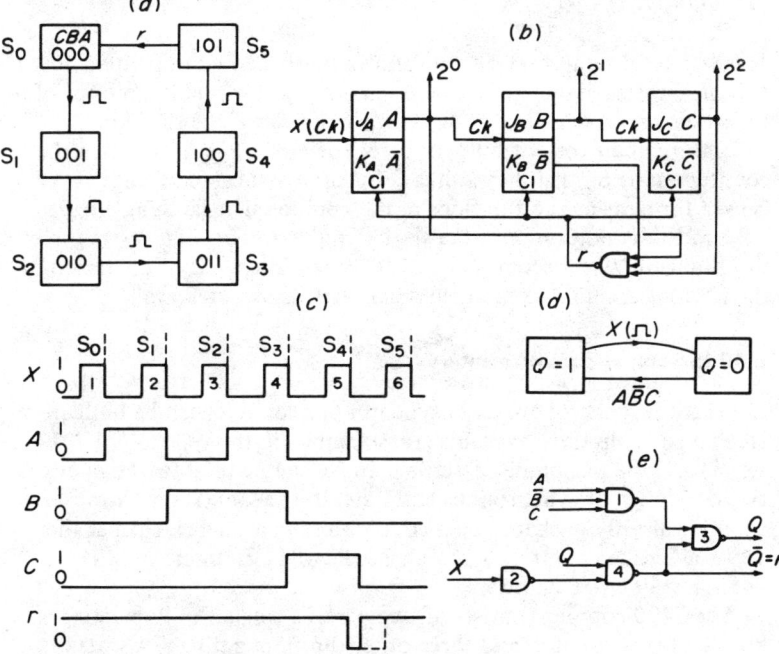

Figure 6.17. Resettable scale-of-five counter. (a) State diagram. (b) Implementation. (c) Timing diagrams. (d) State diagram for the latching circuit. (e) Implementation of the latching circuit

first five states for one clock period, but on entering the sixth state (101) a reset signal $r = \overline{A\bar{B}C}$ is developed by a NAND gate as shown in the circuit implementation in figure 6.17(b). Timing diagrams for the circuit are shown in figure 6.17(c).

The reset times for the individual flip-flops in the counter may well be different. For example, in the circuit described, flip-flop A may reset faster than flip-flop C. The negative-going reset signal will cease to exist when A is cleared and is simply not wide enough to reset flip-flop C.

This problem can be overcome by latching the reset signal until the beginning of the sixth clock pulse, as indicated by the dotted lines in figure 6.17(c). A suitable state diagram for the latching circuit is shown in figure 6.17(d).

The turn-on set of $Q = A\bar{B}C$, and the turn-off set of $Q = X$. Hence, using the NAND sequential equation developed in Chapter 5 gives

$$Q = A\bar{B}C + \bar{X}Q$$

This two-level sum-of-products is shown implemented in figure 6.17(e). It is, in fact, the implementation of an SR flip-flop and the output of the gate marked 4 is the complementary output of the flip-flop.

In this circuit the output Q of gate 3 becomes logical '1' when the counter enters S_5. It follows that $\bar{Q}$ becomes logical '0'. Hence $\bar{Q} = r$ is used for clearing the flip-flops in the counter. The latching circuit stays in this condition until the sixth X pulse arrives. This resets the flip-flop and $\bar{Q} = r$ becomes logical '1' again. The cycle of operation of the latching circuit is completed when $A\bar{B}C$ is detected again.

6.16 Integrated-circuit counters

In practice, synchronous and asynchronous counters can be built from standard JK flip-flops available (for example, in the type 54/74 TTL series) and the techniques described above can be used to design either ripple-through or synchronous counters. In the same series, however, counters already packaged on TTL IC chips are available, such as the 7490, a decade counter; the 7492, a scale-of-12 counter; and the 7493, a scale-of-16 counter.

The 7490 counter consists of two parts: a single flip-flop acting as a scale-of-two counter, and three other flip-flops acting as a scale-of-five counter. When the output of the scale-of-two counter is connected to the input of the scale-of-five counter, a decade counter is produced.

In order to use the 7490 chip, for example, it is not essential to have a detailed knowledge of the circuit or even, for that matter, the logic diagram of the circuit. However, the engineer must be familiar with the function of the chip connections and, in order to use the chip intelligently, must understand the basic principles of counting. For the

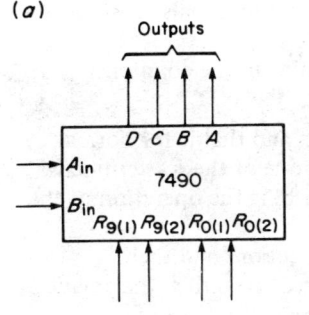

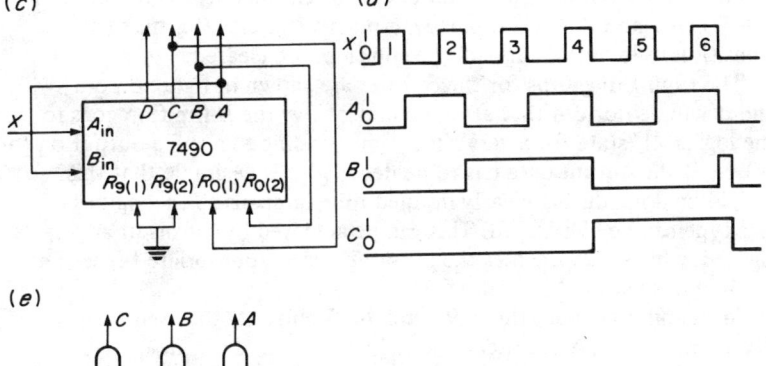

Figure 6.18. (a) Chip connections for the 7490. (b) Truth table for the reset inputs. (c) The 7490 connected as a scale-of-six counter. (d) The 7490 scale-of-six counter timing diagrams. (e) Elimination of spikes with the enable signal, E.

7490 chip the important chip connections are shown in figure 6.18(a). They are:

(1) the four outputs D, C, B and A, where D is the most significant bit;

(2) the input terminal A_{in} to which the input signal is attached;

(3) the input terminal B_{in} which is connected to output A when the counter is operating in the decade mode, otherwise the input signal can be attached here when operating in the scale-of-five mode.

(4) $R_{0(1)}$ and $R_{0(2)}$ which are the direct clear terminals. Both of these must be held at logical '1' to clear all flip-flops.

(5) $R_{9(1)}$ and $R_{9(2)}$ which set a count of nine in the counter if they are both held at a logical '1'.

One other operating rule must be observed, and that is for normal counting at least one of the R_0 terminals and one of the R_9 terminals must be held at logical '0'. A truth table describing the operation of the reset terminals is shown in figure 6.18(b).

Having become familiar with the chip connections and their functions, and armed with the basic principles of counting, the engineer is now in a position to use the chip. For example, if a scale-of-six counter is required, then the chip can be connected to perform this function as shown in figure 6.18(c). In this configuration the chip is acting as a resettable ripple counter and when the output combination $B = 1, C = 1$ and $A = 0$ is reached, terminals R_{01} and R_{02} make a transition to logical '1' and all the flip-flops are cleared.

The timing diagrams for the counter are shown in figure 6.18(d) and it will be noticed that after a count of five the B flip-flop goes to the logical '1' state for a very short time, leading to a spike output on the B line. If the output data has to be decoded it is desirable that this should be done during clearly defined time intervals so that spikes of this type can be eliminated. This can be achieved by means of an enable signal E which only enables the output gates at appropriate times. The method is illustrated in figure 6.18(e).

In a similar fashion, the 7492 and 7493 chips can be used for designing scale-of-N counters. The 7493 consists of a scale-of-two counter succeeded by a scale-of-eight counter and an example of the use of this chip as a scale-of-13 resettable ripple counter is illustrated in figure 6.19.

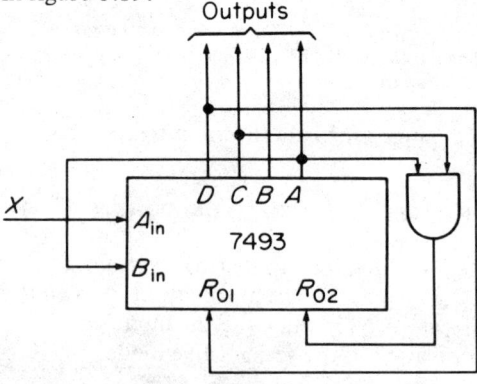

Figure 6.19. The 7493 used as a scale-of-13 ripple counter

6.17 Cascading of IC counter chips

If two counter chips such as the 7490 and the 7493 are cascaded, and a frequency of 320 kHz is applied at the input terminal of the 7490, as shown in figure 6.20, then the frequency of the signal appearing at the output of the 7493 will be at a frequency of 2 kHz. When frequency division by a large number is required the only practical way of achieving this is to use a cascade of counter chips.

It is also possible to construct a scale-of-N counter by cascading counter chips, as shown in figure 6.20. For example, if a scale-of-92 counter is required this can be achieved by cascading two 7490s as

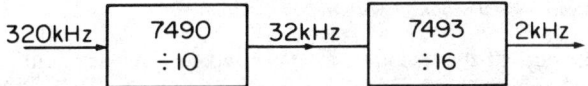

Figure 6.20 Two 7490 chips in cascade used to divide the frequency by 160

shown in figure 6.21. The most significant digit output D_0 of the first chip is connected through to the second chip and acts as the clock signal for it. For every ten X pulses there is one D_0 pulse and on the tenth X pulse the chip labelled 10^0 makes a transition from 1001 to 0000 and the chip labelled 10^1 makes a transition from 0000 to 0001. The counter is allowed to count up to 92 when the signal representing

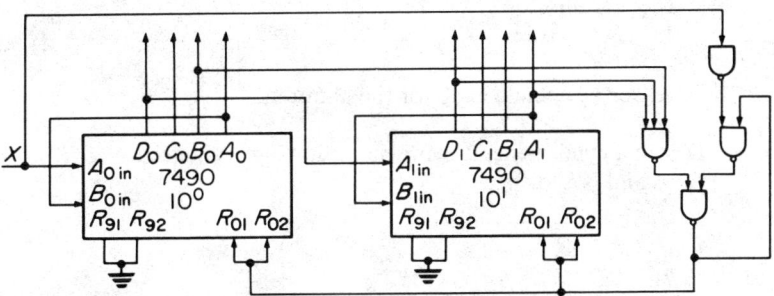

Figure 6.21. Two 7490 chips in cascade form a scale-of-92 counter

this number is fed back via the latch circuit to the clear inputs of the two chips. The latching circuit eliminates problems that may be caused by the flip-flops having different resetting times.

When three 7490s are connected in cascade, as shown in figure 6.22 they can be used as a three-decade NBCD counter. If a decimal display is required the NBCD output from each chip needs to be

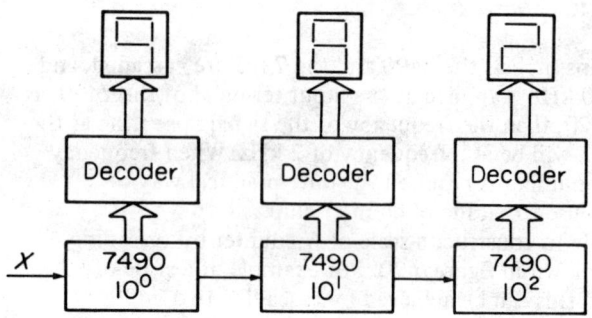

Figure 6.22. Three-decade NBCD counter with decimal display

decoded and the outputs of the decoders are then fed to seven-segment display circuits, as shown in the diagram.

Problems

6.1 Design a synchronous modulo-12 counter using NAND gates and

  (a) T flip-flops,
  (b) SR flip-flops,
  (c) JK flip-flops,
  (d) D flip-flops.

  Develop the decode logic for the counters.

6.2 Design a cyclic generator for the following sequence using JK flip-flops and NAND gates:

Clock Pulse	C	B	A
1	0	0	1
2	1	0	0
3	0	1	0
4	1	0	1
5	1	1	0
6	0	1	1

Examine the behaviour of this circuit in its unused states and show that one of the unused states is a 'lock-in' state. Suggest a way of avoiding a 'lock-in'.

6.3 Convert the binary code in the tabulation shown below to its corresponding Gray code, and design a counter using JK flip-flops and NAND gates to generate this code.

D	C	B	A
0	0	0	0
0	1	1	1
0	1	1	0
0	1	0	1
0	1	0	0
1	0	1	1
1	0	1	0
1	0	0	1
1	0	0	0
1	1	1	1

6.4 The operational characteristics of a PQ flip-flop are as follows:

$PQ = 00$ the next state of the flip-flop is 1, irrespective of its present state.

$PQ = 01$ the next state of the flip-flop is the complement of the present state, irrespective of its present state.

$PQ = 10$ the next state of the flip-flop is the same as the present state, irrespective of its present state.

$PQ = 11$ the next state of the flip-flop is 0, irrespective of its present state.

Using the above information, obtain the steering table of the PQ flip-flop and develop the input equations for a scale-of-eight binary counter which uses this flip-flop.

6.5 The circuit shown in figure P6.5 is to be used to generate an output pulse Q having a time duration equal to 14 clock periods. Draw a timing diagram showing the principal circuit waveforms

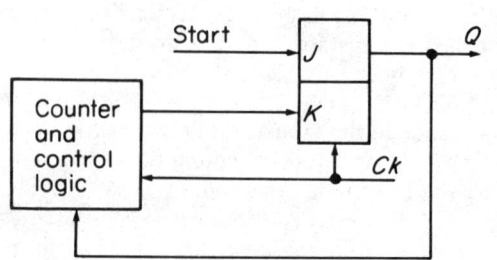

Figure P6.5

and design the counter and control logic using master/slave *JK* flip-flops and NAND gates.

6.6 A five-stage ripple-counter uses flip-flops having a delay time of 30 ns and a decode time of 50 ns. Determine the maximum frequency of operation of the counter. If the counter is operating at this frequency, draw a timing diagram for each of the flip-flops as the count advances from 01111 to 10000.

Assuming that the counter is operated now at a frequency of 8·33 MHz, draw timing diagrams showing the behaviour of the flip-flops between the fifteenth and sixteenth clock pulses.

6.7 Draw the timing diagrams for the following asynchronous counters:

(*a*) a four-bit binary down-counter
(*b*) a four-bit binary up-counter

assuming that the flip-flops used in the counting array trigger on the leading edge of the pulse applied to the clock terminal.

6.8 Design a scale-of-882 counter using 7490 chips and include in the design the latching arrangements required for eliminating the problems raised by the different reset times of the flip-flops.

7

Shift register counters and generators

7.1 Introduction

A shift register is a cascade connection of a number of single-bit storage elements contained in a single IC package. The storage elements are flip-flops and it is arranged that the output of each flip-flop is connected to the input of the next flip-flop in the cascade.

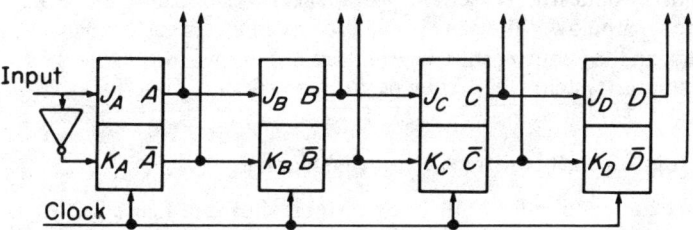

Figure 7.1. *A four-stage shift register using JK flip-flops*

A simple four-stage shift register is shown in figure 7.1. Information is moved into this register, one bit at a time, synchronously with the clock. As one bit is moved into the first flip-flop, other bits stored in the register all move on one place, so that the bit stored in flip-flop A is now stored in flip-flop B, and so on down the cascade, whilst the bit originally stored in flip-flop D is lost.

Shift registers can be classified into four distinct groups:

(*a*) Serial-in/serial-out, in which data can be moved in and out of the register, one bit at a time.

(*b*) Serial-in/parallel-out, in which the register is loaded serially one bit at a time and when an output is required the data stored in all the flip-flops can be read simultaneously.

(c) Parallel-in/serial-out, in which all the flip-flops in the register are loaded simultaneously and when an output is required the stored data is removed from the register one bit at a time under clock control.

(d) Parallel-in/parallel-out, in which all the flip-flops in the register are loaded simultaneously and when an output is required the flip-flops are read simultaneously.

It is also possible to classify shift registers according to their input arrangements:

(a) Double-rail input. For this type of register there are two input terminals which feed either $J-K$ or $S-R$ inputs.

(b) Single-rail input, as illustrated in figure 7.1. Here the first flip-flop in the cascade has been converted into a D-type flip-flop by placing an inverter between the J and K input lines.

In the same way there can be double-rail output, in which the true and complementary outputs of the last flip-flop in the chain are brought out to pins, or alternatively, there can be single-rail output in which only the true output of the last flip-flop is made available at a pin.

Shift registers have a wide range of applications. They can be used for temporary data storage, serial-to-parallel conversion, and vice versa. As will be seen below, they can also be used as counters and sequence generators, and, of course, shift registers are also found in the CPU of a microcomputer system where they perform a variety of functions.

7.2 The four-bit shift register with parallel loading

A typical example of a four-bit shift register is shown in figure 7.2. It has both parallel and serial loading facilities, and the output is taken off serially from the last flip-flop in the cascade. The parallel loading is achieved by a preset operation from one of two sources and it is essential before a parallel loading operation that all the flip-flops should be cleared since a preset operation can only set a flip-flop to logical '1'. Data in this shift register can be shifted to the right only.

7.3 The four-bit shift-left, shift-right register

A typical example of a versatile shift register is the 7495 and its logic diagram is shown in figure 7.3(a). It has facilities for parallel loading and parallel output, serial loading and serial output, and, additionally, it has shift-left and shift-right facilities. This is, in effect, a universal shift register which can operate in all four modes described in §7.1, besides having the facility of bi-directional shifting.

The mode-control (MC) input controls whether data inputs are serial or parallel. With MC = 0, the AND gates marked 1 and the clock gate marked 1 are enabled. In this mode the data is serially entered under the control of clock 1. Alternatively, with MC = 1 the AND gates marked 2 and the clock gate marked 2 are enabled. In this case the input is parallel under the control of clock 2. The operation of the mode control is asynchronous in that changes should not take place when either of the clocks is high.

Data can be loaded in parallel into the shift register and then, by shifting right, can be extracted in serial form. Alternatively, the shift register can be loaded serially and then the data can be read from the parallel output lines in parallel form. Bi-directional shifting is achieved by making the connections shown in figure 7.3(b).

7.4 The use of shift registers as counters

An alternative method of designing digital counters or sequence generators is to use a shift register chip. A typical shift register counter configuration is shown in figure 7.4. The individual flip-flops form part of an N-stage shift register and the connections between individual flip-flops are internal to the chip. The output of each stage and its complement are both available and they may be used to drive combinational feedback logic which provides the J and K inputs to

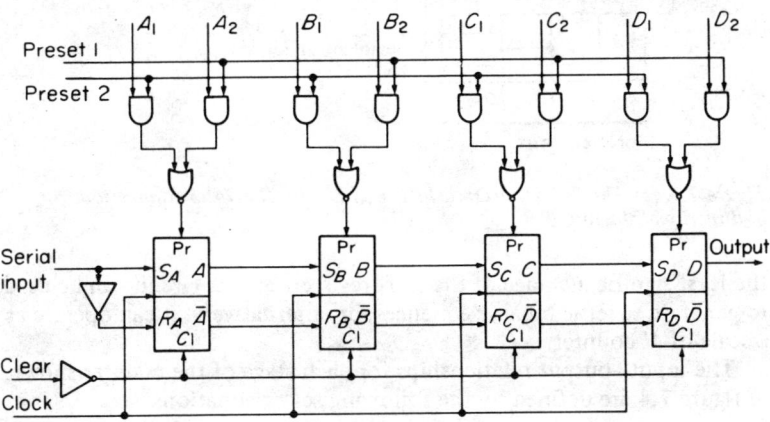

Figure 7.2. A four-bit shift register with both parallel and serial loading

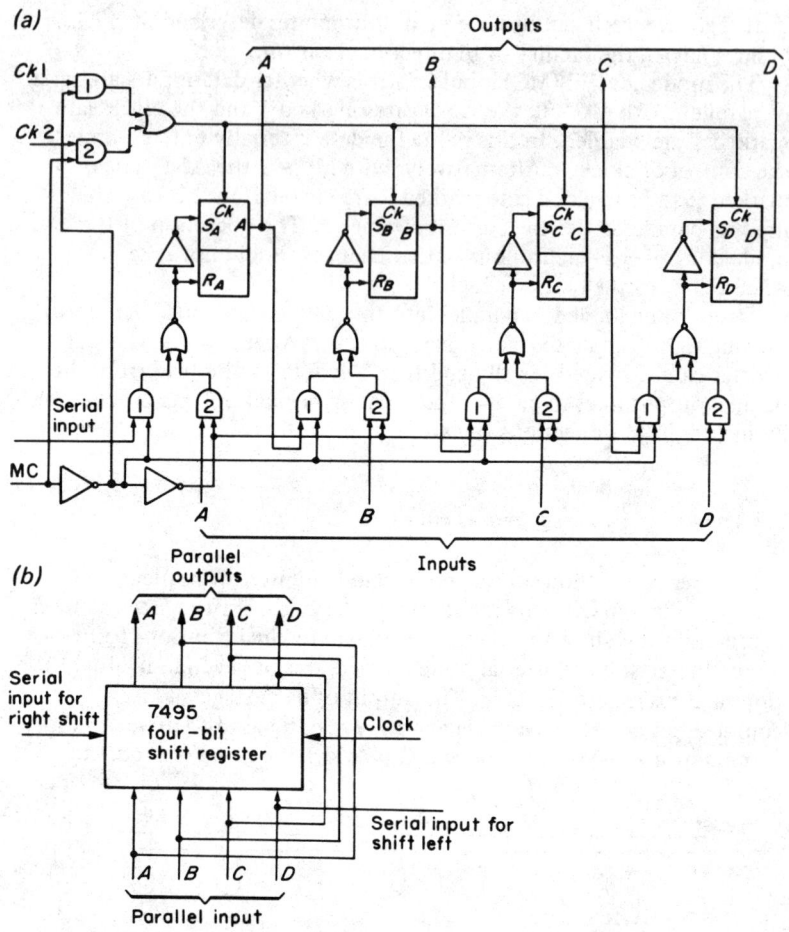

Figure 7.3. (a) The 7495 universal shift register. (b) The 7495 connections for bi-directional shifting

the least significant stage of the shift register. Such a circuit can be used to generate specific binary sequences, or, alternatively, it can operate as a scale-of-M counter.

The input–output relationships for each stage of the counter shown in figure 7.4 are defined by the following set of equations:

$$A^{t+\delta t} = f(AB \ldots N)^t$$

$$B^{t+\delta t} = A^t$$

. .

. .

. .

$$N^{t+\delta t} = (N-1)^t.$$

The feedback circuit produces either a 1 or a 0 which is fed to the input of flip-flop A where it determines the next state of A on receipt of the next clock pulse. For example, assuming that the N-stage shift

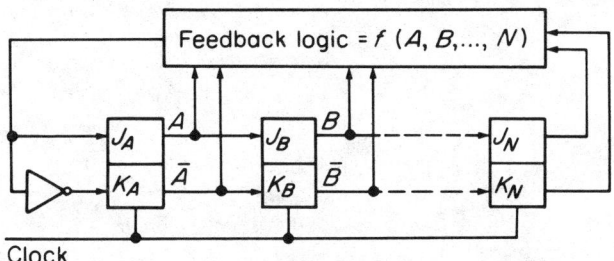

Figure 7.4. *Basic configuration of a feedback shift register*

register is in the state $N \ldots CBA = 0 \ldots 001$, the next state of the shift register will be either $0 \ldots 010$ or $0 \ldots 011$, depending upon whether the feedback logic provides a 0 or a 1 at the J-input of flip-flop A.

7.5 The universal state diagram for shift registers

The transition table for a two-stage shift register is shown in figure 7.5(*a*). If the shift register is initially in the state 00 then there are two possible next states. These are 00 if the J-input to the first flip-flop is a 0, or 01 if the J-input is a 1. Similarly, if the initial state of the shift register is 01 then the two possible next states are either 10 or 11.

The transition table of figure 7.5(*a*) can be translated into the universal state diagram shown in figure 7.5(*b*), alternatively called the de Bruijn diagram. It will be noticed that the shift register is permanently 'locked' in the state 00 if the feedback signal is a 0 and, similarly, it is locked in the state 11 if a 1 is provided by the feedback logic.

Shift register counters and generators

(a)

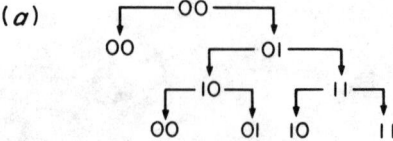

(b)

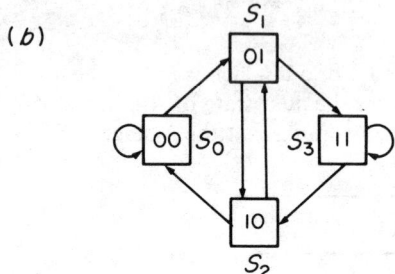

Figure 7.5. Two-stage shift register. (a) Transition table. (b) Universal state diagram

(a)

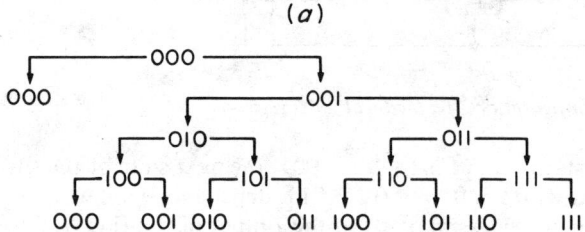

(b)

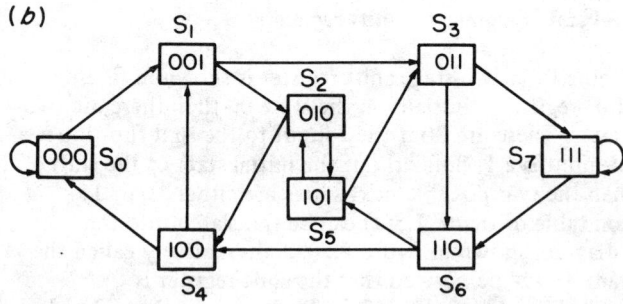

Figure 7.6. Three-stage shift register. (a) Transition table. (b) Universal state diagram

A similar transition table can be developed for a three-stage shift register, and this can be translated into a universal state diagram as shown in figures 7.6(a, b). The universal state diagram for a four-stage shift register may be developed in the same way, and is shown in figure 7.7. Clearly, the complexity of this type of diagram increases rapidly with the number of stages in the shift register.

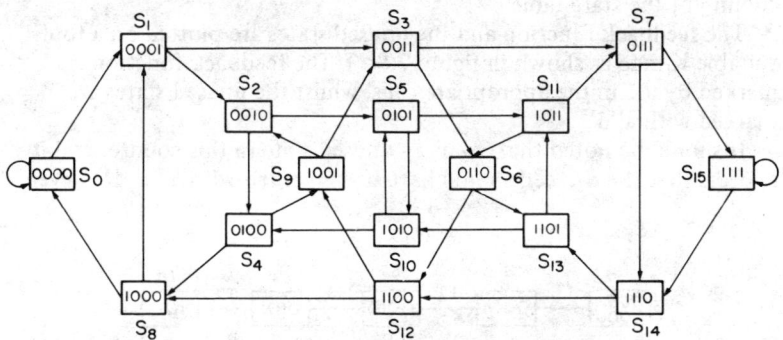

Figure 7.7. *Universal state diagram for a four-stage shift register*

The universal state diagram is a departure from the kind of state diagram that defines a specific sequence, as seen in the previous chapter on counters. All possible internal states of the shift register and all possible transitions between states are shown on the universal state diagram. The logic designer may now choose a suitable sequence of states on the diagram and design the feedback logic so that the shift register will cycle through the chosen sequence of states.

7.6 The design of a decade counter

The first step is to choose a ten-state sequence on the universal state diagram for a four-stage shift register. Figure 7.7 reveals the following three possible sequences:

(a) $S_0 - S_1 - S_2 - S_5 - S_{11} - S_7 - S_{15} - S_{14} - S_{12} - S_8 - S_0$

(b) $S_0 - S_1 - S_3 - S_7 - S_{15} - S_{14} - S_{13} - S_{10} - S_4 - S_8 - S_0$

(c) $S_0 - S_1 - S_2 - S_5 - S_{11} - S_6 - S_{13} - S_{10} - S_4 - S_8 - S_0$

There are, of course, other ten-state sequences besides these. The state diagram for the third of the above sequences is shown in figure 7.8(a).

The second step in the design is to draw up the state table shown in figure 7.8(b) and to determine the logical value of the feedback function for each transition. For example, in going from S_0 to S_1, the state of flip-flop A must change from 0 to 1, and hence the required input to this flip-flop, $J_A = 1$. This is the logical value of the feedback function required for this transition, and it is entered in the final column of the state table.

The feedback function and the unused states are plotted on a four-variable K-map as shown in figure 7.8(c). The feedback function is marked by a 1 in the appropriate cells, whilst the unused states are marked with a 'd'.

It should be noted that S_{15} is an unused state in this counter and it appears that the S_{15} cell should have been marked with a 'd'. However,

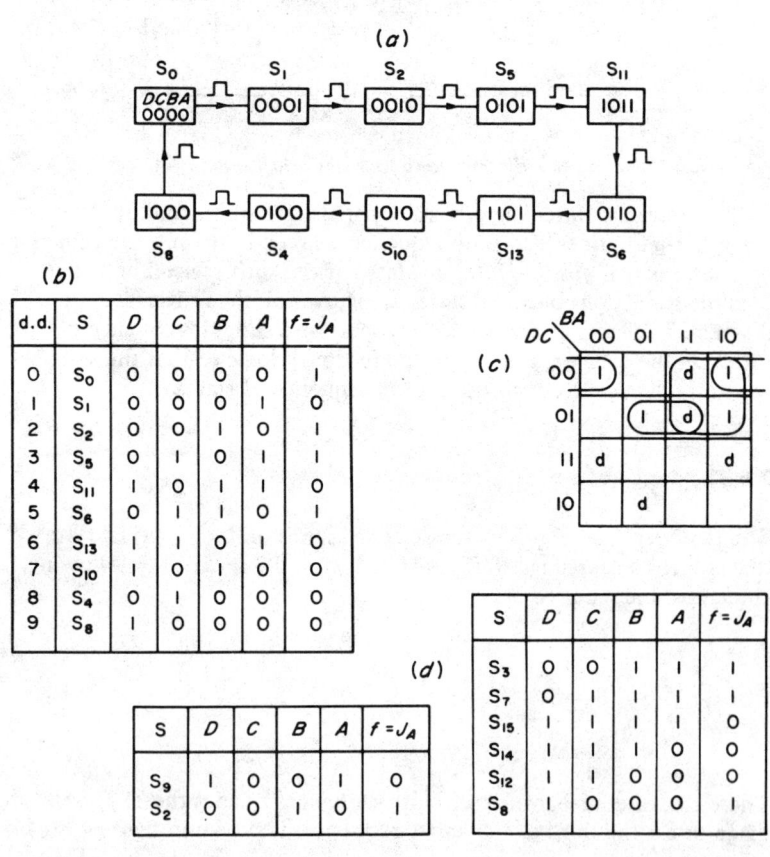

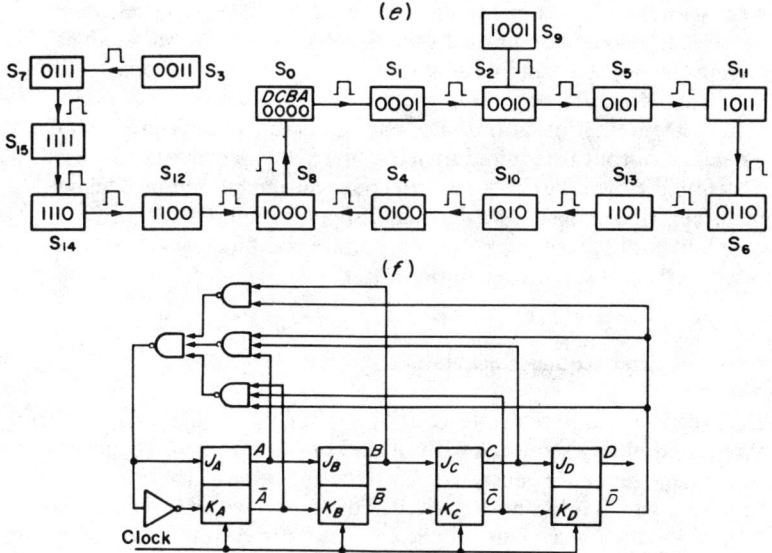

Figure 7.8. (a) State diagram for the decade counter. (b) State table for the decade counter. (c) K-map for the feedback function $f = J_A$. (d) Behaviour of the counter after entering an unused state. (e) Full state diagram for the counter. (f) Implementation of a shift register decade counter

a perfectly general rule that should be observed when designing this type of counter is that the entry in the S_{15} cell on the K-map should always be a 0 and that in the S_0 cell, should always be a 1, irrespective of whether these two states are in the counting sequence. This ensures that the counter will never be locked in either the 0000 or 1111 states.

Using the normal minimisation techniques, the feedback function is found to be

$$f = J_A = B\bar{D} + AC\bar{D} + \bar{A}\bar{C}D$$

If the circuit enters an unused state the logic of the unused states

$$f_u = AB\bar{D} + \bar{A}CD + A\bar{B}\bar{C}D + ABCD$$

can be used to stop the counter, raise an alarm, and reset all the flip-flops to zero.

If the counter enters an unused state due to faulty circuit operation, and if the logic of the unused states is not utilised, it will return to the correct sequence after a maximum of five clock pulses. The behaviour of the circuit when in an unused state is described by the state tables shown in figure 7.8(d) and a full state diagram, including the unused states, is shown in figure 7.8(e).

The implementation of the basic counter is shown in figure 7.8(f). If a decimal output is required then a 4-to-16 line decoder can be used. For example, when $A\bar{B}CD = 1$ the decoder output should be decimal digit three. Alternatively, if a decimal display is required then the counter, in conjunction with the appropriate combinational logic, can be used to drive a seven-segment indicator.

7.7 Shift register sequence generators

A shift register counter with feedback logic can be modified to produce any required binary sequence with the aid of output logic. The length of the binary sequence generated, l, will be the same as that of the count from which it has been derived. For an N-stage shift register the length of the binary sequence is $l \leqslant 2^N$. The basic configuration of such a sequence generator is shown in figure 7.9.

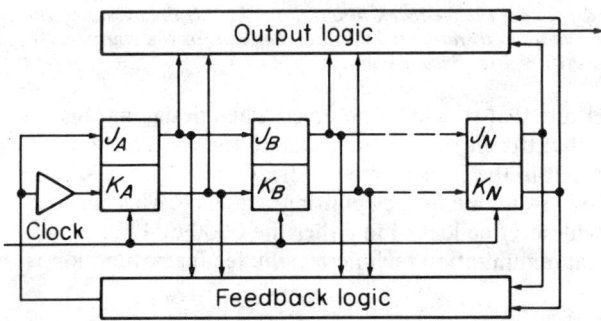

Figure 7.9. *Basic configuration of a binary sequence generator*

As an example, consider the design of a circuit that will generate the binary sequence 1–0–1–1–0–1–1–0. Since there are eight bits in this binary sequence a three-stage shift register will be required. The eight three-bit combinations required to generate this sequence are tabulated in figure 7.10(a), where the binary digits in the column headed q_a are the required sequence, whilst those in columns q_b and q_c are the same sequence delayed in time with respect to the sequence in column q_a, by one and two clock pulses, respectively.

Shift register counters and generators

(a)

Clock pulse	q_c	q_b	q_a
1	1	0	1
2	0	1	0
3	1	0	1
4	0	1	1
5	1	1	0
6	1	0	1
7	0	1	1
8	1	1	0

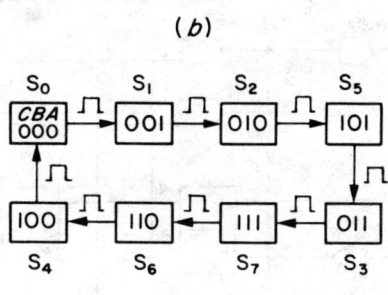

(b)

(c)

S	C	B	A	Feedback function f
S_0	0	0	0	1
S_1	0	0	1	0
S_2	0	1	0	1
S_5	1	0	1	1
S_3	0	1	1	1
S_7	1	1	1	0
S_6	1	1	0	0
S_4	1	0	0	0

(d)

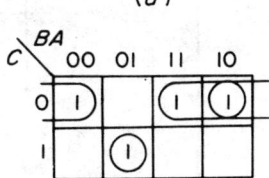

(e)

S	C	B	A	q_a
S_0	0	0	0	1
S_1	0	0	1	0
S_2	0	1	0	1
S_5	1	0	1	1
S_3	0	1	1	0
S_7	1	1	1	1
S_6	1	1	0	1
S_4	1	0	0	0

(f)

C \ BA	00	01	11	10
0	1			1
1		1	1	1

(g)

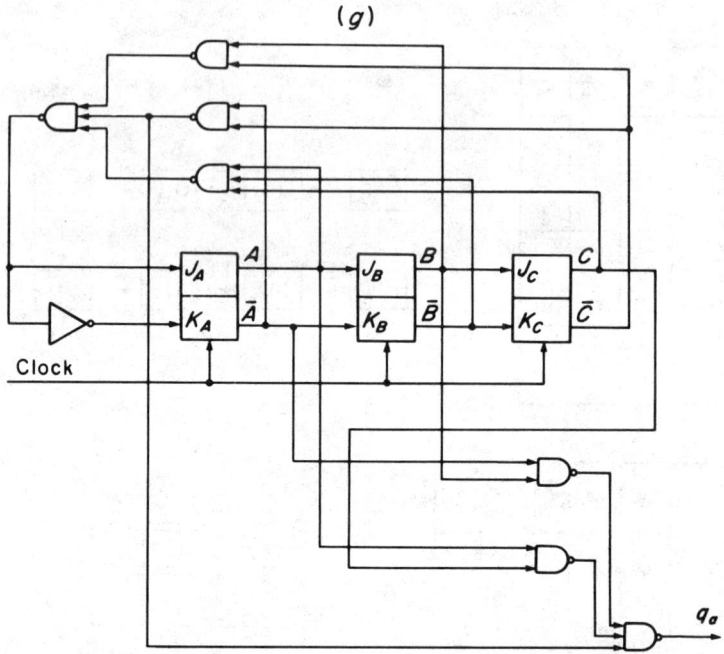

Figure 7.10. (a) The eight three-bit combinations generated from the binary sequence. (b) State diagram for the eight-state sequence. (c) State table for the feedback function. (d) K-map plot of the feedback function. (e) State table for the output function. (f) K-map for the output function q_a. (g) Implementation of a binary-sequence generator

It will be observed that the states 101 and 110 both occur more than once in the tabulation. In the case of the state 101 an ambiguity exists since in one case the next state is 010, and in the other two cases it is 011. Consequently, these eight combinations cannot be generated by the method of direct feedback logic employed in the design of the decade counter, and output logic has to be used to produce the required sequence.

The method employed is to develop an eight-state sequence using direct-feedback logic, the required binary sequence then being derived from this eight-state sequence with the aid of output logic.

An eight-state sequence is obtained from the de Bruijn diagram for a three-stage shift register shown in figure 7.6(b). Such a sequence is:

$$S_0 - S_1 - S_2 - S_5 - S_3 - S_7 - S_6 - S_4 - S_0,$$

and the state diagram for the sequence is shown in figure 7.10(b).

Shift register counters and generators

The state table is drawn up as shown in figure 7.10(c) and the value of the feedback function, either 0 or 1, is determined for each transition as described previously.

The feedback function is plotted on a K-map as shown in figure 7.10(d). Using normal minimisation techniques the feedback function is found to be

$$f = B\bar{C} + \bar{A}\bar{C} + A\bar{B}C$$

The output binary sequence required, q_a, is now tabulated with the eight-state shift register sequence in figure 7.10(e), and the minimal form of the output function

$$q_a = \bar{A}\bar{C} + \bar{A}B + AC$$

is obtained from the K-map shown in figure 7.10(f).

There are other output sequences such as q_b and q_c in figure 7.10(a) which are merely displaced in time from the sequence in column q_a of figure 7.10(e), and the logic for these sequences can be examined to see which requires the least hardware. The implementation of the sequence generator is shown in figure 7.10(g).

7.8 The ring counter

The simplest type of shift register counter is the ring counter, where feedback is provided from the last stage and feeds the inputs of the first stage, as shown in figure 7.11(a). In this circuit there are ten stages and it can be used as a decimal ring counter, since the number of stages is equal to the number of states. The information contained in each stage is shifted to the next stage on the receipt of a clock pulse and the counter circulates a 1 which is initially preset in the first stage, all other stages being simultaneously cleared to 0. The counting sequence of the register is shown in the state table of figure 7.11(b).

The circuit of figure 7.11(a) can be modified so that it becomes self-starting, as shown in figure 7.11(c). The input is

$$J_A = \overline{ABCDEFGHI}$$

and this can only be a 1 provided that $A = B = C = D = E = F = G = H = I = 0$. Clearly, if any section of the counter except the last one

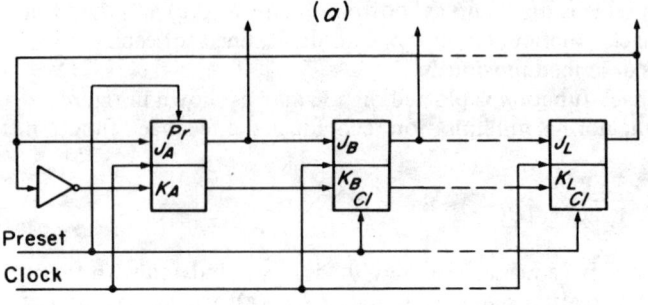

(a)

(b)

Clock pulse	L	I	H	G	F	E	D	C	B	A
0	0	0	0	0	0	0	0	0	0	1
1	0	0	0	0	0	0	0	0	1	0
2	0	0	0	0	0	0	0	1	0	0
3	0	0	0	0	0	0	1	0	0	0
4	0	0	0	0	0	1	0	0	0	0
5	0	0	0	0	1	0	0	0	0	0
6	0	0	0	1	0	0	0	0	0	0
7	0	0	1	0	0	0	0	0	0	0
8	0	1	0	0	0	0	0	0	0	0
9	1	0	0	0	0	0	0	0	0	0

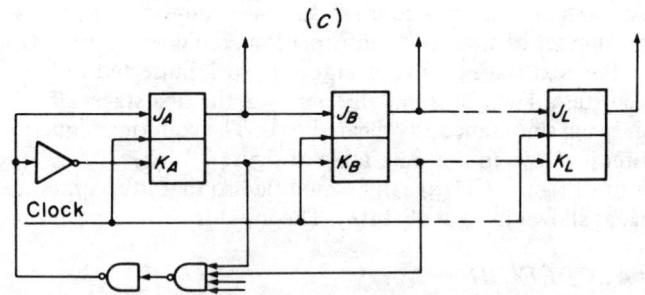

Figure 7.11. (a) The basic ring counter. (b) State table for a ten-stage ring counter. (c) Self-starting ring counter

contains a 1, $J_A = 0$, and the counter will now enter the required sequence within a maximum of ten clock pulses.

The ten outputs of this counter can be used directly as decimal outputs without the need of decoding networks, or, alternatively, the circulating 'one' can be used to enable a group of circuits sequentially. The number of stages required in the latter case will be equal to the number of circuits that have to be enabled.

An obvious advantage of the decimal ring counter is its simplicity and since it requires no feedback logic or decoding circuits it uses fewer components. It does, though, have the disadvantage of not having a binary readout and its counting sequence is radically changed if through faulty circuit operation it enters one of the many unused states.

A binary counter having ten stages will have $2^{10} = 1024$ counting states, and it can in fact count up to 1023_{10}, whereas the decimal ring counter only has ten counting states. It is clear, therefore, that the decimal ring counter has $2^{10} - 10 = 1014$ unused or forbidden states and therefore makes very uneconomic use of the flip-flops. If the counter, for some reason, enters one of these unused states it falls into a forbidden counting sequence, of which there are many, and it will never re-enter the correct counting sequence unless forced to do so.

The two basic faults that can arise are that a section other than that containing the circulating 1, is also set at 1, due to some circuit fault, or, alternatively, the circulating 1 is accidentally set to 0. However, it is not difficult to introduce simple logical networks which detect the presence of additional 1's. A three-stage ring counter having this facility is shown in figure 7.12(a). Similarly, it is not difficult to introduce a logical network which will indicate whether all the sections of the shift register contain 0's. A circuit providing this facility is shown in figure 7.12(b).

A function which can be used for the detection of additional 1's in the three-stage shift register is

$$f_1 = (A + B)C$$

This will not necessarily give an instantaneous indication since it depends where the 1's are situated when the fault arises. If they are held in flip-flops A and B, a one clock pulse delay would take place before an indication occurred. Instantaneous indication would be given by the function

$$f_{1I} = AB + AC + BC$$

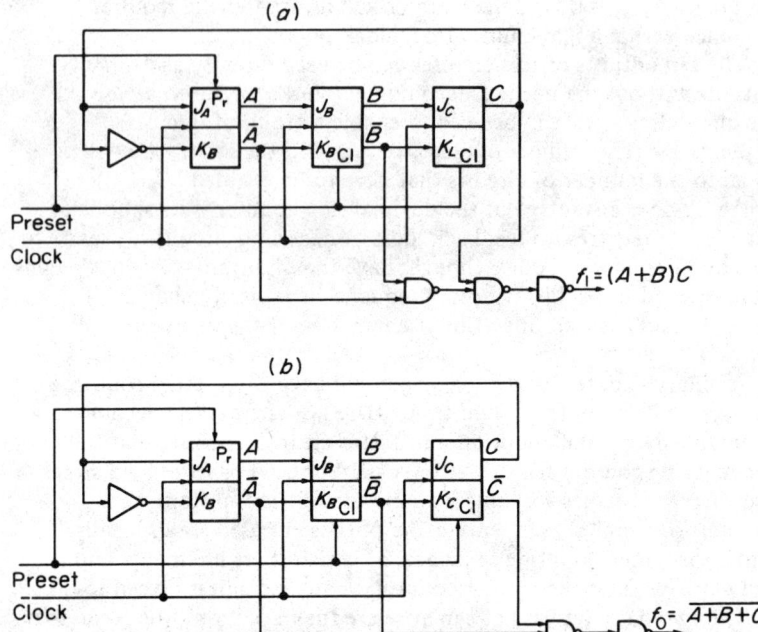

Figure 7.12. *Three-stage ring counter. (a) Circuit for detecting additional 1's. (b) Circuit for detecting all 0's*

Unfortunately, as the number of stages in the counter increases, the function for giving instantaneous detection requires increasingly large amounts of hardware. For example, with four stages the instantaneous function would be

$$f_{1\text{I}} = AB + AC + AD + BC + BD + CD$$

and consequently it might be more appropriate to use the function

$$f_1 = (A+B)(C+D)$$

which would give a fault indication with a maximum of one clock pulse delay.

The function which can be used for detecting all the 0's is given by

$$f_0 = \overline{A+B+C} = \bar{A}\,\bar{B}\,\bar{C}$$

Shift register counters and generators 147

If a ring counter circulating a 0 is required, rather than a 1, it is necessary on starting to preset $(N - 1)$ of the flip-flops in the N-stage shift register to 1 and clear the remaining one. Alternatively, if the complementary outputs are available in the standard ring counter these will provide a circulating 0.

7.9 The twisted ring or Johnson counter

As the name implies, the difference between the twisted ring counter and the ordinary ring counter is that the feedback connections are reversed and in this case the complementary output of the last stage is connected to the J input of the first stage, whilst the inverted form of this signal is fed to the K input. If all the flip-flops are initially preset to the same state, either 0 or 1, then the number of different states in the desired count sequence is equal to twice the number of stages in the shift register. Hence a decade counter can be constructed from a five-stage shift register as illustrated in figure 7.13(a). The counting sequence of the circuit, assuming that initially all the flip-flops are cleared to zero, is given in figure 7.13(b).

This is a ten-state sequence which could have been selected from the universal state diagram of a five-stage shift register. The feedback logic could have been developed by first tabulating the required value of the feedback function in the column headed f in the table of figure 7.13(b) and then plotting this function in conjunction with the unused states on a K-map, as shown in figure 7.13(c). Simplifying, using the normal techniques, gives

$$f = J_A = \overline{E}$$

For this circuit decoding logic is required to obtain a decimal count. This logic is obtained from a five-variable K-map on which the decimal equivalent (corresponding to the clock pulse numbering in figure 7.13(b)) for each of the states in the counting cycle has been marked, as illustrated in figure 7.13(d).

The simplifying adjacencies for decimal 0 and 1 have also been marked on the map and if the reader cares to continue the process of simplification, it will be seen that it is always possible to combine seven unused states with each of the decimal entries.

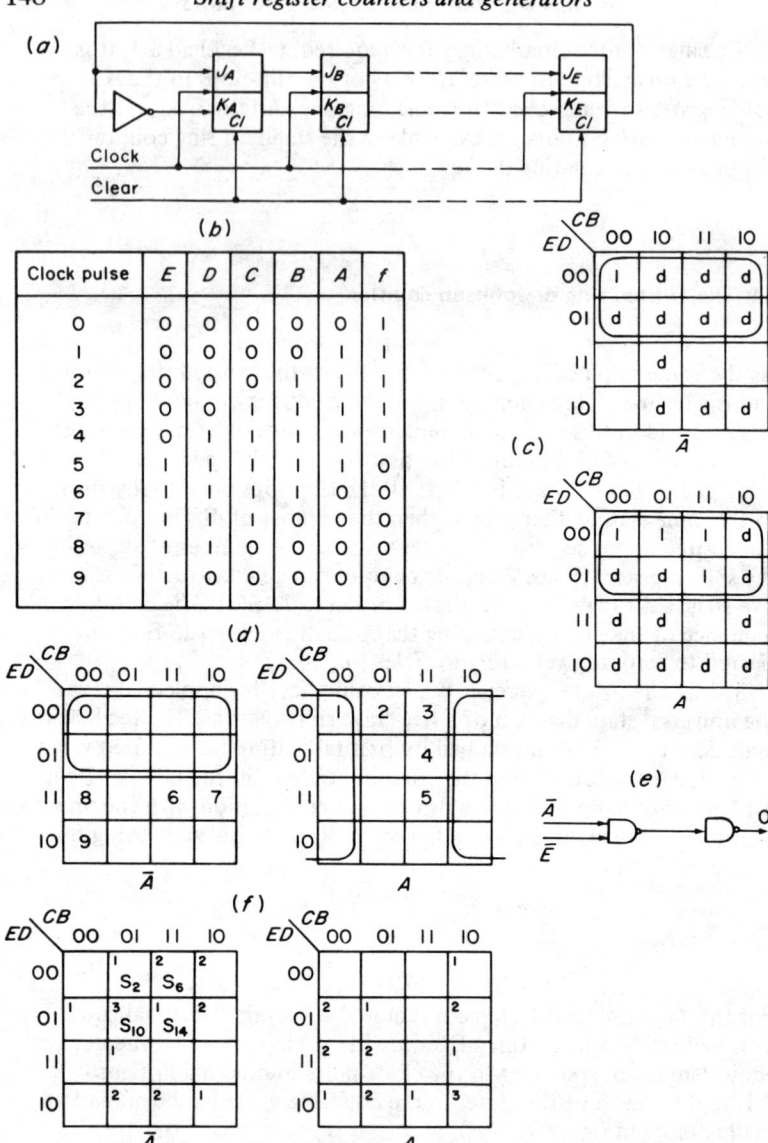

Figure 7.13. (a) Twisted ring decade counter. (b) Counting sequence of a twisted ring counter. (c) K-map plot for determining the feedback function of the twisted ring counter. (d) K-map plot for determination of decimal decode logic. (e) Decoding logic for decimal digit 0. (f) K-map plot for determination of the error-correcting or self-starting function

The resulting decode logic after simplification is tabulated below:

$0 = \bar{A}\bar{E}$ $\quad 5 = AE$
$1 = A\bar{B}$ $\quad 6 = \bar{A}B$
$2 = B\bar{C}$ $\quad 7 = \bar{B}C$
$3 = C\bar{D}$ $\quad 8 = \bar{C}D$
$4 = D\bar{E}$ $\quad 9 = \bar{D}E$

Implementation of the decoding function for decimal digit 0 is shown in figure 7.13(e).

There are also three other undesired and independent count sequences for the Johnson counter:

(1) $S_2-S_5-S_{11}-S_{23}-S_{14}-S_{29}-S_{26}-S_{20}-S_8-S_{17}-S_2$

(2) $S_4-S_9-S_{19}-S_6-S_{13}-S_{27}-S_{22}-S_{12}-S_{25}-S_{18}-S_4$

(3) $S_{10}-S_{21}-S_{10}$.

If the counter should enter any one of these sequences due to faulty circuit operation or when switching on, it will remain in that sequence unless arrangements are made to return the counter to the required sequence.

The unwanted sequences are shown plotted on the K-map in figure 7.13(f), cells marked with a 1 being in sequence 1, and so on. It will be observed that the four adjacent states S_2, S_6, S_{10} and S_{14} are all in one of the three unwanted sequences. If the Boolean function that represents these four states

$$f = \bar{A}B\bar{E}$$

is used to clear the five stages of the counter, then within a maximum of ten clock pulses the counter will enter the desired sequence.

The Johnson counter has an even-numbered cycle length of $2N$, where N is the number of stages in the shift register. However, with a suitable modification of the feedback it is possible to achieve an odd-numbered cycle length of $(2N - 1)$. For example, if the state 00000 is omitted, the counting cycle becomes that shown in the table of figure 7.14(a) and the values of the new feedback function required to produce this sequence are tabulated in the last column of this table.

Shift register counters and generators

(a)

E	D	C	B	A	f
0	0	0	0	1	1
0	0	0	1	1	1
0	0	1	1	1	1
0	1	1	1	1	1
1	1	1	1	1	0
1	1	1	1	0	0
1	1	1	0	0	0
1	1	0	0	0	0
1	0	0	0	0	1

(b)

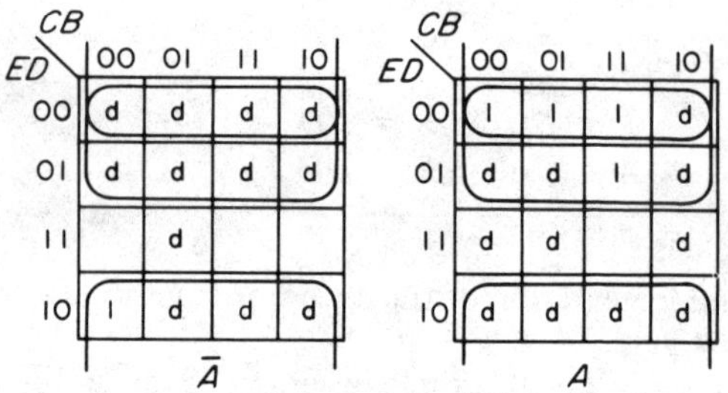

Figure 7.14. (a) Counting sequence of an odd-numbered cycle length Johnson counter. (b) Determination of the feedback function for the Johnson counter of cycle length (2N − 1)

Plotting this function in conjunction with the unused states on the K-map of figure 7.14(b) and minimising, leads to the revised feedback function

$$f = \bar{D} + \bar{E}$$

Alternatively, if the 11111 state is omitted rather than the 00000 state, the revised feedback function can be shown to be $f = \overline{DE}$.

7.10 Shift registers with exclusive-OR feedback

The four-stage shift register shown in figure 7.15(a) has exclusive-OR feedback from stages C and D such that the input to the first stage $J_A = C \oplus D$. To determine the sequence of states for the register, it is assumed initially that the shift register is in the state $D = 0, C = 0$, $B = 0$ and $A = 1$, in which case $J_A = 0 \oplus 0$, and on receipt of the next clock pulse the register enters the state $D = 0, C = 0, B = 1$ and $A = 0$. The complete sequence of states for the register is shown in figure 7.15(b), the value of the feedback function for each state being tabulated in column f.

In all, there are 15 states, and this is the maximum number of states a four-stage register having exclusive-OR feedback can have, so this sequence is termed the maximum-length sequence (MLS). The $S_0 = 0000$ state is not included in the sequence since this is a 'lock-in'

Shift register counters and generators

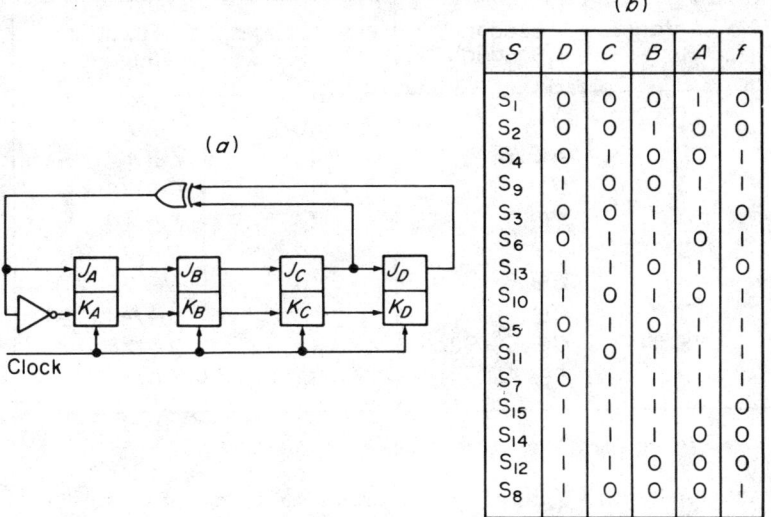

Figure 7.15. (a) Four-stage MLS shift register generator. (b) MLS for the four-stage shift register

state. If the register enters this state $J_A = 0 \oplus 0 = 0$, so that the register is unable to leave this state when the next and subsequent clock pulses arrive. In general, the maximum-length sequence for such a circuit is given by

$$l = 2^N - 1$$

where N is the number of stages in the shift register.

Not all exclusive-OR connections result in a maximum-length sequence. The table in figure 7.16 gives the feedback functions which will give the maximum-length sequence for values of N up to, and including, 18.

Other maximum-length sequences are available with the same register length. For example, if the inverse of the exclusive-OR function $C \oplus D$ is used as feedback, then an alternative maximum-length sequence is obtained and this is shown tabulated in figure 7.17(a).

An examination of the feedback equation column in figure 7.16 shows that one of the digits in the equation is always the Nth digit in the register, and the other digit (or digits) is obtained by looking back down the register. For example, for $N = 4$ the Nth digit is D, and the other digit in the equation, C, is the $(N - 1)$th digit. Two alternative

No of stages, N	Feedback equation	No of stages, N	Feedback equation
1	A	10	$G \oplus J$
2	$A \oplus B$	11	$I \oplus K$
3	$B \oplus C$	12	$F \oplus H \oplus K \oplus L$
4	$C \oplus D$	13	$I \oplus J \oplus L \oplus M$
5	$C \oplus E$	14	$D \oplus H \oplus M \oplus N$
6	$E \oplus F$	15	$N \oplus O$
7	$F \oplus G$	16	$D \oplus M \oplus O \oplus P$
8	$D \oplus E \oplus F \oplus H$	17	$N \oplus Q$
9	$E \oplus I$	18	$K \oplus U$

Figure 7.16. *Feedback functions for maximum-length sequences*

maximum-length sequences for a four-stage register can be obtained by looking forward to the $(N+1)$th digit which, in this case, is A. Hence the other two MLSs are obtained by using the feedback equations $A \oplus D$ and $A \odot D$, and these sequences are shown tabulated in figure 7.17(b).

(a)

S	D	C	B	A	$f = C \odot D$
S_1	0	0	0	1	1
S_3	0	0	1	1	1
S_7	0	1	1	1	0
S_{14}	1	1	1	0	1
S_{13}	1	1	0	1	1
S_{11}	1	0	1	1	0
S_6	0	1	1	0	0
S_{12}	1	1	0	0	1
S_9	1	0	0	1	0
S_2	0	0	1	0	1
S_5	0	1	0	1	0
S_{10}	1	0	1	0	0
S_4	0	1	0	0	0
S_8	1	0	0	0	0
S_0	0	0	0	0	1

(b)

S	D	C	B	A	$f = A \oplus D$
S_1	0	0	0	1	1
S_3	0	0	1	1	1
S_7	0	1	1	1	1
S_{15}	1	1	1	1	0
S_{14}	1	1	1	0	1
S_{13}	1	1	0	1	0
S_{10}	1	0	1	0	1
S_5	0	1	0	1	1
S_{11}	1	0	1	1	0
S_6	0	1	1	0	0
S_{12}	1	1	0	0	1
S_9	1	0	0	1	0
S_2	0	0	1	0	0
S_4	0	1	0	0	0
S_8	1	0	0	0	1

(c)

S	D	C	B	A	$f = A \odot D$
S_1	0	0	0	1	0
S_2	0	0	1	0	1
S_5	0	1	0	1	0
S_{10}	1	0	1	0	0
S_4	0	1	0	0	1
S_9	1	0	0	1	1
S_3	0	0	1	1	0
S_6	0	1	1	0	1
S_{13}	1	1	0	1	1
S_{11}	1	0	1	1	1
S_7	0	1	1	1	0
S_{14}	1	1	1	0	0
S_{12}	1	1	0	0	0
S_8	1	0	0	0	0
S_0	0	0	0	0	1

Figure 7.17. *(a) The* MLS *for a four-stage shift register with feedback $C \odot D$. (b) The* MLS *with feedback $A \oplus D$. (c) The* MLS *with feedback $A \odot D$.*

Shift register counters and generators

Clearly, the circuit shown in figure 7.15(a) can be used as a binary sequence generator, the output sequence being taken directly from the output of one of the flip-flops in the register. In this case, the binary sequence appearing at the output of flip-flop D is 0–0–0–1–0–0–1–1–0–1–0–1–1–1–1. This kind of generator is sometimes referred to as pseudo-random binary sequence generator because the digits in the sequence are in apparently random order. However, the randomness repeats itself every $2^N - 1$ clock pulses. For a given clock frequency, the periodicity of the randomness increases very rapidly with the number of stages in the register. If

$$N = 10 \quad 2^N - 1 = 1023$$

and if the clock frequency is 1 MHz the sequence repeats itself every 1·02 ms. If

$$N = 20 \quad 2^N - 1 = 1\,048\,575$$

and the period of the sequence = 1·05 s. If

$$N = 30 \quad 2^N - 1 = 1\,073\,730\,624$$

and the period of the sequence = 1073·73 s.

Non-maximum-length sequences can be generated with a four-stage register if some other exclusive-OR feedback is used. For example, if the feedback function is $B \oplus D$, one of the sequences tabulated in figure 7.18 will be generated. The sequence generated will depend upon the initial state of the register.

The basic MLS generator shown in figure 7.15(a) is not necessarily self-starting, since on switching on, the initial state of the generator

S	D	C	B	A	f	S	D	C	B	A	f	S	D	C	B	A	f
S_1	0	0	0	1	0	S_3	0	0	1	1	1	S_6	0	1	1	0	1
S_2	0	0	1	0	1	S_7	0	1	1	1	1	S_{13}	1	1	0	1	1
S_5	0	1	0	1	0	S_{15}	1	1	1	1	0	S_{11}	1	0	1	1	0
S_{10}	1	0	1	0	0	S_{14}	1	1	1	0	0						
S_4	0	1	0	0	0	S_{12}	1	1	0	0	1						
S_8	1	0	0	0	1	S_9	1	0	0	1	1						

Figure 7.18. *Non-maximum-length sequences generated by a four-stage shift register with feedback $B \oplus D$*

may be 0000. As the circuit stands there is no way in which it can leave this state. However, with a slight modification to the feedback circuit it is possible to make the generator self-starting. The required modification is the logical addition of the term $\bar{A}\bar{B}\bar{C}\bar{D}$ to the feedback equation so that it becomes

$$f = C \oplus D + \bar{A}\bar{B}\bar{C}\bar{D}$$

This function is plotted on the K-map shown in figure 7.19(a), and after simplification it reduces to

$$f = C \oplus D + \bar{A}\bar{B}\bar{D}$$

The implementation of the self-starting generator is shown in figure 7.19(b).

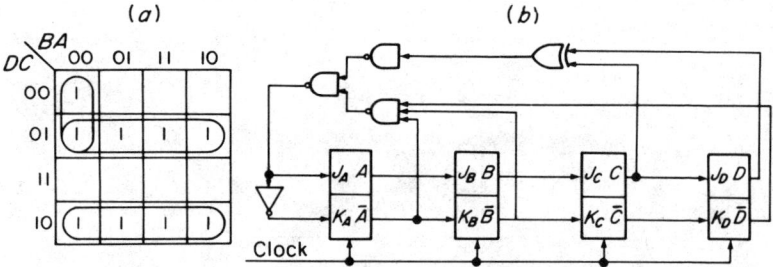

Figure 7.19. (a) K-map plot for a self-starting MLS generator (b).

It is also possible to generate non-maximum-length sequences by using a jump technique. The method of approach is to start with an MLS generator using exclusive-OR feedback and then reduce the length of the sequence by introducing additional feedback. The method will be described for the four-stage shift register generator shown in figure 7.15(a).

It will be assumed that initially the generator is in the state $DCBA = 0011$ (S_3). If, when in this state, the feedback is a 0 then the next state of the generator will be $DCBA = 0110$ (S_6). Alternatively, if the feedback had been 1, then the next state of the generator would have been $DCBA = 0111$ (S_7).

Examination of the state table for the four-stage MLS generator in figure 7.15(b) shows that $C \oplus D = 0$ when the generator is in state S_3 and the next state is therefore S_6. If, however, the feedback is modified to a 1 then the next state of the generator is S_7.

Shift register counters and generators

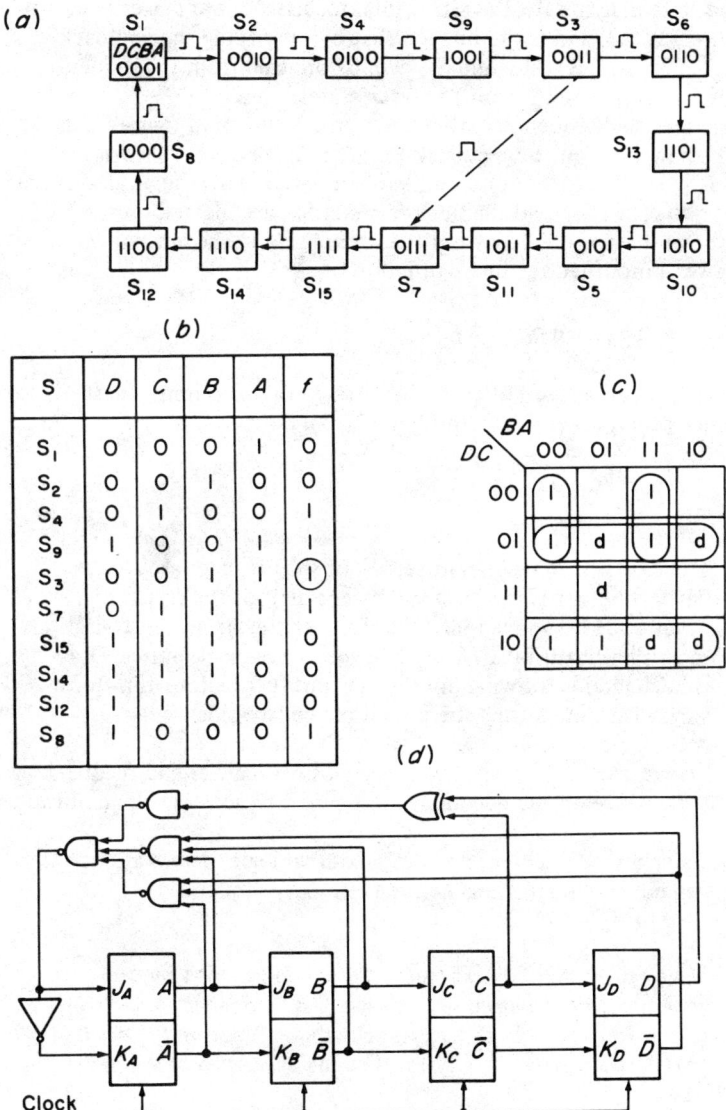

Figure 7.20. (a) State diagram of the four-stage MLS generator with modified feedback showing the jump. (b) Modified MLS sequence. (c) K-map plot of the feedback function f. (d) Implementation of an MLS generator employing the 'jump' technique

Shift register counters and generators

The state diagram for the MLS generator having four stages is shown in figure 7.20(a), and it can be seen that by modifying the feedback, the states S_6, S_{13}, S_{10}, S_5 and S_{11} will be omitted from the sequence, thus reducing its length from 15 to ten states.

The modified sequence for the generator is shown in figure 7.20(b) and the new value of the feedback function in state S_3 is shown encircled. The feedback function, in conjunction with the unused states S_6, S_{13}, S_{10}, S_5, S_{11} and the 'lock-in' state S_0, are plotted on the K-map shown in figure 7.20(c) and then simplified in the normal way. This gives a modified feedback function of

$$f_m = C \oplus D + AB\bar{D} + \bar{A}\bar{B}\bar{D}$$

and the implementation of this self-starting non-maximum-length sequence generator is shown in figure 7.20(d).

Problems

7.1 The contents of a serial-in/serial-out shift register are $DCBA = 0101$, where A is the least significant digit of the register. A serial input 10011 is moved into the shift register, from left to right, most significant bit first, by five successive clock pulses. Draw time diagrams showing how the outputs of the four flip-flops vary with time during the period of the five clock pulses.

7.2 Design a modulo-12 counter using a shift register and feedback logic. Develop the decode logic required to give a decimal output.

7.3 Using a shift register and combinational logic, design a sequence generator, which will generate the binary sequence 0–1–0–0–1–0–1–1–1–0–1.

7.4 Develop the state diagrams for the following shift register generators which employ exclusive-OR feedback:
 (a) A four-stage shift register. Feedback function $f = B \oplus C$.
 (b) A five-stage shift register. Feedback function $f = D \oplus E$.

7.5 A three-stage shift register is to be used to generate two sequences of length 7 and 5, respectively. When a control signal $m = 1$, it generates a sequence of length 7, and when the control signal $m = 0$ it generates a sequence of length 5. Design a shift register

Shift register counters and generators

generator using exclusive-OR feedback to implement the above specification.

7.6 A three-stage shift register ABC having exclusive-OR feedback $B \oplus C$, where A is the least significant stage of the register, is to be used to repeatedly produce a sequence of binary coded decimal digits for e (2·718282) on four output lines P, Q, R and S.

Determine the sequence developed by the generator and develop the combinational logic required to generate the sequence for e.

7.7 Draw a timing diagram for a four-stage twisted ring counter for a period of eight clock pulses. Display the outputs of each of the flip-flops on the timing diagram.

If the counting sequence is to be reduced from eight to seven by the omission of the 1 1 1 1 state, determine the modification of the feedback logic that is required.

8

Clock-driven sequential circuits

8.1 Introduction

In this chapter a design procedure will be established for the design and implementation of clock-driven sequential circuits. Such circuits have many applications in the digital field and consist of both combinational and single-bit memory elements. In practice, the combinational elements will be members of one of the commonly used logic families whilst the single-bit memory elements will be *JK* flip-flops. It is normal to use *JK* flip-flops in clock-driven circuits because they provide the simplest logic implementation.

As previous experience in other fields of electronics shows, circuit analysis is a good deal easier then circuit synthesis and, as a first step towards the establishment of a design procedure, a circuit whose behaviour is known will be analysed. This analysis will allow the identification of those processes which are required for circuit synthesis.

8.2 Analysis of a clocked sequential circuit

The logic diagram shown in figure 8.1(*a*) is that of a clocked sequential circuit. An examination of this diagram shows that the circuit has two inputs, X and clock (⊓), and one output Z. The memory elements used are two *JK* master/slave flip-flops which define the four possible internal states of the circuit, AB = 00, 01, 10 or 11.

An alternative way of representing this circuit is by means of a block diagram such as the one shown in figure 8.1(*b*). This diagram depicts a logic box which contains all the combinational logic, as well as the two

Clock-driven sequential circuits

flip-flops, A and B, whose output combinations define the four internal states of the circuit. The inputs to the box are X and clock, whilst the output from the box is signal Z. The input equations for flip-flops A and B can be obtained directly from figure 8.1(a):

$$J_A = B \qquad J_B = \overline{A}$$

$$K_A = 1 \qquad K_B = \overline{X} + A$$

In § 5.4 the state table for a JK flip-flop was developed, and this is repeated in figure 8.1(c) for convenience. From the table, the plot of the next state function $Q^{t+\delta t}$, shown in figure 8.1(d), is obtained, and after simplification it may be written as

$$Q^{t+\delta t} = (J\overline{Q} + \overline{K}Q)^t$$

By substituting J_A, K_A and J_B, K_B in this equation the next state functions of the two flip-flops A and B are obtained:

$$A^{t+\delta t} = (\overline{A}B)^t$$

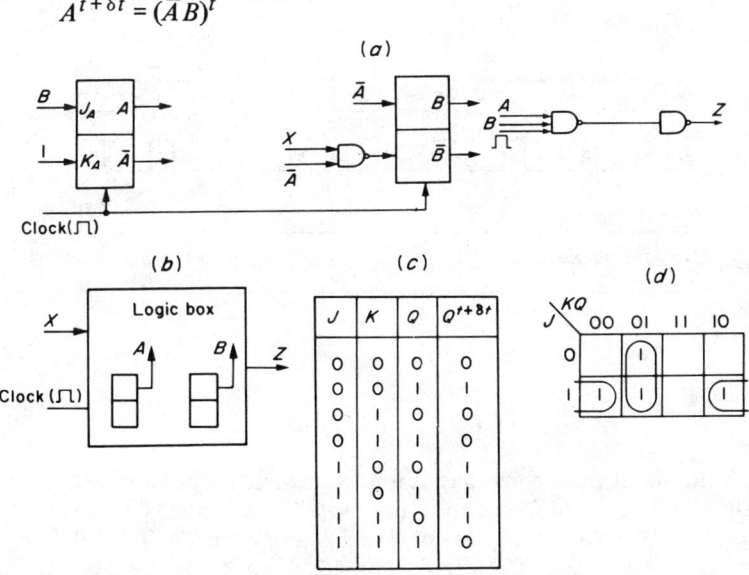

Figure 8.1. (a) Sequential circuit to be analysed. (b) Block diagram for the circuit. (c) State table for a JK flip-flop. (d) K-map for the JK flip-flop. (e) State table for the circuit to be analysed. (f) State diagram of the circuit. (g) Generation of the output signal Z. (h) Timing diagram for the circuit

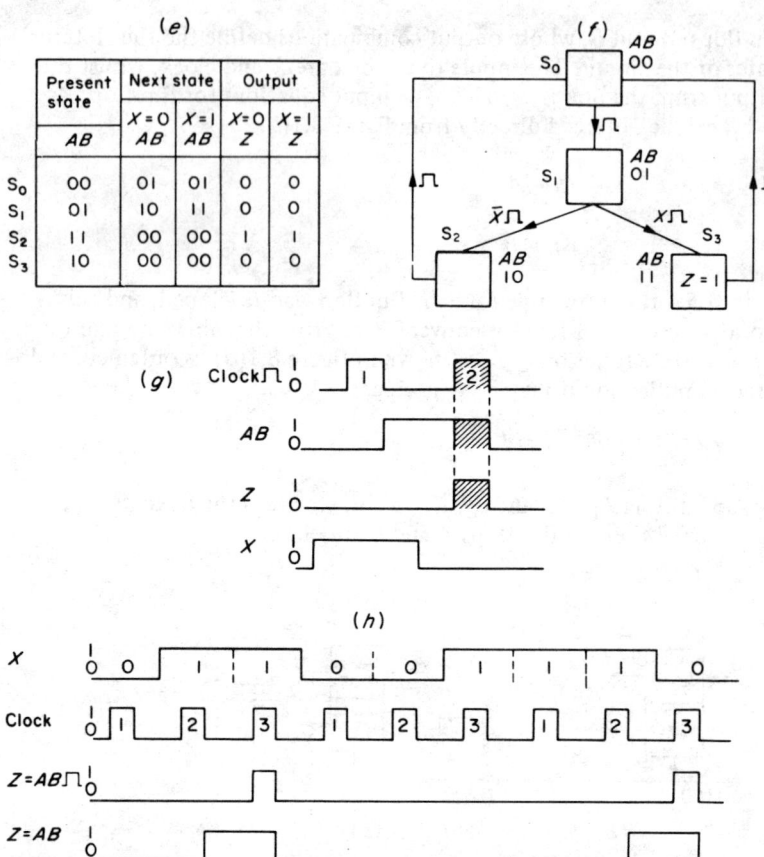

Figure 8.1. Cont.

and

$$B^{t+\delta t} = (\overline{A}BX + \overline{A}\,\overline{B})^t$$

With the aid of these equations, it is now possible for given present state values of A and B, and for a given value of the input signal X, to determine the next state values of A and B. For example, if $A = 0$, $B = 0$ and $X = 0$, then $A^{t+\delta t} = 0$. Alternatively, if $A = 0, B = 0$ and $X = 0$, then $B^{t+\delta t} = 1$.

In the same way as the next states of flip-flops A and B have been obtained, it is possible to determine the output Z of the circuit for all possible combinations of X, A and B. This requires a knowledge of the output equation, which is obtained directly from figure 8.1(*a*) and is

$$Z = AB \sqcap$$

The interpretation of this equation is that the output $Z = 1$, when the present internal state of the circuit is $A = 1$, $B = 1$, and if either $X = 0$ or $X = 1$ is received at the input in conjunction with a clock signal. For all other combinations of A and B the output $Z = 0$, irrespective of the value of X or the presence of the clock. Further, the equation indicates that the time duration of $Z = 1$ at the output, can never be greater than the time duration of the clock pulse.

It is now possible to construct a table showing the present state of the circuit, the next state of the circuit and the output. This table, shown in figure 8.1(*e*), may be regarded as the state table of the circuit. The reader will notice that in this table the internal states have been designated as S_0, S_1, etc.

With the aid of this state table the internal state diagram can now be constructed and it is shown in figure 8.1(*f*). It consists of four rectangles, each of which represents an internal state of the circuit. A transition from one state to the next is represented by a straight line between the two states with an arrowhead indicating the direction of the transition. By the arrowhead is placed the transition signal. For instance, in order to make a transition from S_1 to S_3 the circuit needs to receive X and clock ($X \sqcap$).

Assuming that the flip-flops used are master/slave JK flip-flops, then the transition always takes place on the trailing edge of the clock pulse. For example, the transition from S_1 to S_2 takes place on the trailing edge of the clock pulse which forms part of the transition signal $X \sqcap$.

The output $Z = 1$ has been entered in the rectangle marked S_3. This output allocation should be interpreted as follows. $Z = 1$ if the circuit is in internal state S_3, i.e. $AB = 11$, and if a clock pulse is received. Hence $Z = AB \sqcap$, as stated above.

The generation of the output signal Z is illustrated in figure 8.1(*g*). The circuit enters state S_3 from S_1 on the trailing edge of the clock pulse marked 1. In order for this transition to take place X must be present, as shown in the diagram. The circuit remains in the state S_3 until the trailing edge of the clock pulse marked 2 in the diagram, which initiates the transition from S_3 to S_0. The output $Z = AB \sqcap$ is formed by the ANDing of AB and the clock pulse marked 2. Both of these signals are logical 1 in the shaded region shown on the diagram, hence in this region the output $Z = 1$. It should be noticed that the output is independent of the logical value of X, i.e. $Z = 1$ irrespective of whether $X = 0$ or 1 during the period of the shaded region.

What can be deduced about the function of this circuit from the state diagram? Initial observations indicate that there are two transition paths through the state diagram starting with S_0. The first one is via S_3 and back to S_0, and the second one is via S_2 and back to S_0. Secondly, it is clear that no matter which of these paths is taken from S_0 there are always three transitions before returning to S_0. This infers that strings of three binary digits arriving on the X line are being examined by the circuit.

Certain combinations of the three digits will result in an output on the third clock pulse. This will occur if the path taken through the state diagram is via S_3. Other combinations of the three digits result in the path via S_2 being selected, and in this case there is no output on the Z line during the third clock pulse.

The first transition in the sequence of three, from S_0 to S_1 is initiated by a clock pulse and takes place irrespective of whether the first X digit is a 0 or a 1. Taking the path via S_3 the second digit on the X line must be a 1 if a transition from S_1 to S_3 is to take place. The third transition, from S_3 to S_0 is also initiated by a clock pulse and it occurs irrespective of whether $X = 0$ or $X = 1$.

The diagram below shows that there are four combinations of three digits that will lead to an output of $Z = 1$. They are 010, 011, 110 and

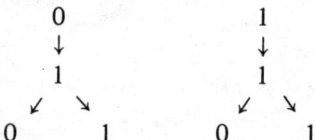

111. Clearly, the remaining four combinations of three digits will be associated with the alternative path through the state diagram, and for this path $Z = 0$ on the third digit.

The timing diagram for the circuit is shown in figure 8.1(h) for three different strings of digits, namely 011, 001 and 110. The X signal is synchronised to the clock and it is assumed that changes in this signal always take place between clock pulses. In the case of the combinations 011 and 110 there is a Z output which lasts for the duration of the clock pulse marked 3, whilst for the 001 combination the output $Z = 0$.

The last waveform shown in the diagram is for $Z = AB$; in other words, the clock signal has been removed from the output equation and as a consequence the output commences on the trailing edge of clock pulse 2 and terminates on the trailing edge of clock pulse 3. It is interesting to note that in the case where $Z = AB$ the output goes high before the third digit has arrived. This is satisfactory in this case since having once recognised what the second digit is by entering the state S_3, it is irrelevant whether the third digit is 0 or 1, i.e. it does not have to be recognised. However, in the case where $Z = AB\sqcap$ the output does not go high until the leading edge of the third clock pulse. The circuit has to recognise this clock pulse before an output can occur.

8.3 The design procedure for clocked sequential circuits

The analysis of the clocked sequential circuit in § 8.2 identifies all of the processes required for circuit design with the exception of one, state reduction, which will be dealt with below. The steps in the design process are summarised in the block diagram shown in figure 8.2.

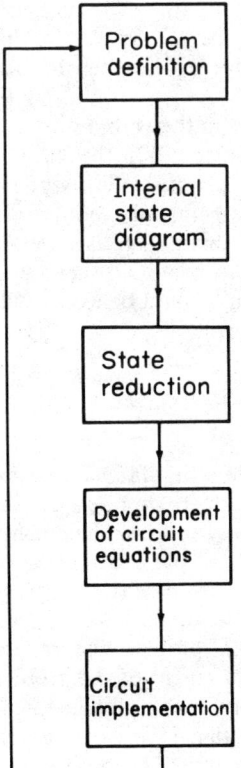

Figure 8.2. Block diagram of the design process for a clock-driven circuit

Step 1: Problem definition.

It is essential that the designer should have detailed knowledge of the external characteristics of the circuit which is to be designed. In the first instance, this will usually consist of a verbal statement of the problem, and, in particular, details of the inputs available and the outputs required. This information is best displayed on a block diagram such as the one shown in figure 8.1(b), if necessary, in conjunction with time diagrams of the input and output waveforms. A typical example of the kind of waveform information that might be supplied is shown in figure 8.1(h).

The statement of the problem in completely unambiguous terms is a difficult task and may require several discussions between designer and customer. If the ambiguities are not resolved at this stage the design process will continue, a circuit implementation will be obtained which will not satisfy the customer's requirements, and there will by necessity be a loop back from step 5 to step 1. In practice, it may well be that several repetitions of the design process will be required before the customer's specification has been satisfied.

For the student of the subject, ambiguities in the verbal statement of the problem very often raise real doubts in the mind. In the learning stage such doubts should not be regarded too seriously. The student should clearly decide on a solution to the problem and should accompany the answer with an attempt at an unambiguous statement of his interpretation of the problem. The solution should then be implemented with hardware, and its behaviour should be examined to see if it verifies his interpretation of the problem.

Step 2: The internal state diagram.

In this step the verbal statement of the problem should be expressed in terms of the internal states of the circuit. There are no rules for constructing internal state diagrams, and the ability to draw them can only be acquired by experience. For example, the inexperienced designer will almost certainly not, in the first instance, produce the internal state diagram shown in figure 8.1(f) for the circuit analysed in § 8.2. To construct the internal state diagram for that problem the designer might have been given the following verbal statement of the problem.

A logic circuit receives binary information, serially on a line X, which is synchronised with an external clock signal. Non-overlapping strings of three successive binary digits are examined by the logic circuit and if the combinations 000, 001, 110 and 111 are detected, a 1 output will appear on the Z line. The output will occur during the third digit of the string and will have a time duration equal to that of the clock pulse. For all other combinations of three binary digits the output Z will be 0.

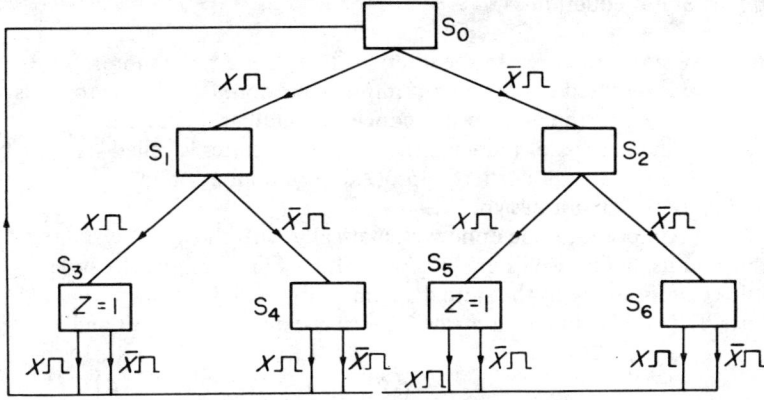

Figure 8.3. *Internal state diagram for a combination detector*

In practice, the inexperienced designer may well develop the tree-like structure of states shown in figure 8.3. The method of approach used to arrive at this diagram would be to commence in a state such as S_0. This internal state of the circuit can be left by two separate transition paths, one associated with the transition signal $X\sqcap$ leading to S_1, and the other associated with the transition signal $\bar{X}\sqcap$ leading to S_2. Each of the states S_1 and S_2 can be left by two transition paths, one associated with the transition signal $X\sqcap$ and the other with the transition signal $\bar{X}\sqcap$. These four paths lead to the states S_3, S_4, S_5 and S_6. For each of these four states there are two exit paths, but the next transition is the third one and consequently all eight exit paths must lead back to the starting state S_0.

The path $S_0 - S_1 - S_3 - S_0$ deals with the combinations 111 and 110, hence the output $Z = 1$ in state S_3. Similarly, the path $S_0 - S_2 - S_5 - S_0$ deals with the combinations 011 and 010 and consequently the output in state S_5, $Z = 1$. The other two paths through the state diagram deal with those combinations that do not have to be detected.

In developing this diagram no short-cuts have been taken. Each combination of three binary digits appears explicitly on the diagram. However, this internal state diagram has eight states in comparison with the four states of girue 8.1(*f*). In terms of hardware, this means that the circuit implementation developed from figure 8.3 would require three *JK* flip-flops and an additional amount of combinational logic.

Step 3: State reduction.

The more states there are in the internal state diagram the more hardware is required for the circuit implementation. For this reason it is in the interests of the designer to reduce the number of states if possible. The process of reducing the number of states is called state reduction and corresponds to the process of minimisation in combinational circuit design.

State reduction can be done systematically with the aid of a state table and using Caldwell's merging procedure. For example, the state table corresponding to the state diagram of figure 8.3 is shown in figure 8.4(*a*). This table has a row for every state of the circuit and a

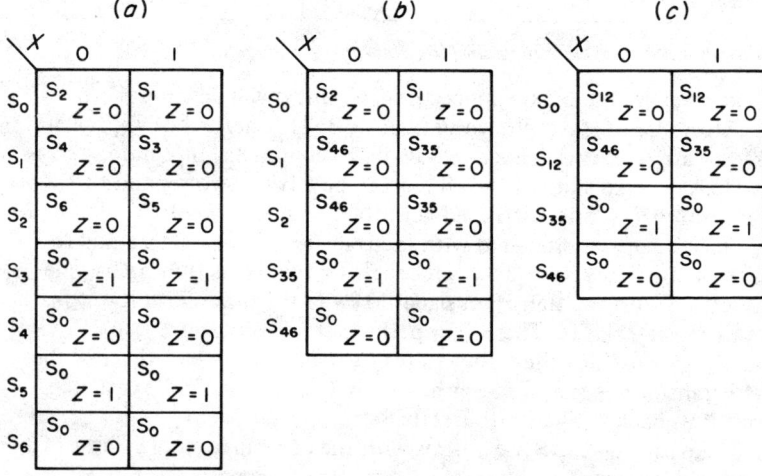

Figure 8.4. Combination detector. (a) State table. (b) Reduced state table. (c) Minimal state table

column for every combination of the input signals. In this case there is only one input signal, and hence there are only two columns. In each cell of the table there are two entries, the first one being the next state assumed by the circuit when it is in a state corresponding to the row heading and it receives an input signal defined by the column heading, the second one being the circuit output.

Caldwell's merging rules state that two rows in the state table can now be merged if the next state entries and outputs in the corresponding columns of each row are alike. In the table of figure 8.4(a) the rows labelled S_4 and S_6 satisfy these two conditions as do the rows labelled S_3 and S_5. After states S_4 and S_6 have been merged the new state is designated S_{46} and everywhere S_4 and S_6 appear in the state table they are replaced by S_{46}. Similarly, S_3 and S_5, when merged, form a new state S_{35}.

Using Caldwell's merging procedure, the state table of figure 8.4(a) can be reduced to that shown in figure 8.4(b) which also has two rows, those headed S_1 and S_2, which can be merged to form a new state S_{12}. Hence the table of figure 8.4(b) can be reduced to that shown in figure 8.4(c), and no further reduction is now possible. The reduced state diagram that can now be constructed from this table is identical to that shown in figure 8.1(f) and is repeated here in figure 8.5 using the state designation of figure 8.4(c).

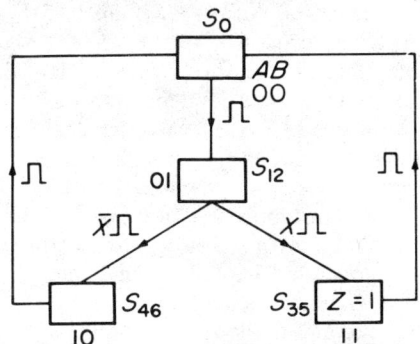

Figure 8.5. State diagram for the combination detector

The best situation to have in practice is one in which the number of states is equal to some power of two, such as 2^n. There is little point in reducing the number of states below 2^n unless it is to a lower power of two, since this would lead to the creation of unused states. For example, if N is the number of states and $2^{n-1} < N < 2^n$, then the number of unused states is $2^n - N$.

Unused states create additional difficulties for the logic designer. Any unused state can be entered at 'power on' or, alternatively, due to faulty circuit operation. It is the responsibility of the logic designer to specify the behaviour of the circuit if it should enter an unused state otherwise a 'lock-in' may occur. This means that if there is no exit path from an unused state, the circuit, on entering it, will remain there for an indefinite period. It should be stressed that unused states are not 'can't happen' states, and for this reason they should not be used for simplification of the circuit equations.

Step 4: Development of circuit equations.

Having obtained the minimum state diagram the next step the designer must take is to allocate secondary variables to the various states. The number of secondary variables required to define a state is governed by the total number of states in the diagram. For example, there are four states in figure 8.5 and therefore two secondary variables are required to define each state uniquely.

One allocation of the secondary variables is shown in figure 8.5. Clearly, there are other possible allocations of these variables and consequently there are a number of different circuit solutions to this problem some of which may lead to more economical circuitry than others. However, it is rarely worth while to search for the minimal solution since this is a very time-consuming process.

The number of secondary variables needed to define a state is related to the number of flip-flops to be used in the circuit implementation. In the state diagram of figure 8.5 there are two secondary variables A and B, and consequently two flip-flops will be required for implementation.

There are at least two techniques for determining the logical equations of the circuit. The first consists of tabulating the J and K inputs for every transition on the state table with the aid of the steering table for a JK flip-flop which was developed in Chapter 5. The J and K inputs for the various transitions are then mapped on a K-map and simplified in the normal way. For the second method, a more direct approach makes use of the process of algebraic reduction described earlier.

For the first method the state table of figure 8.4(c) is tabulated in terms of the secondary variables, as shown in figure 8.6(a). Alongside the state table are eight columns in which the flip-flop input signals J_A, K_A, J_B and K_B are tabulated. The entries in these columns are obtained from the steering table for a JK flip-flop which is reproduced for convenience in figure 8.6(b). In effect, this table gives the J and K signals required to produce the four possible transitions from Q^t to $Q^{t+\delta t}$. The values of J and K that can appear in this tabulation are 0, 1 and $\emptyset$ (don't care).

The entries in the flip-flop input columns of the state table are

Clock-driven sequential circuits

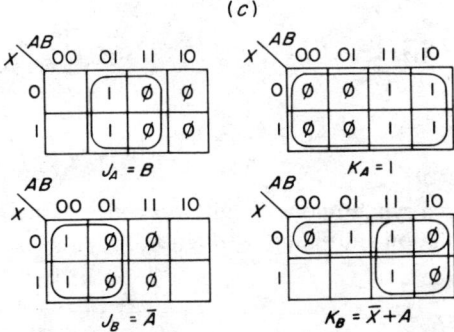

Figure 8.6. (a) State table used for determining the J and K inputs of the flip-flops of the combination detector. (b) Steering table for a JK flip-flop. (c) K-map plots for determining the J and K inputs for the combination detector

obtained in the following manner. If the present state of $AB = 00$ and $X = 0$, then the next state is $AB = 01$. Hence $A^t = 0$ and $A^{t+\delta t} = 0$. Thus from the steering table $J_A = 0$ and $K_A = \emptyset$. Similarly $B^t = 0$ and $B^{t+\delta t} = 1$; consequently from the steering table $J_B = 1$ and $K_B = \emptyset$. All other entries in these eight columns can be obtained by repeating the process just described.

Maps can now be plotted for each of the flip-flop input signals. These are shown in figure 8.6(c) and the equations derived from these maps are as follows:

$$J_A = B \qquad J_B = \overline{A}$$

$$K_A = 1 \qquad K_B = \overline{X} + A$$

170   *Clock-driven sequential circuits*

As might be expected, these equations are identical to the flip-flop input equations for the circuit shown in figure 8.1(*a*).

The output equation is taken directly from the state table in figure 8.6(*a*) and is

$$Z = AB\sqcap$$

In the second method the turn-on and turn-off equations for the individual flip-flops are obtained directly from the state diagram (figure 8.5). Optional products that can be used for minimisation are utilised, if possible, to reduce these equations and then with the aid of the equations $S_Q = J_Q \bar{Q}$ and $R_Q = K_Q Q$ developed in Chapter 5, the J and K input equations for each of the flip-flops are obtained as follows:

$$S_A = S_{12}\, X\sqcap + S_{12}\, \bar{X}\sqcap = S_{12}\sqcap = \bar{A}B\sqcap$$

Hence

$$J_A = B\sqcap$$

Since clock is always present in a synchronous circuit, it is conventional to leave it out of the above equation and simply write it as

$$J_A = B$$

$$R_A = S_{35} + S_{46} + (S_0)$$

where S_0 is an optional product to be used if possible for minimisation. The transition from S_0 to S_{12} as far as the A flip-flop is concerned is $0 \to 0$, and in such a case it does not matter whether $R_A = 0$ or 1. Hence,

$$R_A = AB + A\bar{B} + (\bar{A}\bar{B}) = A + (\bar{A}\bar{B})$$

and

$$K_A = 1$$

In this case there is no point in using $\bar{A}\bar{B}$ to simplify R_A, since use of the relation $R_A = K_A A$ reduces K_A to 1, its simplest possible form.

$$S_B = S_0 + (S_{12} X) = \bar{A}\bar{B} + (\bar{A}BX)$$

Clock-driven sequential circuits

No simplification is possible with the optional product, hence

$$J_B = \bar{A}$$

In this case the optional product $S_{12} X$ originates from the $1 \rightarrow 1$ transition of the secondary variable B when moving from S_{12} to S_{35}, and in such a case it does not matter whether J_B is 1 or 0.

$$R_B = S_{12}\bar{X} + S_{35} + (S_{46})$$
$$= \bar{A}B\bar{X} + AB + (A\bar{B})$$
$$= B\bar{X} + AB$$

Hence

$$K_B = \bar{X} + A$$

and the equations obtained by this method are identical to those obtained by the alternative technique described previously.

Step 5: Circuit implementation.

It is a simple matter now to obtain a circuit implementation using the equations developed in step 4. In this case the circuit implementation will be identical to that shown in figure 8.1(a).

8.4 The design of a sequence generator

Step 1: Problem definition.

Serial binary data is received on the X input line of a logic circuit, each bit of the input data being synchronised with a clock pulse on the clock line. An output signal is generated on the Z line each time the sequence 101 is detected. A block diagram of the circuit is shown in figure 8.7(a) in conjunction with a stream of input data X and the corresponding output Z.

Step 2: The internal state diagram.

A suitable state diagram is shown in figure 8.7(b). In order to detect the sequence 101 only three states, S_0, S_1, and S_2 are required. To

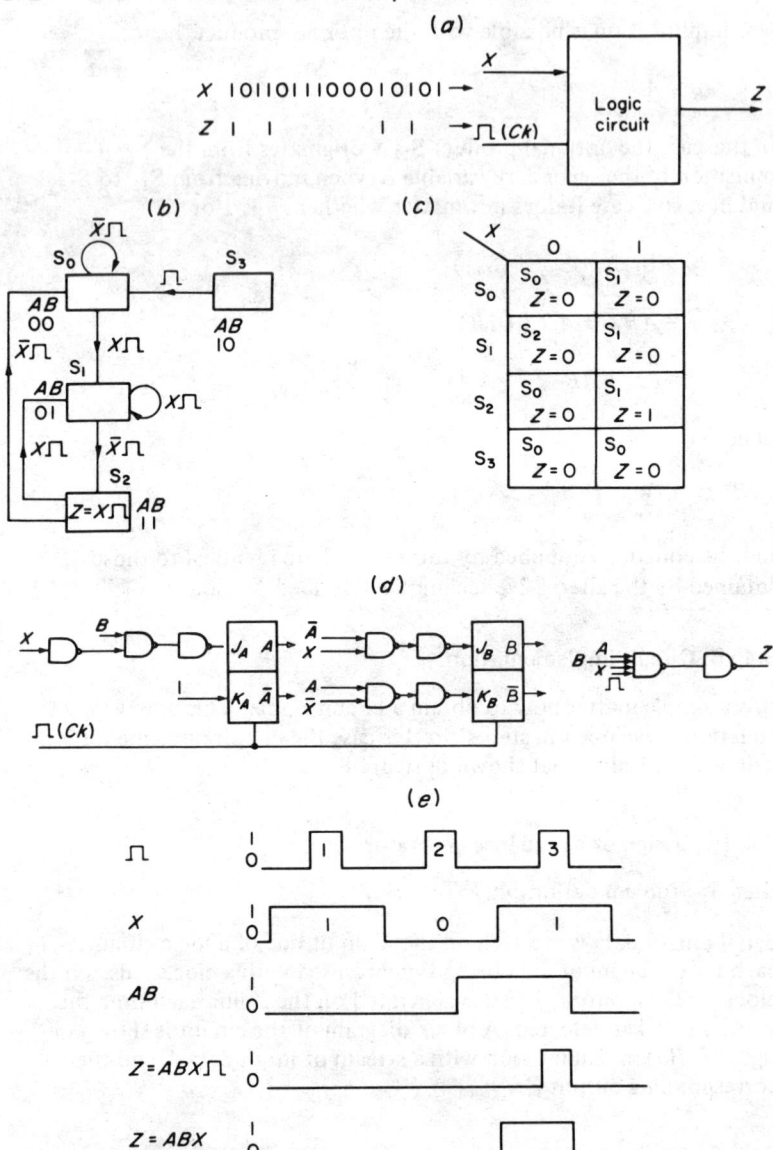

Figure 8.7. The sequence detector. (a) Block diagram. (b) Internal state diagram. (c) The state table for the sequence detector. (d) Circuit implementation. (e) Timing diagrams

define these states, two secondary variables A and B are required. Since there are four combinations of these variables there is one unused state S_3. If, through faulty operation or at 'power-on', the circuit should enter this state it would be desirable to return to the main sequence of states as rapidly as possible. This is achieved by returning S_3 to S_0 via a transition which is initiated by the first clock pulse that occurs after the entry into S_3.

Step 3: State reduction.

Examination of the state table shown in figure 8.7(c) shows that no state reduction is possible.

Step 4: Development of the circuit equations.

$$S_A = S_1 \bar{X} = \bar{A}B\bar{X} \qquad \qquad \therefore J_A = B\bar{X}$$

$$R_A = S_2 + S_3 + (S_0 \bar{X}) + (S_1 X)$$

$$= AB + A\bar{B} + (\bar{A}\bar{B}\bar{X}) + (\bar{A}BX)$$

$$= A + (\bar{A}\bar{B}\bar{X}) + (\bar{A}BX) \qquad \qquad \therefore K_A = 1$$

$$S_B = S_0 X + (S_2 X) + (S_1)$$

$$= \bar{A}\bar{B}X + (ABX) + (\bar{A}B) \qquad \qquad \therefore J_B = \bar{A}X$$

$$R_B = S_2 \bar{X} + (S_3) = AB\bar{X} + (A\bar{B}) \qquad \qquad \therefore K_B = A\bar{X}$$

and from the state diagram, $Z = ABX\sqcap$.

Step 5: Circuit implementation.

The circuit implementation for the detector is shown in figure 8.7(d) and the timing diagrams for a 101 sequence of digits is shown in figure 8.7(e). Assuming that master/slave JK flip-flops are being used, then the circuit enters state AB on the trailing edge of the clock pulse numbered 2, and leaves on the trailing edge of clock pulse 3. If the output is defined as $Z = ABX$, then it goes high when the circuit recognises the leading edge of the X signal associated with the third clock pulse. On the other hand, if the output is defined as $Z = ABX\sqcap$ it does not go high until the leading edge of the third clock pulse has been recognised.

8.5 Moore and Mealy state machines

There are two types of clocked sequential circuits: the first has an output which depends only on the state of the flip-flops and is referred to as a Moore state machine. In the second type the output depends on both the state of the flip-flops and on the logical value of the inputs. Such a circuit is referred to as a Mealy state machine.

The circuit developed in figure 8.7(d) is a Mealy machine, since the output $Z = ABX\sqcap$ depends on the state of the flip-flops and the value of the two inputs X and $\sqcap$. A slight modification to the state diagram for the Mealy machine shown in figure 8.7(b) will convert the final circuit from a Mealy to a Moore representation. This modification is illustrated on the state diagram shown in figure 8.8(a). An additional state S_3 has been introduced and this is the output state of the circuit

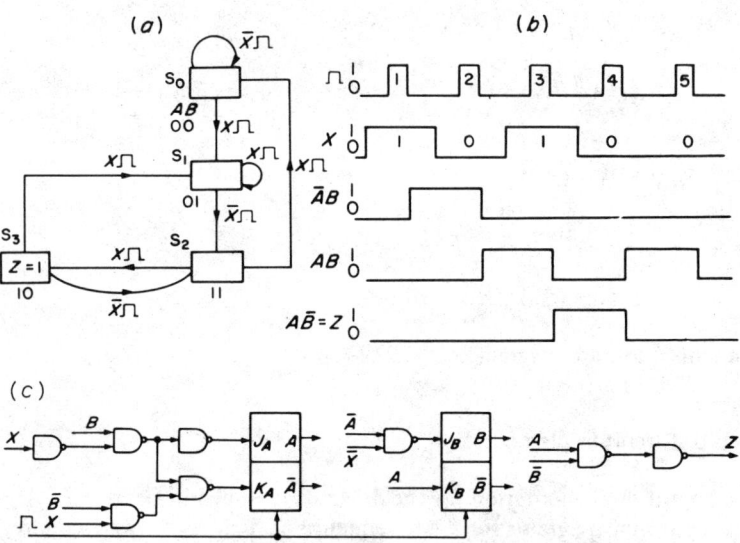

Figure 8.8. *The sequence detector. (a) Moore representation state diagram. (b) Timing diagrams for the Moore representation. (c) Implementation of the Moore representation*

Clock-driven sequential circuits 175

so that $Z = A\bar{B}$. The timing diagrams corresponding to the state diagram are shown in figure 8.8(b).

The circuit equations obtained for the Moore representation of the sequence detector are obtained in the usual way:

$$J_A = B\bar{X} \qquad J_B = A + X \qquad Z = A\bar{B}$$

$$K_A = B\bar{X} + \bar{B}X \qquad K_B = A$$

The Moore form of the sequence detector is shown implemented in figure 8.8(c).

As a further example of the Mealy and Moore representations, consider the following problem.

A logic circuit receives binary information serially on a line X which is synchronised with an external clock signal. Non-overlapping strings of three successive digits are examined by the circuit. If the last digit of the combination is a 1 then an output of 1 will appear on the Z line.

The block diagram for the problem is shown in figure 8.9(a) and the tree-like structure of states developed in figure 8.3, with minor modifications, is a suitable internal state diagram. The modified state diagram is shown in figure 8.9(b). For this problem the entry in states S_3, S_4, S_5 and S_6 is $Z = X\Pi$, indicating that an output will occur if the last digit of the three-bit combination is a 1. The state table is shown in figure 8.9(c) and it is apparent on inspection that states S_3, S_4, S_5 and S_6 can be merged to form one state S_{3456}. After merging, the reduced state table has the form shown in figure 8.9(d).

The reduced internal state diagram constructed from the information presented in the reduced state table is shown in figure 8.9(e). This state diagram will lead to a Mealy-type machine since the output $Z = A\bar{B}X\Pi$ depends on the present state and the input signals. To convert the state diagram to one which will lead to a Moore-type machine, the state S_0 has to be split into two states S_{0A} and S_{0B}. The output Z now appears in the state S_{0A} and is dependent on the state only, so that $Z = \bar{A}\bar{B}C$. It is worth noticing in this case that the conversion increases the number of states from four to five and, as a consequence, the Moore circuit will require an additional flip-flop.

8.6 Pulsed synchronous circuits

The problems dealt with above have assumed that the input is a synchronous level signal as illustrated in figure 8.10. For level signals

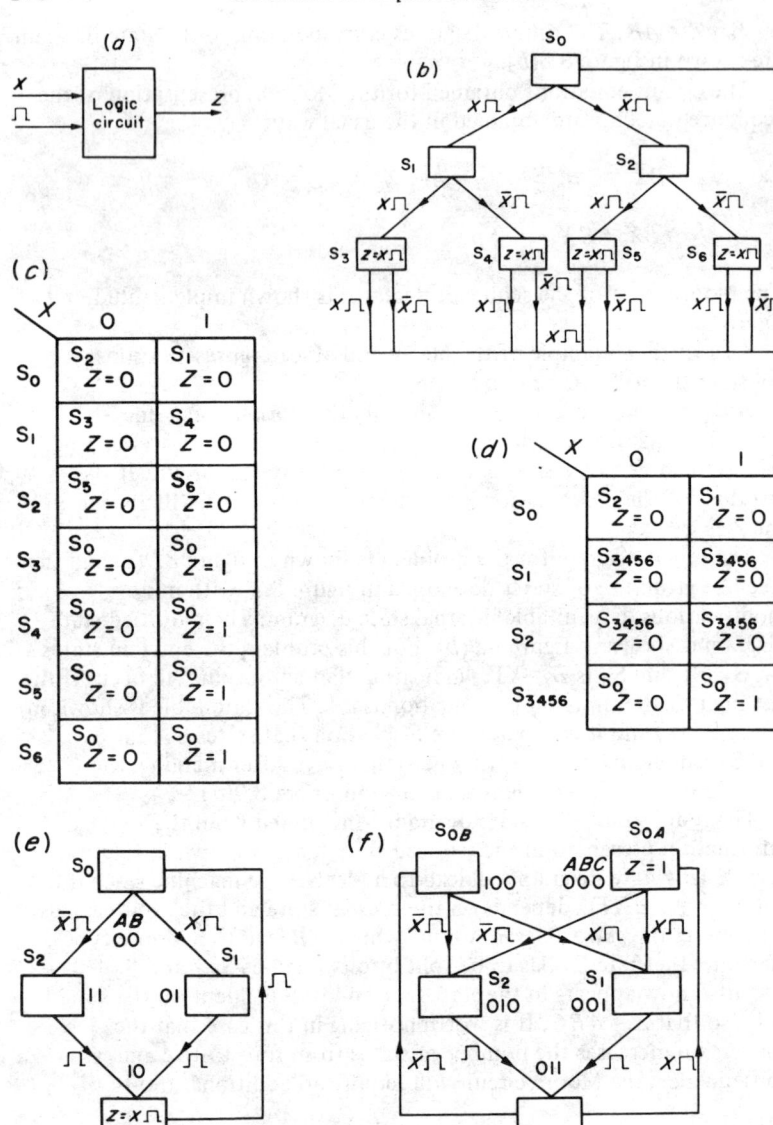

Figure 8.9. (a) Block diagram, (b) basic state diagram, (c) state table, and (d) reduced state table for the word scanner. (e) State diagram for the Mealy-type machine. (f) State diagram for the Moore-type machine

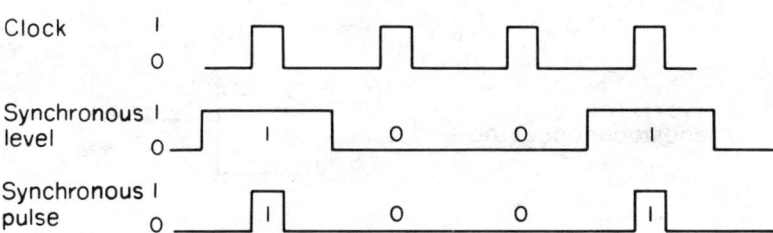

Figure 8.10. *Input signals for synchronous circuits*

of this type changes take place at some time in the interval between clock pulses and the input sequence is defined in terms of the value of the input signal during the time period when the clock is high.

There is an alternative type of input signal which can be applied to a sequential circuit. This is referred to as a synchronous pulse signal and it is illustrated in figure 8.10. It will be observed that the time duration of the input pulses is identical to that of the clock pulses, and again the input sequence is defined in terms of the value of the input signal during the time period when the clock is high.

As an example of the design of a pulsed synchronous circuit, consider the problem in which serial binary data is received on line X, in pulse form, each bit of the input data being synchronised with the incoming clock pulses. The circuit is required to recognise the input sequence $X = 111$, including overlapping sequences, as indicated in figure 8.11(a).

The internal state diagram is shown in figure 8.11(b). S_0 may be regarded as the initial state of the circuit and the machine changes state every time it receives $X = 1$, until it reaches S_2. For values of $X = 0$ the machine always returns to S_0 and if, for some reason, it enters the unused state S_3 it stays there until the next time $X = 1$, when it makes a transition to S_3.

The state table for the circuit, coupled with a tabulation of the J and K input signals for the two flip-flops A and B, is shown in figure 8.11(c), and the K-map plots for the J and K signals are shown in figure 8.11(d). From these plots the following input equations are obtained:

$$J_A = BX \qquad K_A = B\bar{X} + \bar{B}X$$
$$J_B = X \qquad K_B = \bar{X}$$

and the equation of the output signal is

$$Z = ABX$$

178   Clock-driven sequential circuits

(a)

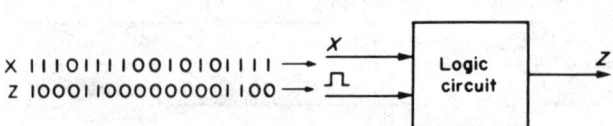

X 1110111100101011111
Z 100011000000001100

(b)

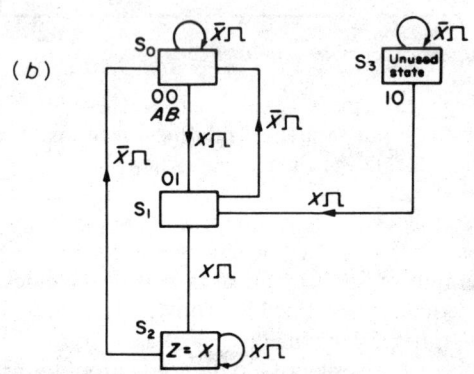

(c)

Present state	Next state		Flip-flop inputs							
	$X=0$	$X=1$	$X=0$		$X=1$		$X=0$		$X=1$	
AB	AB	AB	J_A	K_A	J_A	K_A	J_B	K_B	J_B	K_B
00	00	01	0	ϕ	0	ϕ	0	ϕ	1	ϕ
01	00	11	0	ϕ	1	ϕ	ϕ	1	ϕ	0
11	00	11	ϕ	1	ϕ	0	ϕ	1	ϕ	0
10	10	01	ϕ	0	ϕ	1	0	ϕ	1	ϕ

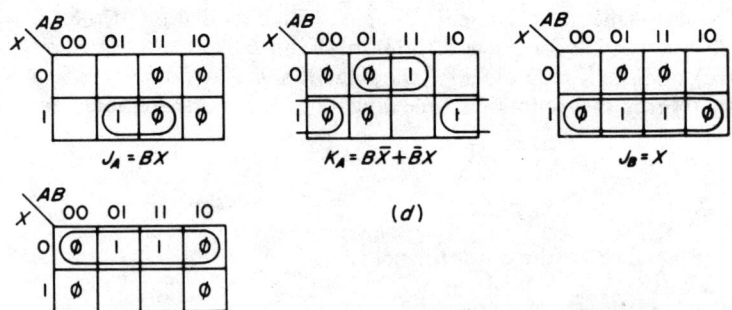

(d)

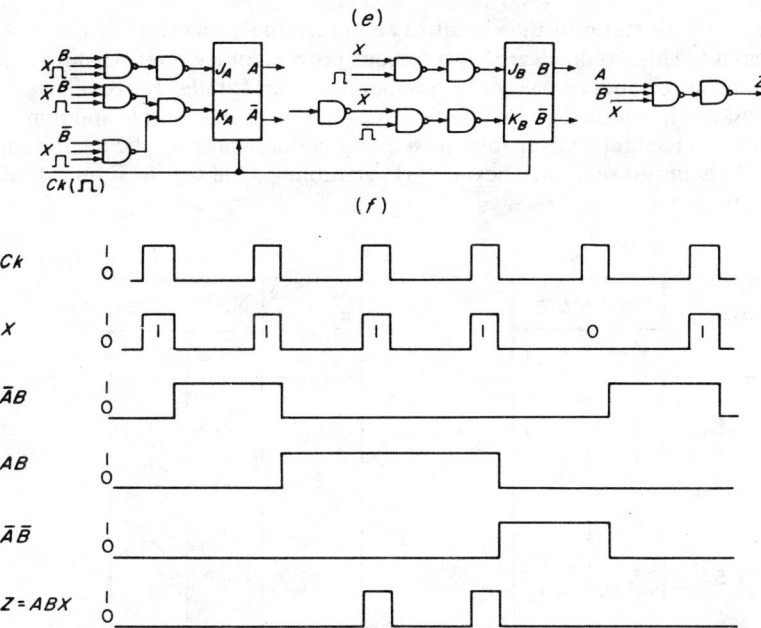

Figure 8.11. (a) *Block diagram,* (b) *internal state diagram, and* (c) *state table for the 111 sequence detector.* (d) *K-maps for determining the J and K input signals.* (e) *Circuit implementation, and* (f) *timing diagrams for the 111 sequence detector*

The circuit implementation of the detector is shown in figure 8.11(*e*). Notice that the clock signal has been ANDed with the J and K inputs of both flip-flops in order to ensure synchronous operation. A typical set of timing diagrams for the circuit is illustrated in figure 8.11(*f*).

8.7 State reduction

A method of state reduction by inspection has already been introduced in § 8.3. In practice, all methods of state reduction depend on the principles of equivalence as defined earlier in Caldwell's merging rules. However, two states, S_p and S_q of a synchronous sequential circuit may also be deemed to be equivalent if, and only if, every possible input sequence produces identical output sequences, irrespective of whether S_p or S_q is the initial state. One method of determining state equivalence would therefore be to apply all possible input sequences to the circuit and tabulate the corresponding output sequences, assuming

180 Clock-driven sequential circuits

each of the states of the circuit to be in turn the initial state of the circuit. This would clearly be a tedious process for a circuit having a number of input signals and a number of internal states. Fortunately, besides the method of inspection, there are two other simple and non-tedious techniques available for state reduction which will be dealt with briefly in this section. They are (a) partitioning, and (b) the implication table.

(a)

Present state	Next state	
	$X=0$	$X=1$
S_0	S_0 $Z=0$	S_4 $Z=1$
S_1	S_4 $Z=1$	S_2 $Z=0$
S_2	S_0 $Z=1$	S_3 $Z=1$
S_3	S_5 $Z=0$	S_6 $Z=1$
S_4	S_1 $Z=1$	S_2 $Z=0$
S_5	S_5 $Z=0$	S_4 $Z=1$
S_6	S_0 $Z=1$	S_3 $Z=1$

(b)

Present state	Next state	
	$X=0$	$X=1$
S_{05}	S_{05} $Z=0$	S_{14} $Z=1$
S_{14}	S_{14} $Z=1$	S_{26} $Z=0$
S_{26}	S_{05} $Z=1$	S_3 $Z=1$
S_3	S_{05} $Z=0$	S_{26} $Z=1$

Figure 8.12. (a) State table for the partitioning example. (b) Reduced state table after partitioning

It will be assumed that the state table shown in figure 8.12(a) has been obtained from an internal state diagram and relates to a problem in which there is a single input line X and a single output line Z.

A first partition is made by placing all those present states into the same section of the partition if the outputs generated are identical for all possible inputs. For example, if the present state is S_0 then the two possible inputs are $X = 0$ and $X = 1$ for which the outputs are $Z = 0$ and $Z = 1$. Similarly, if the present state is S_3 then for $X = 0$ and $X = 1$, the outputs are $Z = 0$ and $Z = 1$. Hence these two states will appear in the same section of the partition. Repeating this process for the remaining states leads to the following first partition

$$P_1 = (S_0, S_3, S_5)(S_1, S_4)(S_2, S_6).$$

In effect, this partition has been obtained by the application to the circuit of an input sequence of length one.

The second partition P_2 is obtained using the following procedure. In the first section of P_1, the next states for S_0, S_3 and S_5, for $X = 0$ are all in the same section of P_1. However, for $X = 1$ the next states for S_0, S_3 and S_5 are S_4, S_6 and S_4 respectively. The next state of S_3 lies in a different section of P_1 for $X = 1$, and consequently the first section of P_1 is now split into two sections, the first one containing S_0 and S_5, the second containing S_3 only. The procedure is repeated for the second section of the partition, but since the next states of S_1 and S_4 will remain in the same section of P_2, irrespective of the value of X, no split of this section is required. An examination of the third section of the partition shows that no split of this section is required either. Hence

$$P_2 = (S_0 S_5)(S_3)(S_1 S_4)(S_2 S_6)$$

This partition has been obtained by the application of an input sequence of length two.

The procedure described above is used to determine P_3, but $P_3 = P_2$. Hence the individual sections of P_2 contain the equivalent states of the circuit and no further partitioning is possible. The reduced state table for the circuit is shown in figure 8.12(b).

The second method of state reduction available to the designer employs the implication table. A state table for a synchronous sequential machine is shown in figure 8.13(a). An initial examination of this table reveals that none of the states can be merged by inspection.

The next step is to develop an implication table by listing all the states vertically except the first one, and all the states horizontally except the last one, as illustrated in figure 8.13(b). The implication table displays all possible combinations of state pairs, and the individual cells in the table represent the testing ground for the equivalence of a state pair. For example, the top left-hand cell at the intersection of S_0 and S_1 is where these two states are tested for equivalence.

One of the conditions for state equivalence is that the next state outputs of a pair of states must be identical if the two states are equivalent. On the implication table all the cells that cannot possibly be equivalent are marked with a cross. For example, S_0 and S_1 cannot be equivalent states since the next state outputs are 0,0 and 1,0 respectively, and hence the cell situated at the intersection of S_0 and S_1

is marked with a cross. Similarly, cells for the other non-equivalent state pairs are marked with a cross as shown in figure 8.13(c).

The next step is to place in the vacant cells the implications required to make the pair of states associated with a particular cell equivalent. For example, the cell at the intersection of S_0 and S_2 contains the implication that S_0 must be equivalent to S_5, and S_2 must be equivalent to S_5 for S_0 and S_2 to be equivalent. The implication table is shown in figure 8.13(d).

If the pairs implied in any of the cells shown in figure 8.13(d) contain only those states defined by the cell, or, alternatively, if the next states of the two states defining the cell are the same state for a

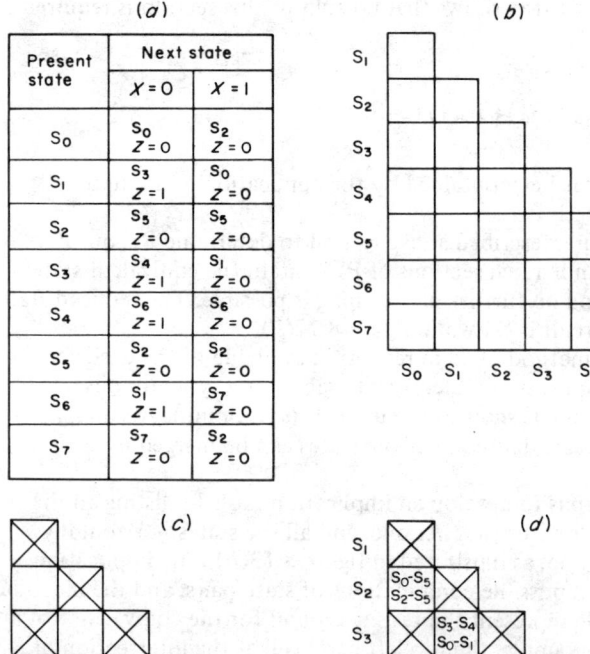

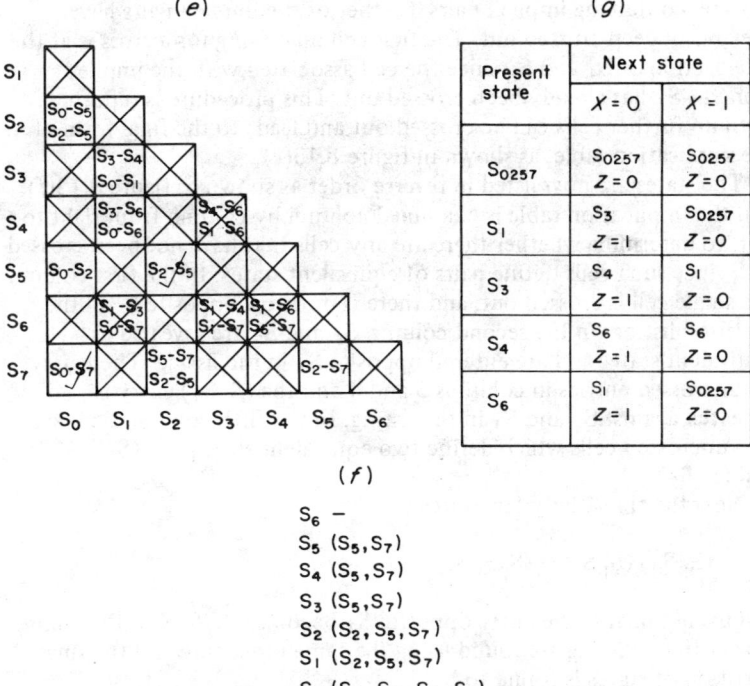

Figure 8.13. (a) State table to be reduced by the implication table (b). (c) Elimination of non-identical outputs. (d) Insertion of implied pairs. (e) Completed implication table. (f) The partition listing. (g) The reduced state table.

given input, the two states defining the cell are equivalent and are identified by a tick. The first section of this rule applies to two cells in figure 8.13(d), first the one defined by S_0 and S_7, and second, the one defined by S_2 and S_5. These two cells have been marked with a tick in figure 8.13(d)

An examination of the state table indicates that S_2 and S_5 are a pair of 'lock-in' states. S_5 can be entered from S_2 on the receipt of a clock pulse and vice versa, but there is no other exit from these states. Consequently, the two states can be merged, and on the receipt of a clock pulse, when in the merged state, the circuit will simply stay there. To leave the merged state it is clear that a reset signal is required. A similar argument can also be applied to states S_0 and S_7.

The next step is to examine the implication table row by row, beginning with the bottom right-hand cell. A cross can be entered into any cell containing implied pairs if either of the implied pairs have previously been crossed out. The first cell qualifying for a cross is at the intersection of S_4 and S_6 since the cell associated with the implied pair S_6-S_7 has already been crossed out. This procedure is repeated until no further cells can be crossed out and leads to the final form of the implication table, as shown in figure 8.13(e).

The states are now listed in reverse order as shown in figure 8.13(f), and the implication table is examined column by column from right to left, to determine whether there are any cells that have not been crossed out, since such cells define pairs of equivalent states. In the first column the single cell is crossed out, and there is no entry opposite S_6 in the partition listing. In the second column S_5 and S_7 are revealed as equivalent states and are entered opposite S_5 in the listing. There are no uncrossed entries in columns 3 and 4 and the ($S_5 S_7$) entry is repeated against S_4 and S_3 in the listing. In the fifth column there are two uncrossed cells which define two equivalent state pairs ($S_2 S_7$) and ($S_2 S_5$).

Now the transitivity law states that

$$(S_p S_q)(S_p S_r) \rightarrow (S_p S_q S_r)$$

and using this rule the entry opposite S_2 becomes ($S_2 S_5 S_7$). Remaining entries in the listing are found using the same procedure and the final partition of states is found to be

$$P = (S_0 S_2 S_5 S_7)(S_1)(S_3)(S_4)(S_6)$$

The reduced state table resulting from this partition is shown in figure 8.13(g).

8.8 State assignment

In all the problems dealt with earlier, a perfectly arbitrary state assignment has been adopted. For example, in the 111 sequence detector developed in § 8.6 the state assignment selected was $S_0 = 00$, $S_1 = 01$, $S_2 = 11$ and $S_3 = 10$. It is clear that alternative state assignments can be selected and they will lead to different circuit solutions. The criterion for a well chosen state assignment is that it should lead to a simple circuit realisation.

The problem associated with state assignment is therefore to allocate secondary variables to the states, which will give, if not the simplest, a

Clock-driven sequential circuits

much simpler circuit solution than a random allocation of secondary variables. Unfortunately, there is no known technique for selecting a state assignment which will give the simplest circuit solution. However, in this section certain rules will be given that will lead to a reasonably simple circuit solution.

Simpler circuits will mean that less gates are required for the circuit implementation, and this in turn may mean that a smaller number of IC chips are required. If the designed circuit is to be manufactured in large numbers there may be a significant reduction in manufacturing costs. A simpler circuit realisation will also result in a reduced number of gate inputs, and hence circuit interconnections, and finally, there may also be a significant saving of space.

The need for a well chosen state assignment will be demonstrated by the following example.

Serial NBCD codes arrive on line X, the most significant bit first, each bit of the four-bit code being synchronised with a clock pulse, as illustrated in figure 8.14(a).

A suitable internal state diagram for this problem is shown in figure 8.14(b). The path $S_0 \rightarrow S_1 \rightarrow S_2 \rightarrow S_3$ is associated with the first eight combinations of the NBCD code, 0000–0111 inclusive, and these are all valid code combinations. A second path $S_0 \rightarrow S_4 \rightarrow S_7 \rightarrow S_3$ is associated with the other two valid code combinations 1000 and 1001. The invalid combinations are covered by paths $S_0 \rightarrow S_4 \rightarrow S_5 \rightarrow S_6$ for code combinations 1100–1111 inclusive and $S_0 \rightarrow S_4 \rightarrow S_6 \rightarrow S_7$ for code combinations 1010 and 1011. The output for the invalid code combinations $Z = 1$ is associated with state S_6.

Three randomly chosen state assignments for this circuit are shown in figure 8.14(c). The input equations for the flip-flops A, B and C, which are required to implement the circuit and the output equation for each state assignment are shown below.

Assignment 1

$J_A = CX + \overline{B}\overline{C}\overline{X}$ $\quad J_B = \overline{C}X + A\overline{C} + \overline{A}C\overline{X}$ $\quad J_C = \overline{A}\overline{B}X$

$K_A = \overline{B}C + B\overline{C}$ $\quad K_B = C\overline{X} + AC + \overline{A}B\overline{C}$ $\quad K_C = \overline{B}\overline{X} + AB$

$$Z = A\overline{B}C\sqcap$$

Assignment 2

$J_A = A + B\overline{C} + \overline{B}C + B\overline{X}$ $\quad J_B = A + C\overline{X}$ $\quad J_C = \overline{A}\overline{B}X$

$K_A = \overline{B} + \overline{C} + X$ $\quad K_B = A\overline{C} + \overline{A}C$ $\quad K_C = \overline{A}B + B\overline{X}$

$$Z = \overline{A}BC\sqcap$$

(a)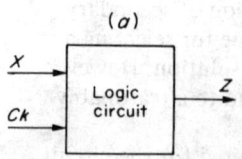

(b)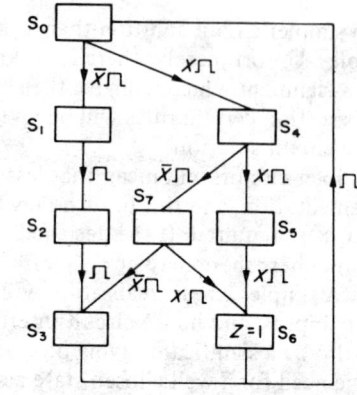

(c)

State	Ass. 1 CBA	Ass. 2 CBA	Ass. 3 CBA
S_0	000	000	000
S_1	001	001	001
S_2	011	010	011
S_3	010	011	010
S_4	110	100	100
S_5	111	101	101
S_6	101	110	111
S_7	100	111	110

(e)

Present state	Next state	
	$X=0$	$X=1$
S_0	S_1	S_4
S_1	S_2	S_2
S_2	S_3	S_3
S_3	S_0	S_0
S_4	S_7	S_5
S_5	S_6	S_6
S_6	S_0	S_0
S_7	S_3	S_6

(d)

Gates NAND	Ass. 1	Ass. 2	Ass. 3
	22	18	16
Chips			
Dual 4 input	1	1	1
Triple 3 input	2	1	1
Quad 2 input	4	4	4
Total chips	7	6	6
Gate inputs	49	39	33

(f)

C \ BA	00	01	11	10
0	S_3	S_6	S_2	S_7
1	S_1	S_4	S_0	S_5

(g)

	Present state	Next state		Flip-flop input signals											
		$X=0$	$X=1$	$X=0$		$X=1$		$X=0$		$X=1$		$X=0$		$X=1$	
	CBA	CBA	CBA	J_C	K_C	J_C	K_C	J_B	K_B	J_B	K_B	J_A	K_A	J_A	K_A
S_0	111	100	101	ø	0	ø	0	ø	1	ø	1	ø	1	ø	0
S_1	100	011	011	ø	1	ø	1	1	ø	1	ø	1	ø	1	ø
S_2	011	000	000	0	ø	0	ø	ø	1	ø	1	ø	1	ø	1
S_3	000	111	111	1	ø	1	ø	1	ø	1	ø	1	ø	1	ø
S_4	101	010	110	ø	1	ø	0	1	ø	1	ø	ø	1	ø	1
S_5	110	001	001	ø	1	ø	1	ø	1	ø	1	1	ø	1	ø
S_6	001	111	111	1	ø	1	ø	1	ø	1	ø	ø	0	ø	0
S_7	010	000	001	0	ø	0	ø	ø	1	ø	1	0	ø	1	ø

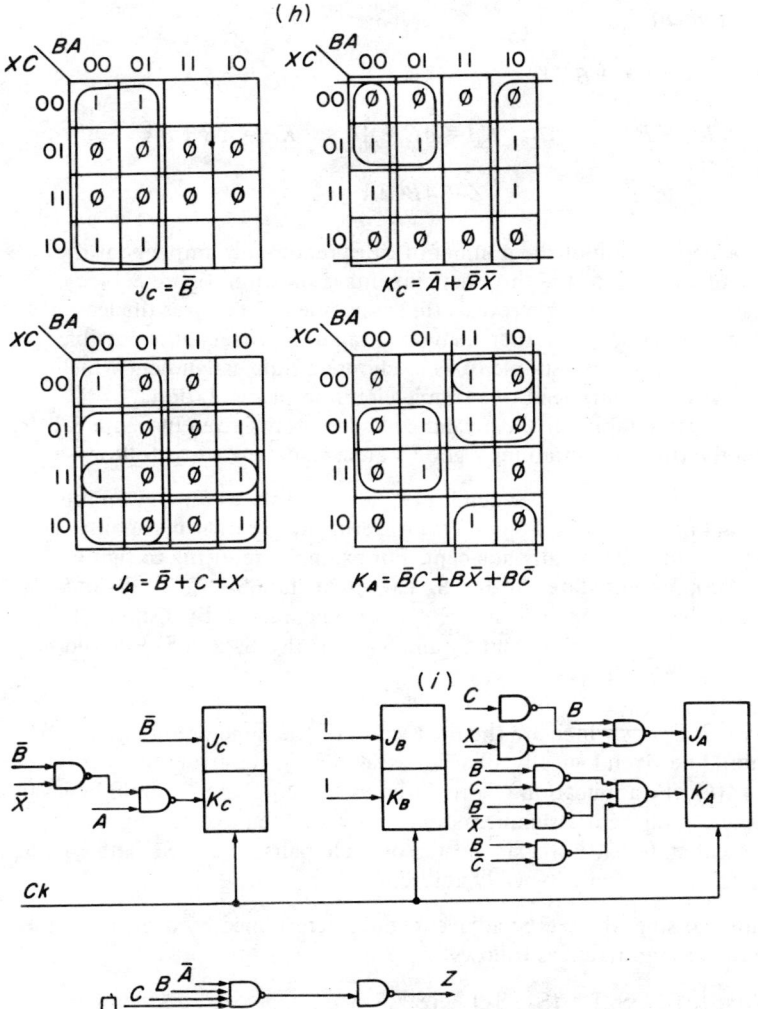

Figure 8.14. (a) Block diagram, (b) internal state diagram, and (c) three possible state assignments for the invalid code detector. (d) Gate and chip comparison for the three randomly selected state assignments. (e) State table for the invalid code detector. (f) State assignment map. (g) State table for the state assignment obtained using rules 1 and 2. (h) K-maps for the invalid code detector. (i) Implementation of the invalid code detector

Assignment 3

$$J_A = CX + \overline{B}\overline{C}\overline{X} \qquad J_B = A + C\overline{X} \qquad J_C = \overline{A}\overline{B}X$$

$$K_A = B \qquad K_B = AC + \overline{A}\overline{C} \qquad K_C = AB + B\overline{X}$$

$$Z = ABC\sqcap$$

A comparison of the number of gates required to implement the circuit for each of the three assignments is shown in figure 8.14(d). Inspection of this table reveals that assignment 3 requires the least hardware. However, rather than use a random process, it is possible with the aid of two simple rules to choose a state assignment which will with certainty lead to a simple circuit implementation.

The state table for the invalid code detector is given in figure 8.14(e) and the rules for obtaining a good state assignment are as follows:

Rule 1: Present states which lead to identical states for a given input should be given state assignments differing in one digit place only, i.e. they should be logically adjacent. For example, referring to figure 8.14(e), present states S_3 and S_6 have next state S_0 for $X = 0$, and should be given logically adjacent state assignments. By a similar line of reasoning, it is clear that S_2 and S_7 (and also S_5 and S_7) should be made logically adjacent.

Rule 2: States which are the next states of the same present state should be given logically adjacent assignments. Referring to figure 8.14(e), the application of rule 2 shows that S_1 and S_4 should be given logically adjacent assignments since they are the next states of the present state S_0. For a similar reason state pairs S_7 and S_5, and S_3 and S_6 should be made logically adjacent.

Summarising, the sets of adjacent states determined by using rules 1 and 2 are tabulated as follows:

Rule 1: (S_3, S_6) (S_2, S_7) (S_5, S_7)

Rule 2: (S_1, S_4) (S_5, S_7) (S_3, S_6)

A suitable state assignment is shown plotted in figure 8.14(f), in which the above adjacencies are satisfied. If it is not possible to satisfy all the adjacencies obtained using these rules without conflict, then the adjacencies obtained from the first rule should have priority.

The state table for the state assignment shown in figure 8.14(f) is illustrated in figure 8.14(g), together with the tabulation of the flip-flop inputs which have been derived from the state table in the normal way. K-maps for simplifying the J and K input signals are shown in figure 8.14(h). Notice that it is not necessary to plot the J_B and K_B signals on maps since all the entries in the J_B and K_B tabulations are 1 or $\emptyset$. Hence $J_B = K_B = 1$.

The flip-flop input signals obtained from the maps are

$$J_C = \bar{B} \qquad J_B = 1 \qquad J_A = \bar{B} + C + X$$

$$K_C = \bar{A} + B\bar{X} \qquad K_B = 1 \qquad K_A = \bar{B}C + B\bar{X} + B\bar{C}$$

$$Z = \bar{A}\bar{B}C\sqcap$$

The implementation of the invalid code detector is shown in figure 8.14(i) and the diagram reveals that ten NAND gates are required for circuit implementation which can be provided by the following chips.

Dual 4 input	1
Triple 3 input	1
Quad 2 input	2
Total chips	4
Gate inputs	23

It is clear that this state assignment requires less hardware than any of the other three randomly selected assignments shown in figure 8.14(c).

Problems

8.1 For the sequential circuit shown in figure P8.1 find

(a) the state table;
(b) the internal state diagram, and
(c) the function of the circuit.

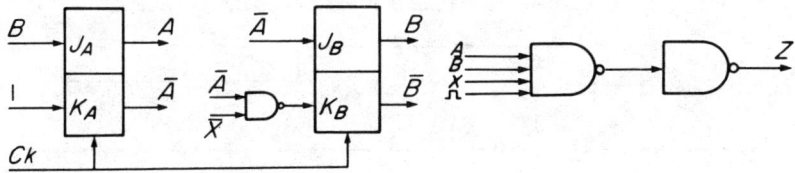

Figure P8.1

8.2 Serial binary data X, synchronised with the clock, is fed to the logic network shown in figure P8.2. An output will occur on the Z line of the network whenever the string of digits 1101 is

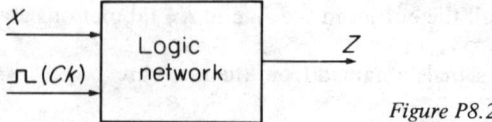

Figure P8.2

received (n.b. An output will occur for overlapping strings). Develop a synchronous sequential circuit to implement the above specification.

8.3 XS3 information is received serially, most significant bit first, and in synchronism with the clock, by the logic network shown in figure P8.3. The function of the network is to generate an

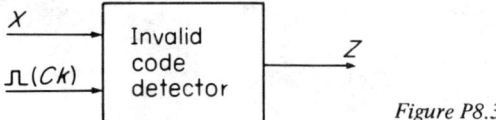

Figure P8.3

output signal $Z = 1$ when an invalid code combination has been received. Using JK flip-flops and NAND gates, develop a synchronous sequential logic circuit that will perform this function.

8.4 A clock signal X is to be gated on and off by a signal m. The gating signal must be arranged so that the circuit produces complete clock pulses only. A timing diagram for the network is shown in figure P8.4. Develop a synchronous sequential circuit for implementing the above specification.

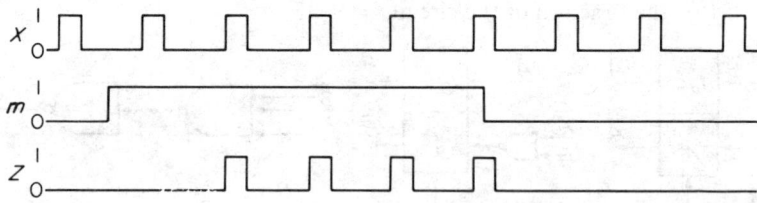

Figure P8.4

8.5 A circuit is to be designed in which a single clock pulse Z is to be selected by a push button control S. The push button is made at random intervals and the time duration for which the push button contact is on, is long in comparison with the periodic time of the clock. A typical timing diagram is shown in figure P8.5. Design and implement a circuit which will satisfy the given specification.

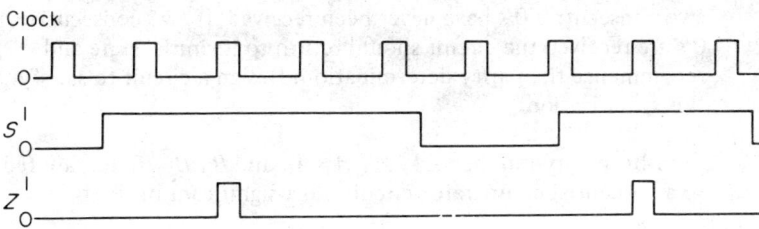

Figure P8.5

8.6 A serially operated full adder is supplied with two binary digits A and B which are fed in, one pair of bits at a time, in synchronism with the clock, and with the carry generated by a previous addition. The adder generates a sum digit and a carry. The carry is stored in the memory box in figure P8.6 until the next pair of binary digits arrive in conjunction with their associated clock pulse.

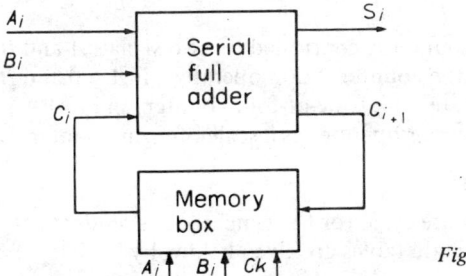

Figure P8.6

Develop the logic for the hardware contents of the memory box using a JK flip-flop and NAND gates. Also draw a timing diagram displaying the important waveforms of the system during the adding process.

8.7 A sequential network has two inputs, X and clock, and one output Z. Incoming data are examined in consecutive groups of four digits and the output $Z = 1$ if any of the three input sequences 1010, 0110 or 0010 should occur. Develop a state diagram and implement the circuit using *JK* flip-flops and NOR gates.

8.8 A sequential logic network is to be used for determining the parity of a continuous string of binary digits. If an even number of 1's has been received the output of the network $Z = 1$ provided two consecutive 0's have never been received. If two consecutive 0's are received the circuit should return to its initial state and recommence the parity determination. Design a circuit to satisfy this specification.

8.9 Four-bit binary numbers, $A_3 \, A_2 \, A_1 \, A_0$ and $B_3 \, B_2 \, B_1 \, B_0$, are fed to a sequential comparator circuit most significant bit first, as

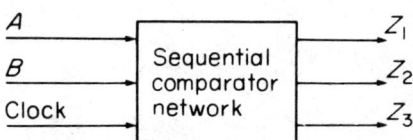

Figure P8.9

shown in figure P8.9. Design the synchronous sequential circuit whose outputs are $Z_1 = 1$, if $A > B$, $Z_2 = 1$ if $A = B$ and $Z_3 = 1$ if $A < B$.

8.10 A synchronous counter is controlled by two signals A and B. If $A = 0$ and $B = 0$ the counter is non-operative, if $A = 0$ and $B = 1$ the counter operates as a scale-of-four counter, and if $A = 1$ and $B = 0$ the counter operates as a scale-of-eight counter. Design this circuit.

8.11 Find a minimal state table for the synchronous sequential machines whose state tables are given below, by

(*a*) Caldwell's merging rules;
(*b*) partitioning; and
(*c*) the implication table.

Present state	Next state X = 0	Next state X = 1	Present state	Next state X = 0	Next state X = 1	Present state	Next state X = 0	Next state X = 1
S_0	S_1 Z = 0	S_2 Z = 0	S_0	S_1 Z = 0	S_8 Z = 0	S_8	S_9 Z = 0	S_{12} Z = 0
S_1	S_3 Z = 0	S_4 Z = 0	S_1	S_2 Z = 0	S_5 Z = 0	S_9	S_{10} Z = 0	S_{11} Z = 0
S_2	S_5 Z = 0	S_6 Z = 0	S_2	S_3 Z = 0	S_4 Z = 0	S_{10}	S_0 Z = 0	S_0 Z = 0
S_3	S_0 Z = 1	S_0 Z = 0	S_3	S_0 Z = 0	S_0 Z = 0	S_{11}	S_0 Z = 0	S_0 Z = 0
S_4	S_0 Z = 0	S_0 Z = 0	S_4	S_0 Z = 0	S_0 Z = 1	S_{12}	S_{13} Z = 0	S_{14} Z = 0
S_5	S_0 Z = 0	S_0 Z = 0	S_5	S_6 Z = 0	S_7 Z = 0	S_{13}	S_0 Z = 0	S_0 Z = 0
S_6	S_0 Z = 0	S_0 Z = 1	S_6	S_0 Z = 0	S_0 Z = 0	S_{14}	S_0 Z = 0	S_0 Z = 1
			S_7	S_0 Z = 0	S_0 Z = 1			

8.12 A synchronous sequential machine has the following state table:

Present state	Next state X = 0	Next state X = 1
S_0	S_1 Z = 0	S_4 Z = 0
S_1	S_0 Z = 0	S_5 Z = 0
S_2	S_3 Z = 0	S_5 Z = 1
S_3	S_2 Z = 0	S_4 Z = 1
S_4	S_1 Z = 0	S_2 Z = 0
S_5	S_0 Z = 0	S_3 Z = 0
S_6	S_1 Z = 0	S_0 Z = 0

Determine the output sequences of the machine for all possible input sequences of length 1, 2, 3, ..., and hence find a partition of the states.

8.13 Using the rules given in this chapter, obtain good state assignments for the state tables developed in answer to problems 8.2 and 8.3.

8.14 Design a synchronous sequential circuit with JK flip-flops whose characteristic equations are

$$A^{t+\delta t} = (\bar{A}\bar{B}X + B\bar{X})^t$$

$$B^{t+\delta t} = (AX)^t$$

and whose output equation is

$$Z = AB$$

where X is the input signal. Also obtain the state table and the state diagram for the circuit.

9

Event-driven circuits

9.1 Introduction

There are many sequential circuits which are driven by events rather than by a train of clock pulses. For example, a digital alarm system will be activated by the event that raises the alarm. In this case it is the event that drives the logic, and since the events are irregular in occurrence such a circuit is referred to as an asynchronous sequential circuit, or alternatively, and perhaps more meaningfully, as an event-driven circuit.

In this chapter a design procedure will be established for the design of event-driven circuits based on the NAND sequential equation, $Q^{t+\delta t} = (S + \bar{R}Q)^t$, developed in Chapter 5 from the state table of an SR flip-flop.

9.2 The museum problem

As an example of an event-driven circuit, a sequential circuit will be developed that counts the number of people entering a museum. The input sensors are two light beams six inches apart, as shown in figure 9.1(a), and it has been arranged that only one person can enter the counting system at any one time. A person is only counted if the first beam X_1 is broken followed by the second beam X_2. Persons leaving the museum break beam X_2 first, then X_1, and are not to be counted, as well as those who hesitate and break one beam only. It will be assumed that the breaking of a beam generates a logical '1' signal.

A block diagram for this problem is shown in figure 9.1(b). The two inputs X_1 and X_2 are generated when a person sequentially breaks the beams at the museum entrance. These input signals are referred to as the primary variables. The circuit has one output Z which for a short period is logical '1' when a person is entering the museum, and this signal could be used as the asynchronous clock input to a multi-decade counter.

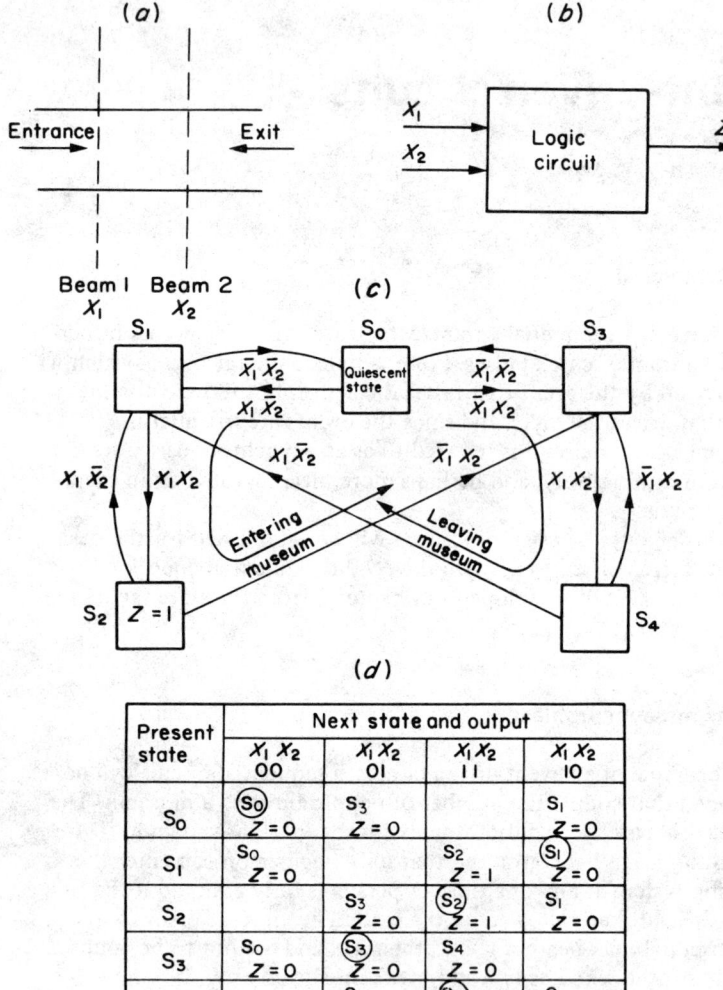

(e)

Present state	Next state and output			
	$X_1 X_2$ 00	$X_1 X_2$ 01	$X_1 X_2$ 11	$X_1 X_2$ 10
S_{012}	(S₀₁₂) Z=0	S_{34} Z=0	(S₀₁₂) Z=1	(S₀₁₂) Z=0
S_{34}	S_{012} Z=0	(S₃₄) Z=0	(S₃₄) Z=0	S_{012} Z=0

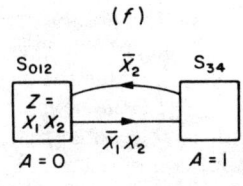

(f)

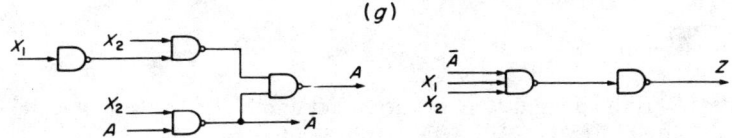

(g)

Figure 9.1. (a) Schematic diagram of the museum entrance. (b) Block diagram for the museum problem. (c) Internal state diagram. (d) State table. (e) Reduced state table. (f) Reduced state diagram. (g) Primitive circuit for the museum problem

The internal state diagram for the problem is shown in figure 9.1(c). S_0 may be regarded as the quiescent state on this diagram in the sense that it can represent the condition of a person having either fully entered or fully left the museum. The path taken through the state diagram if a person enters the museum is $S_0 \rightarrow S_1 \rightarrow S_2 \rightarrow S_3 \rightarrow S_0$, and the path taken through the state diagram if a person leaves the museum is $S_0 \rightarrow S_3 \rightarrow S_4 \rightarrow S_1$ and back to S_0. Allowance is also made on the state diagram for people who hesitate when entering or leaving the museum. For example, the person who breaks beam X_1, only, on entering, and then turns away from the entrance will result in a transition path $S_0 \rightarrow S_1 \rightarrow S_0$ being followed.

A state table can now be drawn up which summarises the information appearing on the internal state diagram. The table has five rows, one for each possible present state, and four columns, one for each of the possible combinations of the input signals, as illustrated in figure 9.1(d). Each of the 20 cells in the table is at the intersection of a row headed with the present state and a column headed with a particular combination of input signals. The entries in any one of these cells consist of the next state, and the output corresponding to the intersecting present state and input combination. For example, in the first row of the table the present state is S_0. If the input signal is

$X_1 X_2 = 01$ then the next state of the circuit is S_3 and the output is $Z = 0$. These are then entered in the second column of the first row.

It will be observed that some of the next state entries in the table have been circled. These states are known as stable states. In the first row of the table the present state is S_0 and for the input combination $X_1 X_2 = 00$ the next state is S_0. Clearly, no transition has been made and a stable state exists. If, however, the present state is S_0 and the input combination is $X_1 X_2 = 10$, the next state is S_1 and a transition, which requires a finite time, takes place from S_0 to S_1. The circuit is not in a stable state during the transition.

The condition for stability is

$$S^t = S^{t+\delta t}$$

where S^t is the present state of the circuit and $S^{t+\delta t}$ is the next state of the circuit. It follows that the condition for instability is

$$S^t \neq S^{t+\delta t}$$

A further examination of the state table shows that some cells do not have an entry at all. For example, in the first row where the present state is S_0 there is no entry in the cell corresponding to the input combination $X_1 X_2 = 11$. The state diagram shows that the state S_0 is entered from either S_1 or S_3 with an input signal $X_1 X_2 = 00$. If it is now required, after entering the state S_0, that the input signal should become $X_1 X_2 = 11$ a simultaneous change of the two primary variables must occur.

The circuits that will be developed in this chapter will only allow the change of one primary variable at any given instant. This constraint is obviously satisfied in the museum problem since the light beams are six inches apart and cannot be broken simultaneously. A circuit designed employing this constraint is said to be operating in the fundamental mode.

State tables which contain vacant cells are referred to as incompletely specified tables. In the state table of figure 9.1(d) the vacancies correspond to forbidden input combinations. These vacancies can be regarded as 'can't happen' conditions, and may enable a simplification of the state table which at first sight did not seem possible. The justification of the allocation of a 'can't happen' condition in a state table is the same as for 'can't happen' conditions in combinational logic problems. If an event cannot happen the designer 'doesn't care' what the circuit would do in response to it.

Event-driven circuits

When simplifying a state table, it is possible to assign a next state and output to a vacancy in such a way as to make the row in which the vacancy occurs identical to a second row. These two identical rows can then be merged using Caldwell's merging rules as described in Chapter 8.

An examination of the state table of figure 9.1(d) shows that the rows headed S_0, S_1 and S_2, and those headed S_3 and S_4 are identical and can be merged using Caldwell's merging rules. S_0, S_1 and S_2 are merged to form the new state S_{012}, and states S_3 and S_4 are merged to form the new state S_{34}. The reduced state table is shown in figure 9.1(e) and the reduced state diagram which is constructed from the information in the reduced state table is shown in figure 9.1(f).

Since there are only two states in the reduced state diagram just one secondary variable A is required to define them. For the state S_{012}, $A = 0$ and for the state S_{34}, $A = 1$. The output Z appears in state S_{012} and will be logical '1' for the input combination $X_1 X_2 = 11$.

The primitive circuit can now be obtained with the aid of the NAND sequential equation

$$Q^{t+\delta t} = (S + \bar{R}Q)^t$$

developed in Chapter 5, where S was defined as the turn-on set of Q and R was defined as the turn-off set of Q. The turn-on and turn-off sets of A can be obtained directly from the reduced state diagram.

Turn-on set of $A = \bar{X}_1 X_2$

Turn-off set of $A = \bar{X}_2$

Hence $\quad A = \bar{X}_1 X_2 + X_2 A$

and the output Z may be written as

$$Z = \bar{A} X_1 X_2$$

The implementation of the primitive circuit is shown in figure 9.1(g).

9.3 Races and cycles

When the secondary variables were allocated to the internal states of a clock-driven machine, the criterion for the allocation was that it should lead to a minimum hardware implementation. It was pointed out that there is no known method for the allocation of secondary variables which will lead to this minimum hardware implementation, although guidelines were presented, which, when used, will lead to a simple, if not the simplest, circuit. The criterion for the allocation of secondary variables in event-driven circuits is somewhat different, and in this section those factors which govern this allocation will be examined.

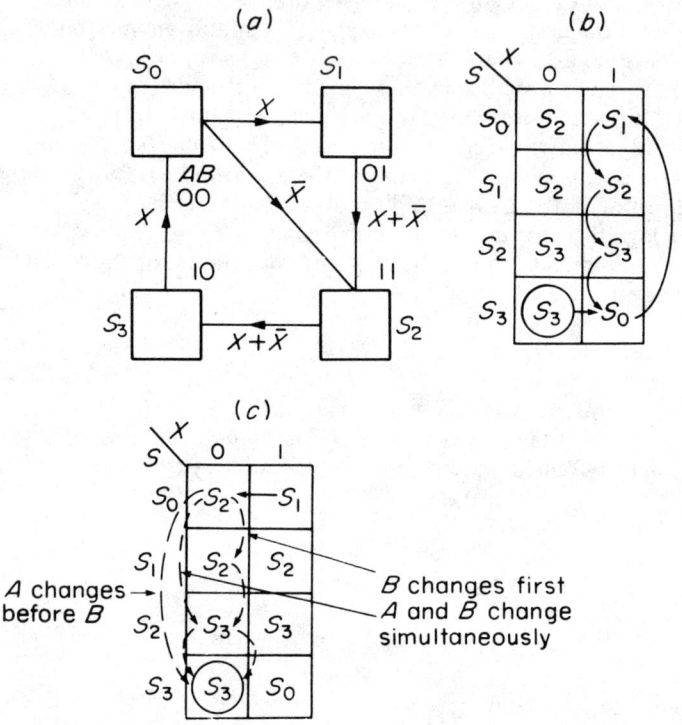

Figure 9.2. (a) State diagram for a circuit exhibiting cyclic behaviour and a non-critical race. (b) State table illustrating a cycle. (c) State table illustrating non-critical races

The internal state diagram of a state machine is shown in figure 9.2(a). The secondary variables A and B have been allocated to the internal states in a purely arbitrary way and the state table for the machine which has been tabulated with the aid of the state diagram is shown in figure 9.2(b).

If it is assumed that the machine starts in the stable state defined by

$AB = 10$ (S_3) and $X = 0$, and that X then changes from 0 to 1, the machine will first make a transition to the unstable state defined by $AB = 00$ (S_0) and $X = 1$. Since $X = 1$ when the state S_0 is entered, a further transition will be made to S_1, and assuming there is no further change in the input variable the circuit passes through S_2 and returns to S_3, thus completing one cycle of the state diagram. As long as $X = 1$ this cycling process will continue.

An examination of the state table shows the circuit first leaving $\widehat{S_3}$ to enter S_0 in the bottom right-hand cell of the state table. Thereafter the circuit cycles in turn through the unstable states tabulated in the $X = 1$ column and will continue to do so until X returns to 0. This type of circuit behaviour is called a cycle.

If now X is changed from 1 to 0 when the machine is in the internal state $AB = 00$, the machine should make a direct transition to the state $AB = 11$, followed by a further transition to the stable state defined by $X = 0$ and $AB = 10$. The circuit can reach this final destination by a number of routes simply because AB can change from 00 to 11 in a number of different ways. For example, A and B may change simultaneously, and in that case the transition sequence will be as described above. However, it may be that B changes before A, in which case a transition is first made to $S_1 (AB = 01)$ and this is followed by further transitions to states $S_2 (11)$ and $S_3 (10)$, the stable state. Alternatively, A may change before B, in which case the circuit will enter the stable state defined by $X = 0$ and $AB = 10$ directly from S_0. Once a circuit enters a stable state it will stay there until a further change in the primary variables occurs. Consequently for A changing before B the circuit enters state S_3 by none of the transition paths shown on the state diagram and it now stays there until such time as X changes from 0 to 1. The various transitions between states for the three conditions described above are illustrated in the state table shown in figure 9.2(c).

From the foregoing remarks it may be concluded that whenever two secondary variables must change in response to a change in a primary variable, a race condition exists. The condition has its origin in the different delays that occur when the A and B signals are generated. In the case described above, the race is called a non-critical race since, irrespective of the transitions made, the circuit always ends up in the same stable state.

Unfortunately, there are races that can occur in event-driven circuits in which the final state reached depends upon the order in which the secondary variables change. Such races are termed critical races. For example, the internal state diagram of a state machine is given in figure

9.3(a) and the corresponding state table is shown in figure 9.3(b). It will be assumed that the circuit is in the state defined by $AB = 11$ and $X = 1$. If X is now changed to 0 the circuit will make a direct transition to the stable state defined by $X = 1$ and $AB = 00$ (S_0) providing A and B change simultaneously. Alternatively, if B changes before A the circuit will make a transition to the state defined by $X = 1$ and $AB = 10$. Since this is a stable state the circuit will remain there, and in fact a

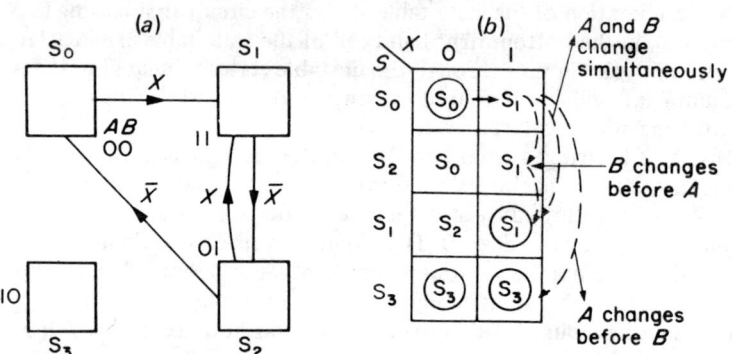

Figure 9.3. (a) State diagram for a circuit exhibiting critical races. (b) State table illustrating a critical race.

quick glance at the state diagram shows that the circuit remains locked in that state indefinitely because of the absence of an exit from the state. However, A may change before B and then the circuit will initially make a transition to the state defined by $X = 0$ and $AB = 01$. This state is unstable and the circuit makes a further transition to the state defined by $X = 0$ and $AB = 00$. The transitions described can clearly lead to faulty circuit operation. A critical race occurs in the circuit because it is possible to end up in one of two stable states, depending upon the order in which the secondary variables change.

The various transitions which can take place in this machine are indicated on the state table shown in figure 9.3(b).

9.4 Race-free assignment for a three-state machine

If critical races are to be avoided, it is necessary to provide a race-free assignment of the secondary variables on the internal state diagram. In effect, this means that when a transition is made from one internal state to the next, only one secondary variable should be allowed to change.

The three-state diagram shown in figure 9.4(a) requires two secondary variables to define the three states. An arbitrary secondary assignment has been made on the diagram, but an inspection of it reveals that on making a transition from S_2 to S_0 both secondary variables must change. It is in fact impossible to find a race-free assignment for a three-state diagram if transitions are required between each pair of states.

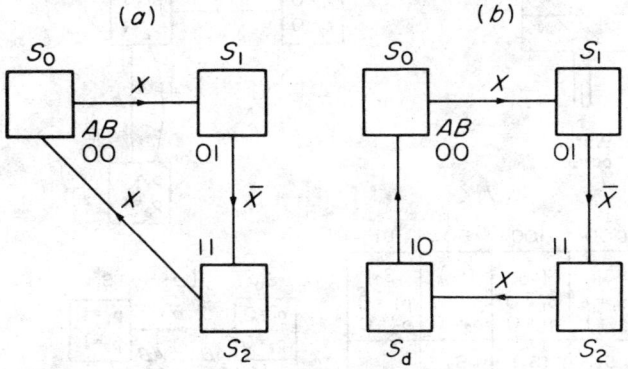

Figure 9.4. (a) State machine requiring race-free secondary assignment. (b) Inclusion of a dummy state to give race-free assignment

Furthermore, two secondary variables can define four states which implies that for the three-state diagram of figure 9.4(a) there is one unused state which has been omitted from the diagram. The presence of an unused state also creates problems for the designer. If the exit from the state has not been specified it can become a 'lock-in' state as described in § 9.3.

These two problems are overcome by incorporating the unused state $AB = 10$ (S_d) in the state diagram (figure 9.4b) and allowing the circuit to return unconditionally from this dummy state to state S_0. An event-driven circuit will now be designed which requires the inclusion of a dummy state.

9.5 The pump problem

Step 1: Problem definition.

Water is pumped into a water tower by two pumps p_1 and p_2. Both pumps are to turn on when the water goes below level 1 and are to

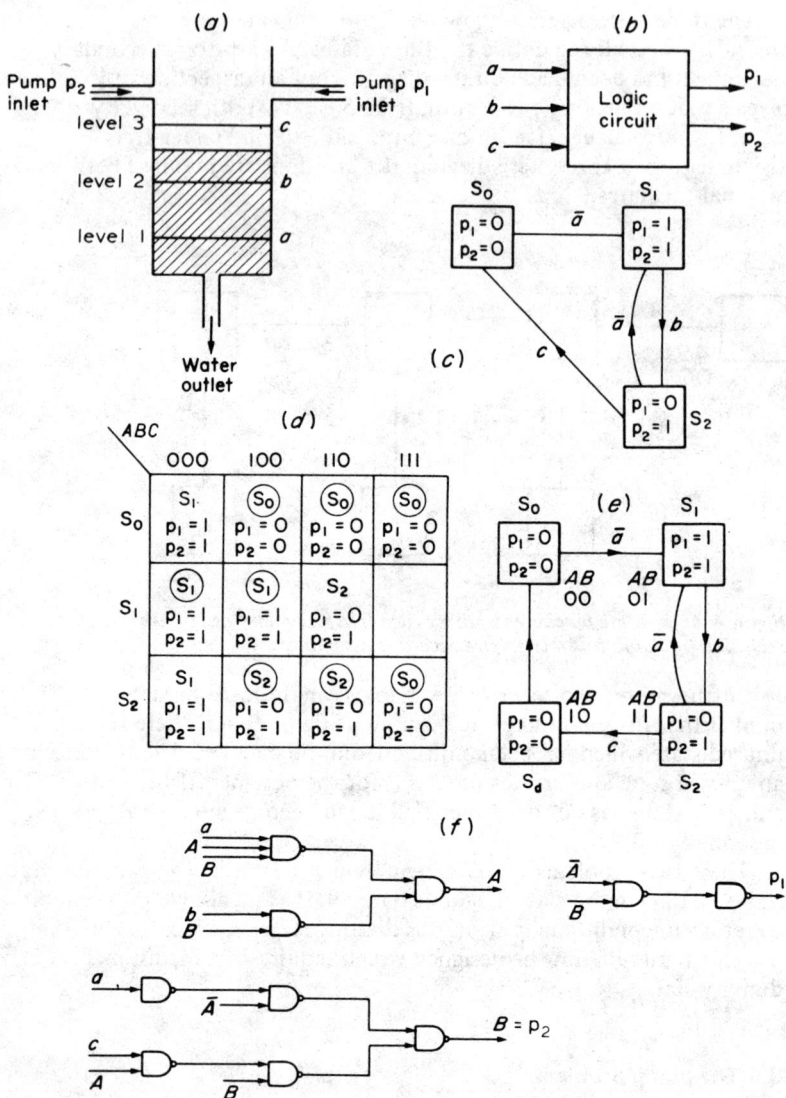

Figure 9.5. (a) Block schematic for the water pump problem. (b) Block diagram of a pump controller. (c) Basic internal state diagram for the pump problem. (d) State stable. (e) Modified state diagram. (f) Circuit implementation of the pump controller

Event-driven circuits

remain on until the water reaches level 2, when pump p_1 turns off and remains off until the water is below level 1 again. Pump p_2 remains on until level 3 is reached, when it also turns off and remains off until the water falls below level 1 again. Level sensors are used to provide level detection signals as follows:

Signal $a = 1$ when the water is at or above level 1, otherwise $a = 0$.
Signal $b = 1$ when the water is at or above level 2, otherwise $b = 0$.
Signal $c = 1$ when the water is at or above level 3, otherwise $c = 0$.

The aim is to develop an event-driven circuit to control pumps p_1 and p_2 according to the specification given above.

A schematic diagram of the water tower is shown in figure 9.5(a) and a block diagram of the circuit is shown in figure 9.5(b).

Step 2: The internal state diagram.

A suitable internal state diagram is shown in figure 9.5(c), in which the state S_0 is related to the condition when the water is above level 'c' and both pumps p_1 and p_2 are off. As the tank empties the water level falls until it is below 'a' and a transition is then made to S_1, since $\bar{a} = 1$. When in state S_1 both pumps are on. If the water continues to rise and reaches level 'b' a transition is made to S_2 and pump p_1 is then turned off. In state S_2, two options are available: if the water level falls below 'a' again a transition will be made back to S_1 on the signal $\bar{a} = 1$. Alternatively, if the water continues rising, when the level 'c' is reached a transition is made to S_0 and both the pumps are turned off.

Step 3: The state table.

The state table for the pump problem is shown in figure 9.5(d). It should be observed that input combinations $abc = 001, 010, 011$ and 101 are missing from the table since these combinations can only exist under fault conditions.

Two secondary variables A and B are required to define three states. Because there are transitions between every pair of states, a race-free assignment of secondary variables is not possible. Consequently, an additional dummy state S_d is added to the state diagram. The modified state diagram is shown in figure 9.5(e).

Step 4: Development of circuit equations.

$$\text{Turn-on set of } A = bB$$

$$\text{Turn-off set of } A = \overline{B} + B\overline{a} = \overline{B} + \overline{a}$$

$$\text{Turn-on set of } B = \overline{a}\overline{A}$$

$$\text{Turn-off set of } B = cA$$

Hence

$$A = bB + (\overline{\overline{B} + \overline{a}})A$$

$$= bB + aAB$$

and

$$B = \overline{a}\overline{A} + \overline{(cA)}B$$

$$= \overline{a}\overline{A} + (\overline{c} + \overline{A})B$$

Also

$$p_1 = \overline{A}B$$

and

$$p_2 = \overline{A}B + AB = B$$

Step 5: Circuit implementation

The circuit implementation of the pump controller is shown in figure 9.5(f).

9.6 Race-free assignment for a four-state machine

The internal state diagram for a four-state machine is shown in figure 9.6. All the transitions on this diagram take place between adjacent states, for example, S_0 to S_1, and S_1 to S_2 etc. Because the state diagram is structured in this way, the individual states can each be defined by two secondary variables A and B, and these can be allocated in such a way that when a transition takes place between adjacent states only one secondary variable changes.

If the state diagram for the machine includes a transition between two states that are not adjacent, for example S_3 to S_1 in the state diagram of figure 9.7(a), then a race-free assignment is not possible with two secondary variables. The state diagram reveals that with the

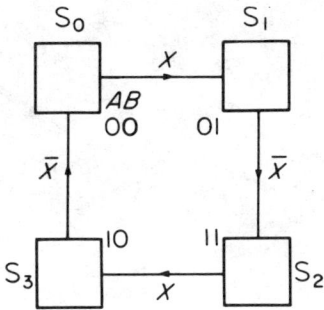

Figure 9.6. State diagram for a four-state machine with transitions between adjacent states

same secondary assignment as the one shown in figure 9.6 there is a double change in secondary variables when a transition is made from S_3 to S_1. No matter how the secondary variables are allocated there will always be at least one transition which will result in a double change of secondary variable, and this implies that a race-free assignment can only be achieved by using three secondary variables.

Race-free assignments can most easily be obtained from a K-map of the three secondary variables, as shown in figure 9.7(b). It is a property of the K-map that adjacent cells differ in one digit position only, and consequently two states allocated to adjacent cells will have secondary assignments that differ in one digit place only.

Four of the internal states of the circuit have been allocated to cells such that S_0 is adjacent to S_1, S_1 to S_2, and S_1 to S_3. However, for a race-free assignment S_2 should be adjacent to S_3 and so should S_0. Examination of the K-map indicates that such adjacencies are impossible, and consequently the transitions from S_2 to S_3 and S_3 to S_0 have been made via the dummy states S_{d1} and S_{d2}, respectively. The modified state diagram consists of six internal states, two of which are dummies, as shown in figure 9.7(c). Each transition on this diagram exhibits one change of secondary variable; hence the assignment is race-free.

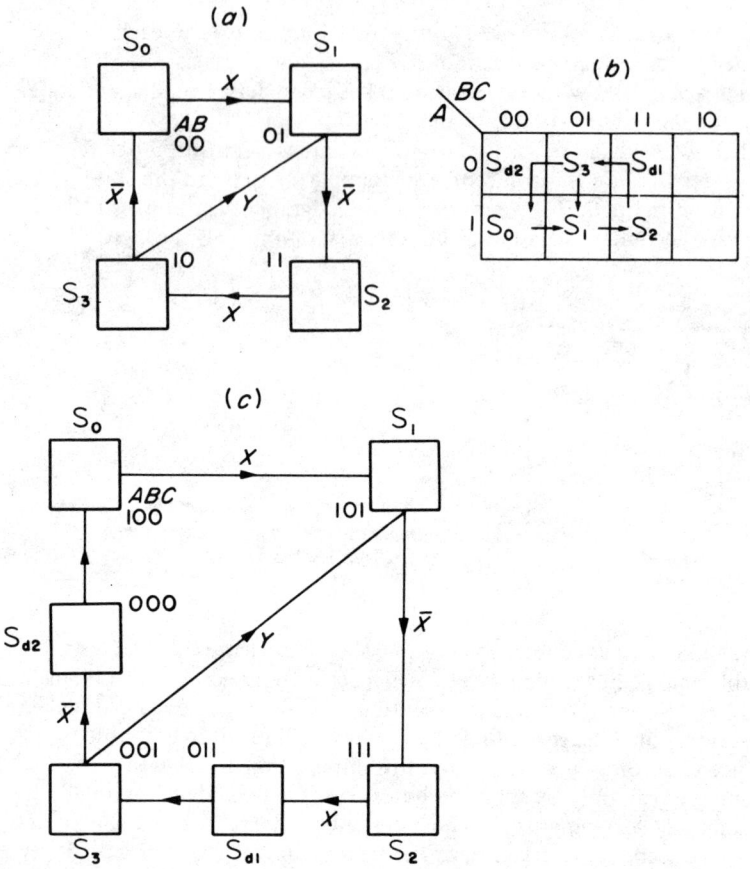

Figure 9.7. (a) State diagram for a four-state machine with one diagonal transition. (b) K-map for determining a race-free assignment. (c) Race-free state diagram for a four-state machine having a diagonal transition

The one additonal complication that can be added to the four-state diagram in figure 9.7(a) is the introduction of the second diagonal transition from S_2 to S_0 as shown in figure 9.8(a). On the K-map (figure 9.8b) the four original states S_0, S_1, S_2 and S_3 are allocated as in the previous example, but an additional transition from S_2 to S_0 is now required, and this is made via the dummy state S_{d3}. The modified state diagram is shown in figure 9.8(c) and this state diagram can now be used to determine the NAND implementation of the machine.

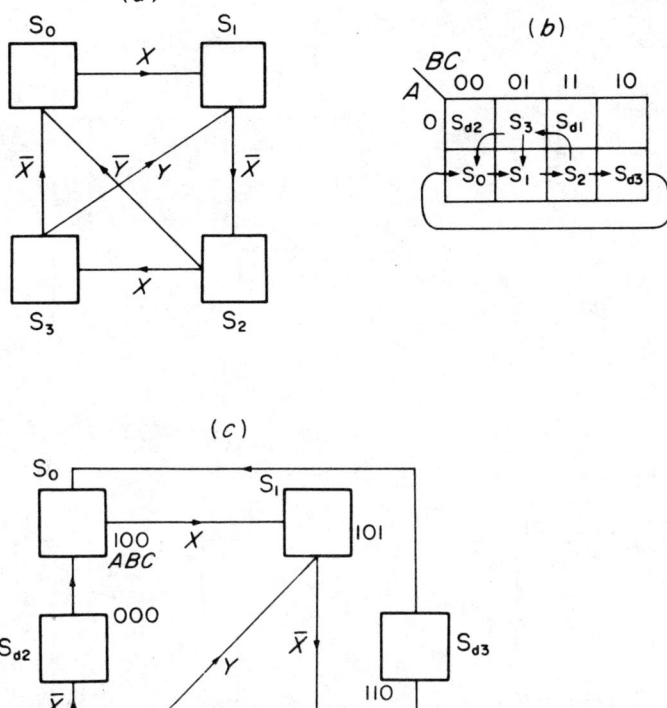

Figure 9.8. (a) A four-state machine with two diagonal transitions. (b) K-map for determining a race-free assignment. (c) Modified state diagram for a race-free assignment

9.7 A sequence detector

Finally, a further example of an event-driven circuit will be studied in order to emphasise some of the problems faced by the designer when developing this type of circuit. The example concerns a sequence detector which has two inputs, X_1 and X_2, and one output, Z, as illustrated in the block diagram of figure 9.9(a), and which will give an

210 Event-driven circuits

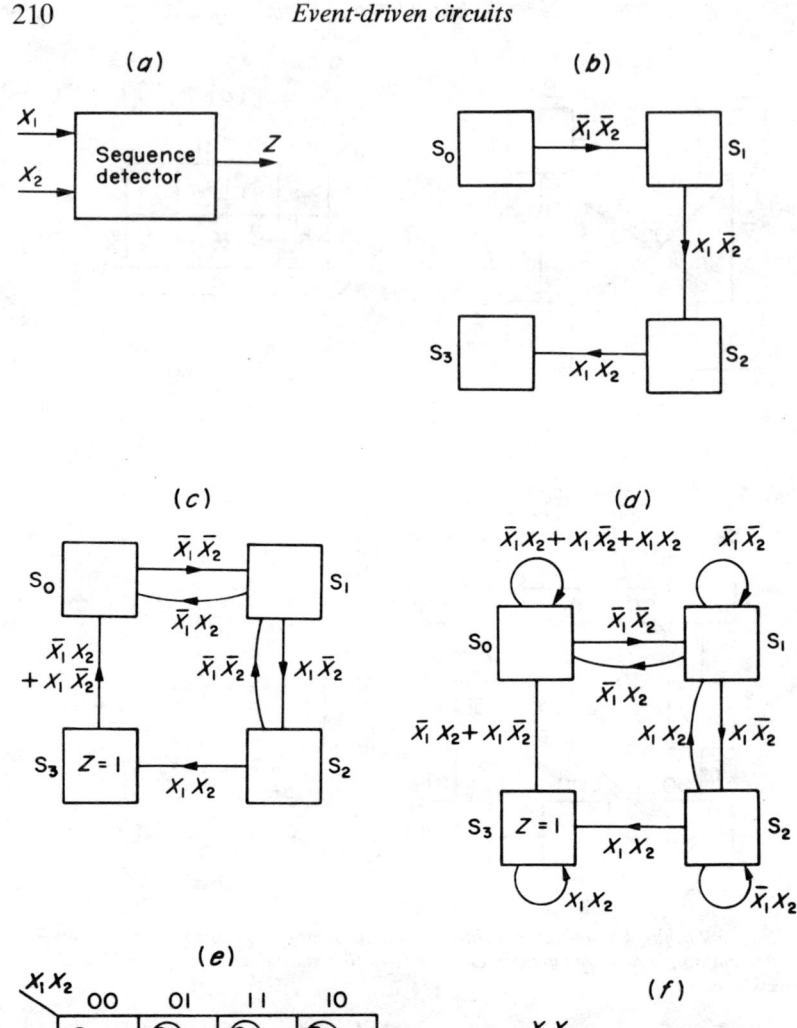

Event-driven circuits

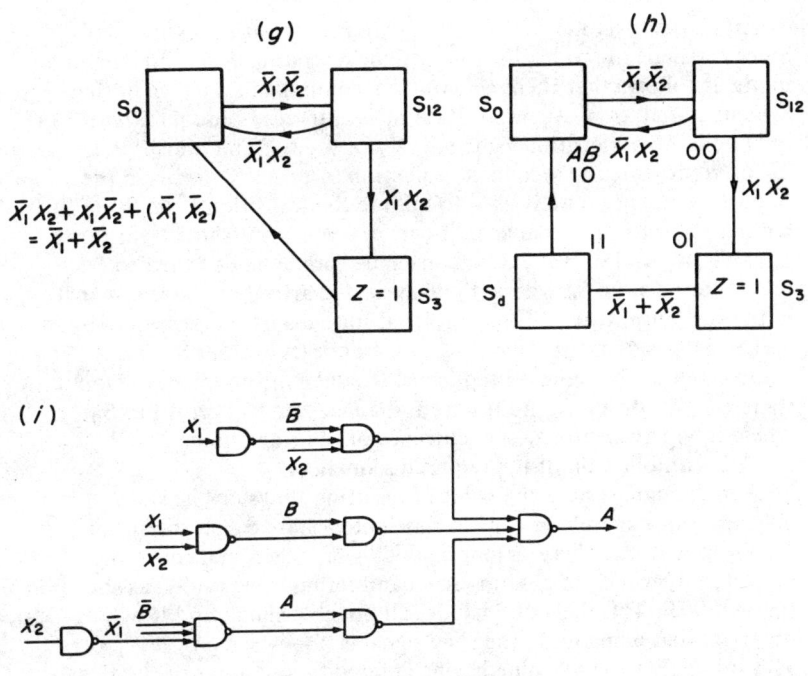

Figure 9.9. (a) Block diagram of a sequence detector. (b) Basic elements of the internal state diagram. (c) Internal state diagram for the sequence detector. (d) Complete internal state diagram including slings. (e) State table. (f) Reduced state table. (g) Reduced state diagram. (h) State diagram including dummy state and secondary assignment. (i) Implementation of the sequence detector

output $Z = 1$ when the sequence of primary signals $X_1 X_2 = 00, 10, 11$ has occurred.

Initially, the designer must develop the internal state diagram. In this type of problem a good beginning to the state diagram is to insert the required sequence as shown in figure 9.9(b). This requires four states connected via three transitions initiated by the transition signals $\bar{X}_1 \bar{X}_2$, $X_1 \bar{X}_2$ and $X_1 X_2$, respectively.

To complete the state diagram it is now only necessary to insert the additional transition paths that may originate at each of the states. For example, the machine enters state S_1 on the transition signal $\bar{X}_1 X_2$. Since the machine to be designed will be operating in the fundamental mode, it follows that there cannot be a simultaneous change in the primary variables $X_1 X_2$ when state S_1 is entered. Consequently, it can only be left on transition signals $X_1 \bar{X}_2$ or $\bar{X}_1 X_2$. The transition signal $X_1 \bar{X}_2$ represents the second combination of primary signals in the required sequence, and is used to initiate the transition from S_1 to S_2. On the other hand, a change in X_2 from 0 → 1 results in a transition signal $\bar{X}_1 X_2$ and in this case the machine should be designed to return to the state S_0 where it will await the arrival of the first signal in the sequence $\bar{X}_1 \bar{X}_2$. The completed state diagram is shown in figure 9.9(c). In this diagram the output $Z = 1$ appears in state S_3 at the completion of the required sequence. If, when in this state, the input signals $\bar{X}_1 X_2$ or $X_1 \bar{X}_2$ are received, the machine will return to S_0, where it will await the next occurrence of the transition signal $\bar{X}_1 \bar{X}_2$, the first combination of the required sequence.

Some designers have the habit of inserting slings on the state diagram. For example, if the machine enters state S_1 on the signal $\bar{X}_1 \bar{X}_2$, it will stay there as long as this signal exists, and this can be indicated by a sling originating from and terminating on S_1, as shown in figure 9.9(d). This diagram includes all possible slings, and it will be observed that when in S_0 the sling signal is $\bar{X}_1 X_2 + X_1 \bar{X}_2 + X_1 X_2$. This infers that the machine having entered S_0 on either of the signals $\bar{X}_1 X_2$ or $X_1 \bar{X}_2$ it would be possible to get a change of primary signal from either $\bar{X}_1 X_2$ to $X_1 X_2$ or, alternatively, from $X_1 \bar{X}_2$ to $X_1 X_2$. If such a sequence of events occurs, the machine will remain in state S_0 and will only leave this state if the primary signals X_1 and X_2 change in either of the following two sequences:

(a) 11 → 01 → 00, or
(b) 11 → 10 → 00.

The state table is constructed from the information given on the state diagram and is shown in figure 9.9(e). Examination of the table shows that rows S_1 and S_2 are mergeable, and hence the table can be reduced to three rows. At first sight this may appear to be a disadvantage, for two reasons: firstly, it leads to the presence of an unused state, and secondly, since the state diagram will now only consist of three states, a race-free secondary assignment is not possible. However, the unused state can be re-introduced as a dummy state having an unconditional transition to the next state. The presence of an unconditional transition on the state diagram will lead to simpler turn-on and turn-off conditions, and hence to a simpler logic implementation.

The reduced state table is shown in figure 9.9(*f*), and it will be noticed that there is one unoccupied cell on this diagram on the row headed S_3. This effectively is a 'can't happen' condition. If the present state is S_3 then a transition signal $\bar{X}_1 \bar{X}_2$ is forbidden. Since this signal cannot occur when the machine is in state S_3 it may be used as an optional term added into the Boolean equation for the $S_3 \to S_0$ transition, as shown in the reduced state diagram of figure 9.9(*g*). In this case the optional term leads to a simplification of the transition signal.

Finally, the state diagram including the dummy state and with a suitable secondary assignment is shown in figure 9.9(*h*). The turn-on and turn-off equations are taken directly from this diagram:

Turn on set of $A = \bar{B}\bar{X}_1 X_2 + B(\bar{X}_1 + \bar{X}_2)$

Turn off set of $A = \bar{B}\bar{X}_1 \bar{X}_2$

$$A = \bar{B}\bar{X}_1 X_2 + B(\bar{X}_1 + \bar{X}_2) + (\overline{\bar{B}\bar{X}_1\bar{X}_2})A$$

$$= \bar{B}\bar{X}_1 \bar{X}_2 + B(\bar{X}_1 + \bar{X}_2) + (B + X_1 + X_2)A$$

Turn-on set of $B = \bar{A} X_1 X_2$

Turn-off set of $B = A$

$$B = \bar{A} X_1 X_2 + \bar{A} B$$

The output Z is given by $Z = S_3 = \bar{A} B$, and the machine implementation is shown in figure 9.9(*i*).

An alternative method of design for the sequence detector would be to use *SR* flip-flops. The state table compiled from the information given in the state diagram of figure 9.9(*h*) is shown in figure 9.10(*a*) and in this diagram assignment of the secondary variables A and B has been placed alongside the present states. However, it is more convenient to rearrange the rows of this table so that the secondary variables appear in normal K-map order. At the same time the states within the cells are replaced by the secondary variables that define them, as shown in figure 9.10(*b*).

This table may be regarded as a flip-flop excitation map which, in conjunction with the steering table for the *SR* flip-flop shown in figure 9.10(*c*), enables the designer to obtain the K-maps for the *S* and *R* inputs of both flip-flops. For example, when the present state of the

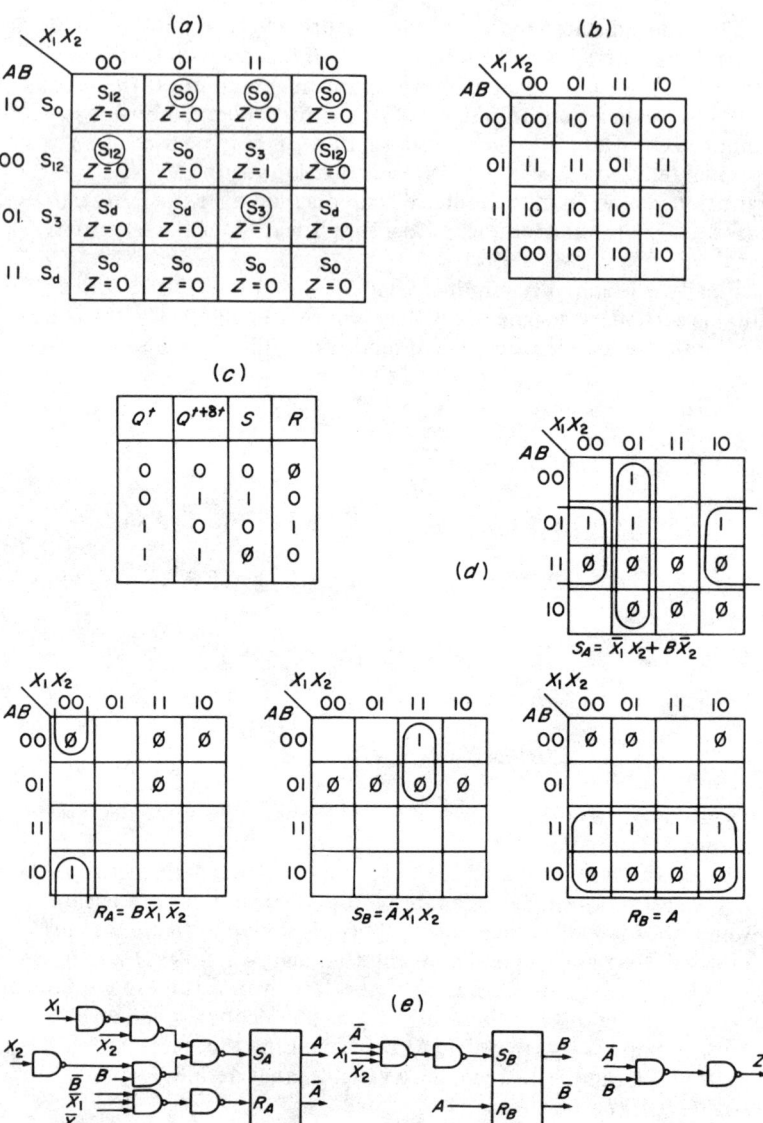

Figure 9.10. (a) State table used for SR flip-flop implementation. (b) Flip-flop excitation table. (c) Steering table for an SR flip-flop. (d) K-maps for the flip-flop input signals. (e) Implementation of a sequence detector using SR flip-flops

machine $AB = 00$ and the input combination $X_1 X_2 = 01$ is received, then the next state is $AB = 10$. Clearly, flip-flop A has made a $0 \to 1$ transition whilst flip-flop B has made a $0 \to 0$ transition.

The K-maps for the flip-flop input signals are shown in figure 9.10(d) and, after simplification, the following equations are obtained for the set and reset signals:

$$S_A = \bar{X}_1 X_2 + B\bar{X}_2 \qquad S_B = \bar{A} X_1 X_2$$

$$R_A = \bar{B} \bar{X}_1 \bar{X}_2 \qquad R_B = A$$

The output is given by $Z = S_3 = \bar{A}B$ and the machine implementation is shown in figure 9.10(e).

Problems

9.1 A double-sequence detector has two inputs, X_1 and X_2, and one output Z. For an input sequence $X_1 X_2 = 00, 10, 11$ the output Z becomes 1, and when the reverse sequence is received the output Z returns to 0. A typical timing diagram for the detector is shown in figure P9.1.

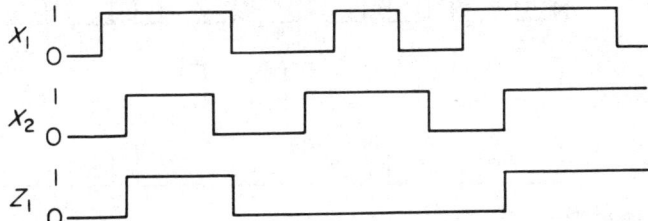

Figure P9.1

Develop a state diagram for the detector and hence obtain a state table. If possible reduce the state table and implement the detector using

(a) NAND gates, and
(b) SR flip-flops and NAND gates.

9.2 The characteristics of a JK flip-flop are specified below:
(a) If $Q = 0$ when $J = 1$, $K = 0$ and $Ck = 1$, the output Q goes from $0 \to 1$ on the trailing edge of the clock pulse.

(b) If $Q = 1$ when $J = 0$, $K = 1$ and $Ck = 1$, the output Q goes from $1 \rightarrow 0$ on the trailing edge of the clock pulse.
(c) If $J = 1$, $K = 1$ and $Ck = 1$, the output Q toggles.

Develop an event-driven circuit to implement the above specification and draw a timing diagram for the flip-flop.

9.3 X_1 and X_2 are the two inputs to an asynchronous circuit which has two outputs Z_1 and Z_2. When $X_1 X_2 = 00$ the output $Z_1 Z_2 = 00$. If a $0 \rightarrow 1$ change in X_1 precedes a $0 \rightarrow 1$ change in X_2, then the output of the circuit is $Z_1 Z_2 = 01$. Alternatively, if a $0 \rightarrow 1$ change in X_2 precedes a $0 \rightarrow 1$ change in X_1, then the output of the circuit is $Z_1 Z_2 = 10$. In both cases the outputs remain at 01 and 10, respectively, until $X_1 X_2 = 00$ again.

9.4 Develop an asynchronous circuit that will give an output clock pulse (Z) after every second data pulse arrives on the X input line. The arrival of the data pulses is purely random and it is to be assumed that the minimum time for a pair of consecutive data pulses is greater than the periodic time of the clock. A typical timing diagram is shown in figure P9.4.

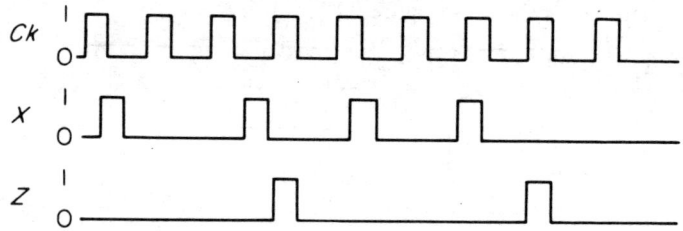

Figure P9.4

9.5 A logic circuit has two asynchronous inputs, X_1 and X_2, and also a synchronous clock signal. The circuit is to be designed so that

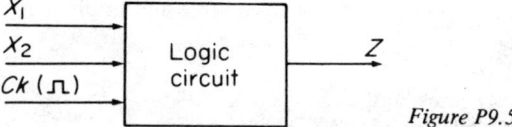

Figure P9.5

the first complete clock pulse that occurs after X_1 and X_2 have become 1, in that order, is output on the line marked Z in the figure P9.5. After the output of the clock pulse the circuit must return to its quiescent state when $X_1 X_2 = 0$.

Design a circuit that satisfies this specification and implement the design using NAND gates.

9.6 Analyse the fundamental mode circuit shown in figure P9.6
  (a) Determine the state table.
  (b) Determine the state diagram.
  (c) Use the state table to determine the output response to the input sequence $X_1 X_2 = 00, 01, 11, 10, 11, 01, 00, 10, 00, 01$. Initial conditions $X_1 = X_2 = A = 0$.

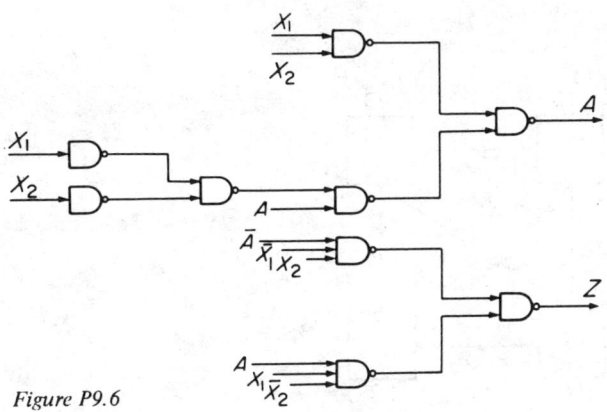

Figure P9.6

9.7 Analyse the circuit shown in figure P9.7:
  (a) Determine the state table.
  (b) Determine the state diagram.

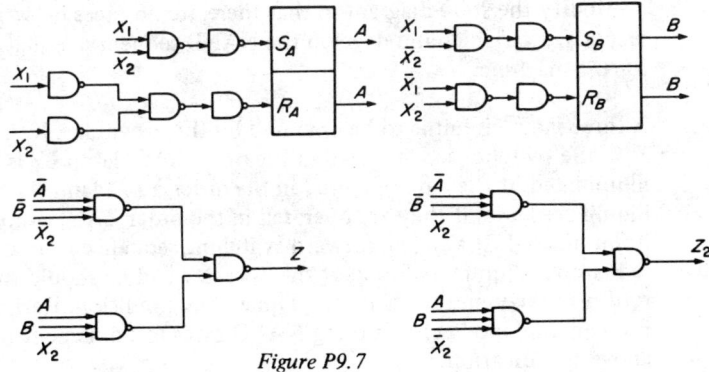

Figure P9.7

(c) Use the state table to determine the output response to the input sequence $X_1 X_2 = 00, 01, 11, 10, 00, 01, 11, 01, 11, 10, 00$. Assume the initial conditions are $X_1 = X_2 = 0$ and $A = B = 0$.

9.8 The internal state diagram for a four-state digital machine is shown in figure P9.8. Construct a state table for the machine and identify all races that will occur if the machine is implemented from the given state diagram, stating whether they are critical or non-critical. For each race, give all the state transitions which may occur.

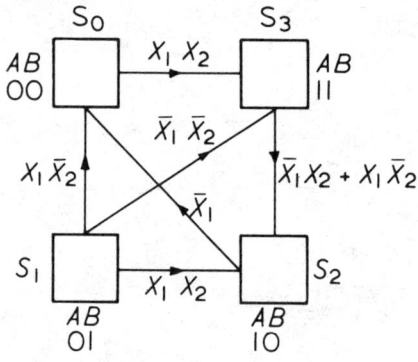

Figure P9.8

Modify the state diagram so that there are no races between secondary variables and develop the NAND sequential equations for the machine.

9.9 A three-lamp circuit is to be operated by three switches X, Y and Z. If the switches are operated in the order XYZ, lamp L_1 is illuminated; if they are operated in the order YZX, lamp L_2 is illuminated, and if they are operated in the order ZXY, lamp L_3 is illuminated. If a wrong forward switching sequence is used, a red lamp is illuminated to alert the operator and he should then return the switches to their initial quiescent condition. Design a fundamental mode circuit using NAND gates to implement the above specification.

9.10 Find a circuit realisation for the given state table (figure P9.10) after choosing a race-free secondary assignment, using SR flip-flops and NAND gates.

	$X_1 X_2$ 00	01	11	10
S_0	(S₀) $Z_1 Z_2$ 0 0	S_1 01	(S₀) 00	S_3 10
S_1	S_0 00	(S₁) 01	(S₁) 01	S_2 10
S_2	S_3 00	(S₂) 01	(S₂) 01	(S₂) 10
S_3	(S₃) 00	S_3 10	S_0 10	(S₃) 10

Figure P9.10

10
Digital design with MSI

10.1 Introduction

In recent years a large number of MSI and LSI circuits have been introduced and these are now available to the logic designer for use in the implementation of digital systems. Integrated circuits for use in digital systems can be divided into three distinct groups:

   (a)   SSI (small-scale integrated) circuits are packages containing discrete gates and flip-flops, etc;

   (b)   MSI (medium-scale integrated) circuits can be regarded as digital sub-systems — for example, adders, multiplexers, comparators, multipliers, etc — and

   (c)   LSI (large-scale integrated) circuits can be regarded as digital systems in their own right — for example, eight-bit microprocessors, DMA (direct memory access) controllers, calculators, etc.

The introduction of MSI circuits is tending to result in the replacement of the old methods of logic design. Traditionally, the design engineer has developed a logic function as the solution to a particular problem. This function has been minimised using the methods described in previous chapters, and then implemented using SSI circuits. However, when implementing logic functions with MSI circuits such as the multiplexer, the Boolean function is expressed in its canonical form, (i.e. each term in the Boolean function contains all the variables in their true or complemented form) and is implemented directly without minimisation.

The cost of a digital system is approximately proportional to the number of ICs in the system, so that to reduce costs the package count should be minimised. The logic designer should therefore be looking for the replacement of a large number of SSI circuits by one or more MSI packages. It is frequently better to use a standard MSI package even if this introduces redundant or unused gates rather than to design with SSI circuits.

10.2 Data selector or multiplexer

The multiplexer selects 1-out-of-n lines, where n is usually 4, 8 or 16. A block diagram of a multiplexer having four input lines D_0, D_1, D_2 and D_3, and two output lines f and $\bar{f}$ is shown in figure 10.1(a). The device also has two control lines A and B and may have an 'enable' line E. When functioning as a data selector, the multiplexer may be regarded as a single-pole switch which selects, say, 1-out-of-4 lines, as shown in figure 10.1(b). Implementation of the multiplexer at the gate level is shown in figure 10.1(c).

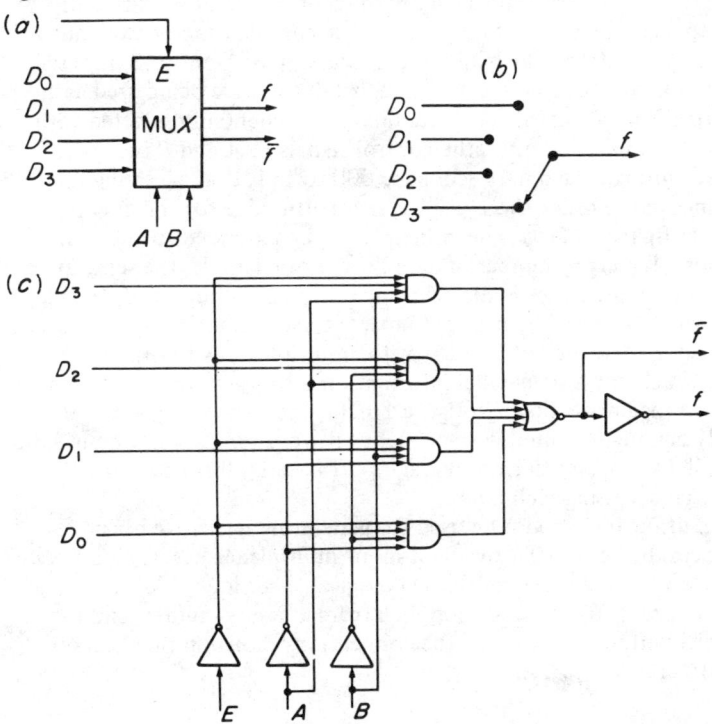

Figure 10.1. (a) Block diagram of a four-input multiplexer. (b) Selection of 1-out-of-4 data lines by means of a single-pole switch. (c) Implementation of a four-input multiplexer

In essence, the circuit is an AND-OR-invert gate having complementary outputs. The Boolean function which represents the output f of this circuit is

$$f = \bar{A}\bar{B}D_0 + \bar{A}BD_1 + A\bar{B}D_2 + ABD_3$$

Data lines can be selected by applying the appropriate binary-coded signal to the control lines A and B. When the control signal $\bar{A}\bar{B} = 1$ the output of the circuit is D_0, and so on.

Some multiplexers are provided with an input enable line as shown in figure 10.1(c). When the input to this line is logical '0' the four AND gates are enabled. With the enable signal equal to logical '1' the operation of the multiplexer is inhibited.

The number of data lines to be selected can be increased either by choosing a multiplexer with a large number of data lines, or, alternatively, by using a combination of multiplexers. A combination of two four-input data selectors, which allows the selection of 1-out-of-8 lines is shown in figure 10.2(a), the enable signal in this case being used as an additional control signal. The data lines are sequentially selected with the aid of a binary counter, the control signals E, A and B being clocked through the binary sequence 000 to 111, thus accessing the data lines in the order D_0, D_1, ..., D_7. A truth table for the circuit is shown in figure 10.2(b). This principle can be extended to allow the selection of a larger number of data lines. For example, the selection of 1-out-of-64 lines can be achieved using nine eight-input multiplexers, as shown in figure 10.3, arranged in two levels of multiplexing.

An alternative way of looking at the multiplexer is to regard it as a device which converts parallel information into serial form. For example, in the arrangement shown in figure 10.2(a) the two multiplexers M_1 and M_2 can be presented with an eight-bit word on the eight input lines in parallel form, and this can be taken off in serial form by using the sequential accessing technique.

A multiplexer can also be used for generating repetitive binary sequences. In figure 10.4 the eight-input multiplexer has its data lines connected to +5 volts or ground, i.e. logical '1' or logical '0'. The control lines A, B, and C are supplied from a binary counter and the repetitive output sequence of the connections shown in the diagram is 10010011.

10.3 The multiplexer as a logic function generator

The equation for a multiplexer having four input lines is

$$f = \bar{A}\bar{B}D_0 + \bar{A}BD_1 + A\bar{B}D_2 + ABD_3$$

Digital design with MSI

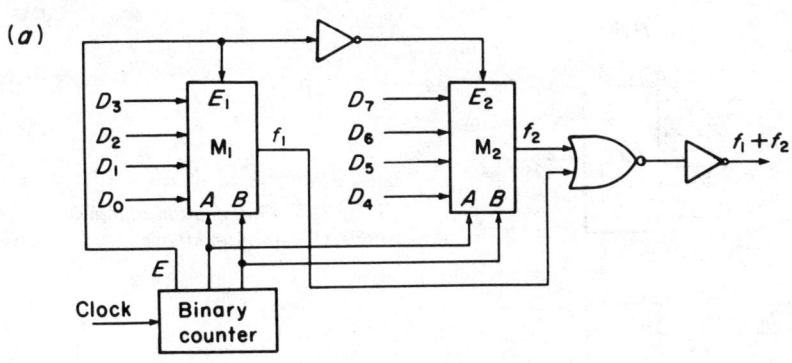

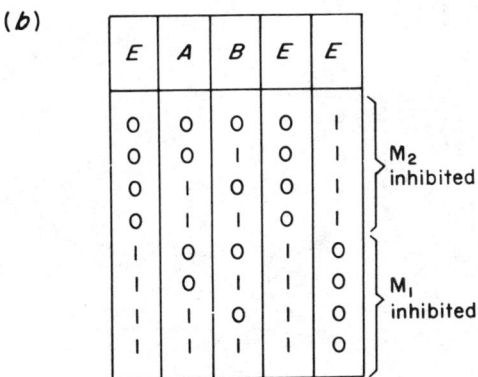

Figure 10.2. (a) Combination of two multiplexers providing a 1-out-of-8 data selector. (b) Truth table for a 1-out-of-8 data selector

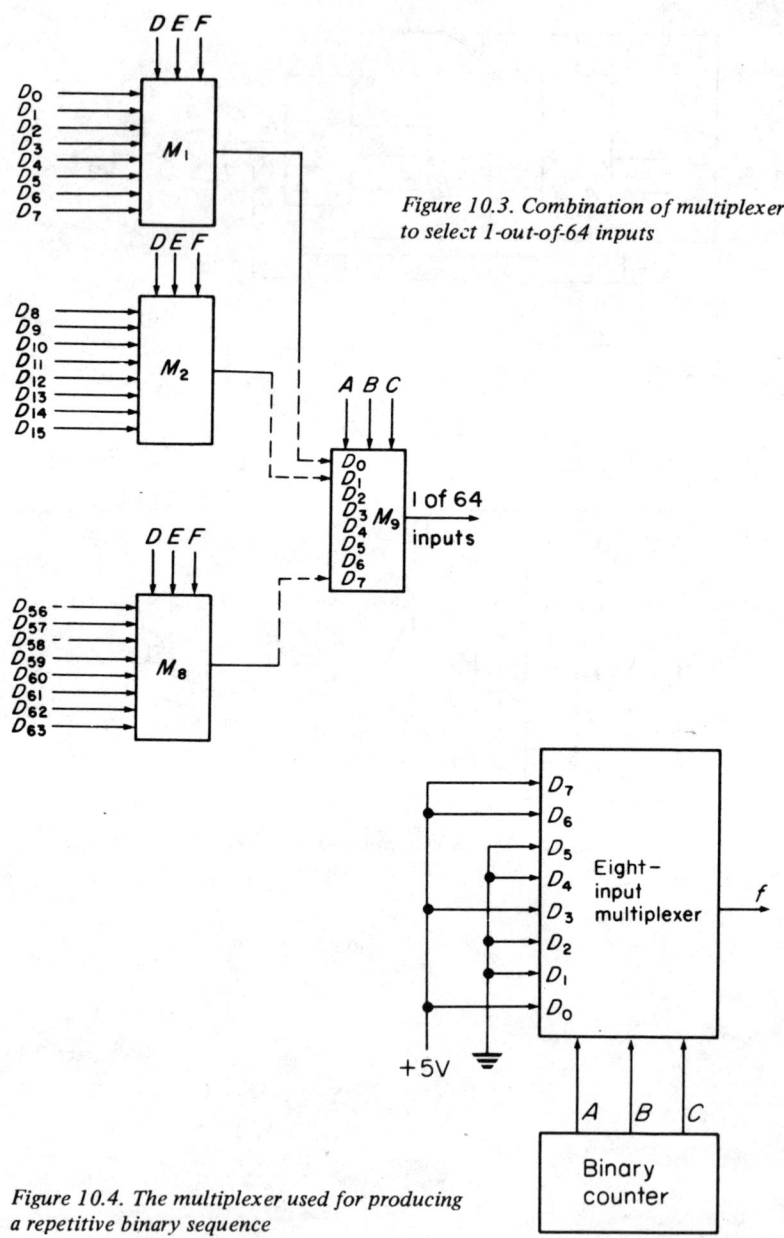

Figure 10.3. Combination of multiplexers to select 1-out-of-64 inputs

Figure 10.4. The multiplexer used for producing a repetitive binary sequence

Digital design with MSI

where the Boolean variables A and B are used as the signals for the control lines. A and B can be factored out of any function of n variables, and the residue functions of $n-2$ variables can then be applied to the data lines. For example, if $n = 3$, four signals of one variable can be applied to each of the data lines. Assuming that the third variable is C, the possible signals that can be applied to the data lines are C, $\bar{C}$, 0 and 1. In all, there are $4^4 = 256$ possible combinations of the four input signals, and a multiplexer with four input lines can generate any one of the 256 possible Boolean functions of three variables.

For the four-input multiplexer, there are three possible choices for the control variables: AB, AC and BC, and these combinations can be associated with individual data lines, as indicated in figure 10.5. For example, with control variables A and B, the input line D_0 is associated with those cells marked $A = 0$ and $B = 0$, that is, the top two left-hand

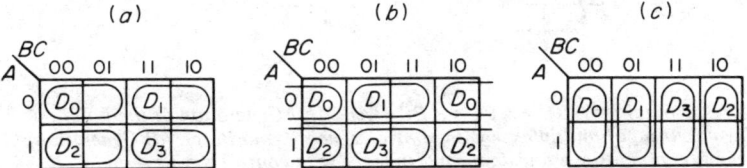

Figure 10.5. Association of data lines with control signals for a four-input multiplexer. The control variables are A and B in (a), A and C in (b), and B and C in (c)

cells on the K-map of figure 10.5(a). In effect, the K-map for three variables has now been split into four two-cell, one-variable K-maps, each of these two-cell maps being associated with a data line.

As an example, the three-variable function

$$f = \bar{A}\bar{B}C + \bar{A}BC + \bar{A}B\bar{C} + ABC$$

will be implemented using a four-input multiplexer. The function is first plotted on a K-map as shown in figure 10.6(a), and an arbitrary choice of control variables, say A and B, is made. The four one-variable functions associated with each data line are simplified. For example, the two cells associated with D_1 are both marked with a 1, and hence the input to data line D_1 is $C + \bar{C} = 1$. The remaining inputs obtained from the K-map in figure 10.6(a) are

$$D_0 = C \quad D_2 = 0 \quad D_3 = C$$

and the implementation of the function is shown in figure 10.6(b). If

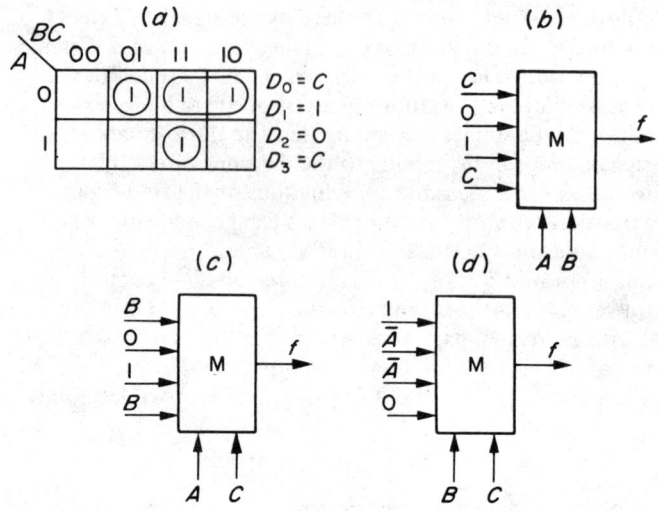

Figure 10.6. (a) $f = \bar{A}\bar{B}C + \bar{A}BC + \bar{A}B\bar{C} + ABC$. (b) Generation of the function using a four-input multiplexer with control signals A and B. (c), (d) Generation of the function using a four-input multiplexer with control signals A and C, B and C, respectively

the choice of control variables had been A and C, the inputs to the multiplexer would have been

$$D_0 = B \quad D_1 = 1 \quad D_2 = 0 \quad D_3 = B$$

and likewise for a choice of control variables B and C, the inputs would have been

$$D_0 = 0 \quad D_1 = \bar{A} \quad D_2 = \bar{A} \quad D_3 = 1$$

The corresponding implementations are shown in figures 10.6(c, d).

Consider now the implementation of the four-variable function

$$f = \sum 0, 1, 5, 6, 7, 9, 10, 14, 15$$

using a four-input multiplexer. Since a four-input multiplexer is to be used, the application of two variables to its control lines will leave residue functions of two variables to be applied to the data lines. There are six possible choices for the control variables: AB, AC, AD, BC, BD or CD; these combinations can be associated with the data lines indicated in figure 10.5. It will be assumed in this example that A and B are chosen as the control variables, and the K-map associating them with the data lines is shown in figure 10.7(*a*). The four-variable K-map has now been divided into four four-cell, two-variable maps, and simplification can only take place within the confines of the two-variable maps.

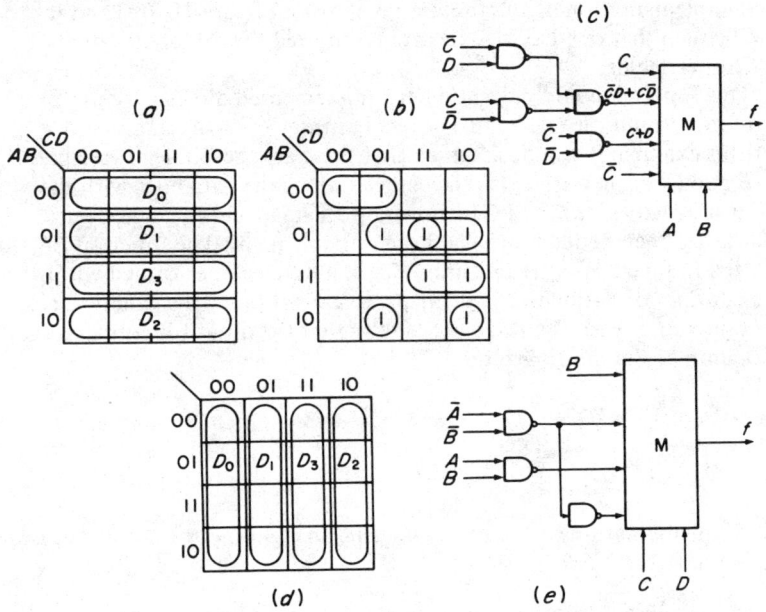

Figure 10.7. (a) Association of data lines with variables A and B. (b) K-map of $f = \Sigma\ 0, 1, 5, 6, 7, 9, 10, 14, 15$. (c) Implementation of the function using a four-input multiplexer. (d) Association of data lines with control variables C and D. (e) Alternative implementation of the function using C and D as the control variables

The K-map plot of the function is shown in figure 10.7(*b*) and the data line inputs obtained from the four rows of this map are

$$D_0 = \overline{C} \qquad D_1 = C + D \qquad D_2 = \overline{C}D + C\overline{D} \qquad D_3 = C$$

The implementation of the function is shown in figure 10.7(*c*).

It should be pointed out that it is useful to examine the various possible choices of control variables to ascertain whether there is a simpler solution. In this case a simpler solution is obtained if C and D are chosen as the control variables. The association of the data lines with the control variables is shown in figure 10.7(d) and the inputs to the data lines are obtained with the aid of the K-map plot shown in figure 10.7(b). They are

$$D_0 = \bar{A}\bar{B} \quad D_1 = \bar{A} + \bar{B} \quad D_2 = A + B \quad D_3 = B$$

The implementation of the function is shown in figure 10.7(e) and it can be seen that one less NAND gate is required for this choice of control variables.

This function could have also been implemented directly by using an eight-input multiplexer. Such a device has three control signals, and for this example it will be assumed that these are the Boolean variables A, B and C. In figure 10.8(a) the association of the data lines with the control variables A, B and C is shown on a K-map and it can be seen that it has been divided into eight one-variable maps. The function is plotted in figure 10.8(b) and simplification has been performed within the confines of the one-variable maps. Examination of the function plot in conjunction with figure 10.8(a) shows that the input functions to the data lines of the multiplexer are

$$D_0 = 1 \quad D_2 = D \quad D_4 = D \quad D_6 = 0$$

$$D_1 = 0 \quad D_3 = 1 \quad D_5 = \bar{D} \quad D_7 = 1$$

The implementation of the function using an eight-input multiplexer is shown in figure 10.8(c).

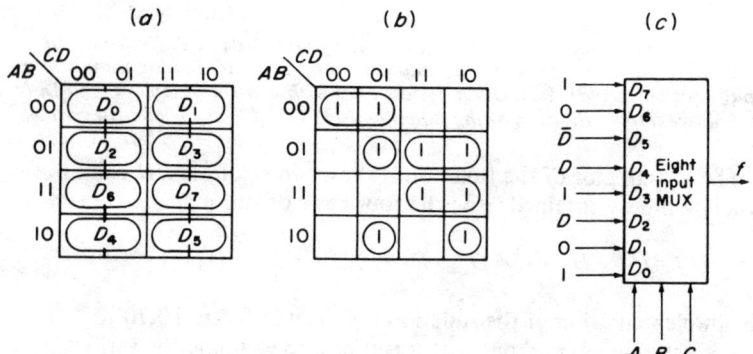

Figure 10.8. (a) Association of data lines with control variables ABC. (b) Simplification of $f = \Sigma\ 0, 1, 5, 6, 7, 9, 10, 14, 15$. (c) Implementation of the function

As the number of variables associated with the Boolean function to be implemented increases, it may become necessary to use more than one level of multiplexing, and this technique is illustrated in the next example, where it is required that the five-variable function

$$f = \sum 0, 1, 3, 6, 7, 8, 14, 15, 17, 18, 20, 21, 22, 24, 27, 28, 31$$

be implemented using four-input multiplexers.

For the first level of multiplexing the control variables D and E have been arbitrarily chosen. The function f is now listed at the left-hand side of the table shown in figure 10.9(a) which contains four columns headed $\overline{DE}$, $\overline{D}E$, $D\overline{E}$ and DE, respectively. In the column headed $\overline{DE}$ are listed all those terms of the three variables (A, B and C) associated with $\overline{DE}$. For example, in the case of the term $\overline{A}\,\overline{B}\overline{C}\overline{D}\overline{E}$ the entry in

(a)

Function		$\overline{D}\overline{E}$	$\overline{D}E$	$D\overline{E}$	DE
$f = \overline{A}\overline{B}\overline{C}\overline{D}\overline{E}$	(0)	$\overline{A}\overline{B}\overline{C}$			
$\overline{A}\overline{B}\overline{C}\overline{D}E$	(1)		$\overline{A}\overline{B}\overline{C}$		
$\overline{A}\overline{B}CDE$	(3)				$\overline{A}\overline{B}C$
$\overline{A}BC\overline{D}\overline{E}$	(6)		$\overline{A}BC$		
$\overline{A}BCDE$	(7)				$\overline{A}BC$
$A\overline{B}\overline{C}\overline{D}\overline{E}$	(8)	$A\overline{B}\overline{C}$			
$\overline{A}BCD\overline{E}$	(14)			$\overline{A}BC$	
$\overline{A}BCDE$	(15)				$\overline{A}BC$
$A\overline{B}\overline{C}\overline{D}E$	(17)		$A\overline{B}\overline{C}$		
$A\overline{B}CD\overline{E}$	(18)			$A\overline{B}\overline{C}$	
$A\overline{B}C\overline{D}\overline{E}$	(20)	$A\overline{B}C$			
$A\overline{B}C\overline{D}E$	(21)		$A\overline{B}C$		
$A\overline{B}CD\overline{E}$	(22)			$A\overline{B}C$	
$AB\overline{C}\overline{D}\overline{E}$	(24)	$AB\overline{C}$			
$AB\overline{C}DE$	(27)				$AB\overline{C}$
$ABC\overline{D}\overline{E}$	(28)	ABC			
$ABCDE$	(31)				ABC

(b)

(c)

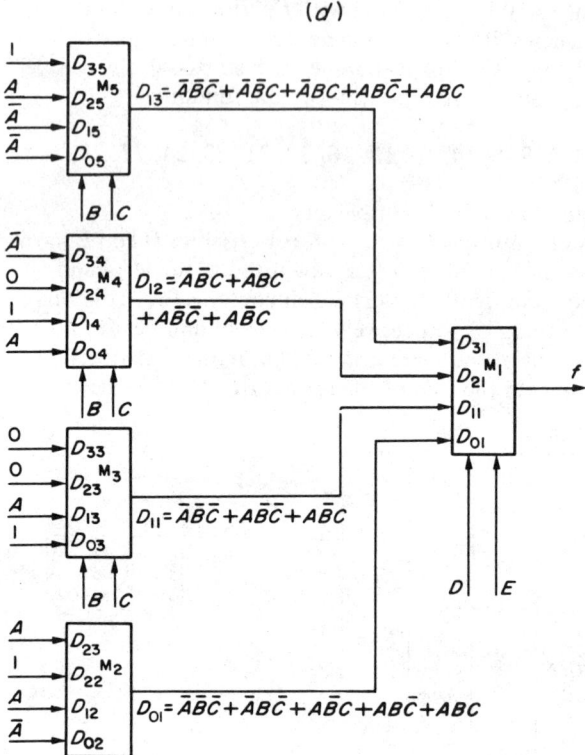

Figure 10.9. (a) Determination of the inputs to the first-level multiplexer. (b) Association of data lines with control variables B and C. (c) K-maps for determining the data input signals for the four second-level multiplexers. (d) Two levels of multiplexing used to generate the function f

the $\overline{D}\overline{E}$ column will be $\overline{A}\overline{B}\overline{C}$. This procedure is repeated for each term in the five-variable function and an entry is made in the appropriate column in each case.

The input functions for the first-level multiplexer are now seen to be

$$D_{01} = \overline{A}\,\overline{B}\overline{C} + \overline{A}\,B\overline{C} + A\overline{B}C + AB\overline{C} + ABC$$

$$D_{11} = \overline{A}\,\overline{B}\overline{C} + A\overline{B}\overline{C} + A\overline{B}C$$

$$D_{12} = \bar{A}\bar{B}C + \bar{A}BC + A\bar{B}\bar{C} + AB\bar{C}$$

$$D_{13} = \bar{A}\bar{B}\bar{C} + \bar{A}\bar{B}C + \bar{A}BC + AB\bar{C} + ABC$$

These three-variable functions can be generated with four-input multiplexers, as described above, at the second level of multiplexing.

For the second level of multiplexing, B and C have been chosen as the control variables, as indicated in the K-map shown in figure 10.9(b). The K-maps for determining the inputs to the data lines of the second-level multiplexers are shown in figure 10.9(c) and from these maps the various input signals are found to be

$$D_{02} = \bar{A}, \quad D_{12} = A, \quad D_{22} = 1, \quad D_{32} = A$$

$$D_{03} = 1, \quad D_{13} = A, \quad D_{23} = 0, \quad D_{33} = 0$$

$$D_{04} = A, \quad D_{14} = 1, \quad D_{24} = 0, \quad D_{34} = \bar{A}$$

$$D_{05} = \bar{A}, \quad D_{15} = \bar{A}, \quad D_{25} = A, \quad D_{35} = 1$$

The implementation of the function is shown in figure 10.9(d).

10.4 Decoders and demultiplexers

A demultiplexer performs the opposite function to that of a multiplexer. A block diagram of the device is shown in figure 10.10(a). A single data

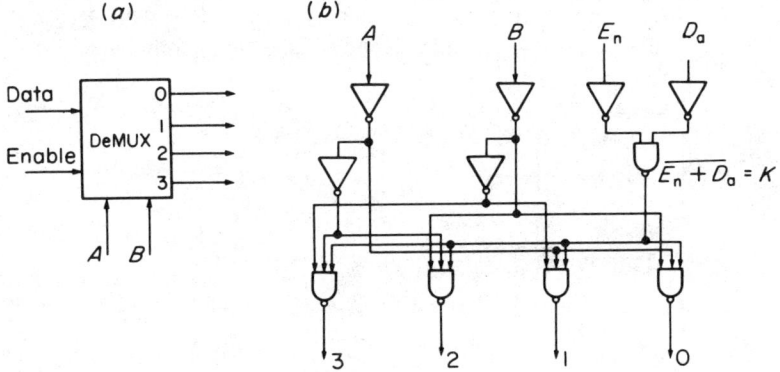

Figure 10.10. (a) A 2-to-4 line demultiplexer. (b) Logic diagram

input line can be connected to one of many output lines by the appropriate choice of signal on the control lines.

With two control lines there are four possible addresses, and hence the maximum number of output lines that can be selected is four, as illustrated in figure 10.10(b). The enable and data lines E_n and D_a respectively, are connected to the output gates by what is, effectively, a NOR gate. This input arrangement allows two modes of operation.

In the first mode, if $E_n = 0$ and $D_a = 0$, $K = 1$, thus enabling all the output gates. For any other values of E_n and D_a, $K = 0$, thus disabling all the output gates, that is, the output lines will be permanently set to logic '1'. In this mode, the 1-to-4 line demultiplexer will act as a decoder allowing each of the four possible combinations of the input signals A and B to appear on the selected line. For example, $\bar{A}\bar{B}$ will give an output on line 0, and so on. Alternatively, the circuit can be operated as a generator of the four canonical terms of two Boolean variables. If $P_1 = \bar{A}B$ is the input to the control lines A and B, then the output on line 1 = $\bar{P}_1$.

For the second mode, if $E_n = 0$, $D_a = 0$, then $K = 1$. For a control signal $f_2 = A\bar{B}$, then the output on line 2 = 0 = D_a. Alternatively, if $E_n = 0$ and $D_a = 1$, then $K = 0$, and for a control signal $f_3 = AB$, the output on line 3 = 1 = D_a. In this mode the data is transferred to the output gate selected by the address applied to the control lines.

10.5 Decoder applications

A commonly used decoder has four input lines and ten output lines, as shown in figure 10.11. If the input signals are $A = 0, B = 0, C = 0$ and $D = 0$ then the output line marked 0 will be at logical '0'. This decoder can be used as a NBCD-to-decimal decoder, or, alternatively, when

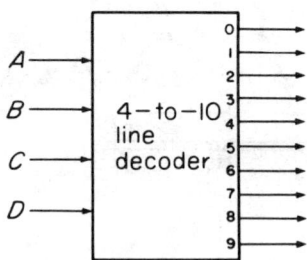

Figure 10.11. Block diagram of a 4-to-10 line decoder

used in conjunction with a certain amount of combinational logic it can be used for code conversion.

As an example, the 4-to-10 line decoder will be used to convert the NBCD code into the 3-out-of-5 code. These two codes are tabulated

(a)

P terms	NBCD				3-out of -5 code				
	A	B	C	D	V	W	X	Y	Z
P_0	0	0	0	0	1	0	0	1	1
P_1	0	0	0	1	1	0	1	0	1
P_2	0	0	1	0	1	1	0	0	1
P_3	0	0	1	1	0	0	1	1	1
P_4	0	1	0	0	0	1	0	1	1
P_5	0	1	0	1	0	1	1	0	1
P_6	0	1	1	0	0	1	1	1	0
P_7	0	1	1	1	1	0	1	1	0
P_8	1	0	0	0	1	1	0	1	0
P_9	1	0	0	1	1	1	1	0	0

(b)

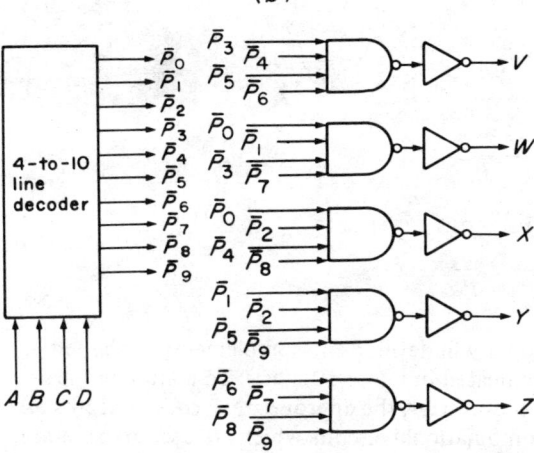

Figure 10.12. (a) Tabulation of 3-out-of-5 and NBCD codes. (b) The NBCD to 3-out-of-5 code converter

in figure 10.12(a). From the tabulation, the equations for the 3-out-of-5 code outputs are written down directly in terms of the P-terms.

$$V = P_0 + P_1 + P_2 + P_7 + P_8 + P_9$$
$$W = P_2 + P_4 + P_5 + P_6 + P_8 + P_9$$
$$X = P_1 + P_3 + P_5 + P_6 + P_7 + P_9$$
$$Y = P_0 + P_3 + P_4 + P_6 + P_7 + P_8$$
$$Z = P_0 + P_1 + P_2 + P_3 + P_4 + P_5$$

Now

$$\overline{V} = P_3 + P_4 + P_5 + P_6$$

hence

$$V = \overline{P_3 + P_4 + P_5 + P_6}$$

and

$$V = \overline{P}_3 \, \overline{P}_4 \, \overline{P}_5 \, \overline{P}_6$$

Similarly

$$\overline{W} = \overline{P}_0 \, \overline{P}_1 \, \overline{P}_3 \, \overline{P}_7$$
$$\overline{X} = \overline{P}_0 \, \overline{P}_2 \, \overline{P}_4 \, \overline{P}_8$$
$$\overline{Y} = \overline{P}_1 \, \overline{P}_2 \, \overline{P}_5 \, \overline{P}_9$$
$$\overline{Z} = \overline{P}_6 \, \overline{P}_7 \, \overline{P}_8 \, \overline{P}_9$$

A 4-to-10 line decoder will develop the complements of the ten required P-terms, as indicated in figure 10.12(b). To complete the implementation of the converter, the appropriate decoder outputs are fed to five separate combinational circuits which are used to generate the V, W, X, Y and Z signals.

The technique used in this example is useful where there are many functions of the same number of variables to be implemented. In

Digital design with MSI 235

comparison, a multiplexer implementation would require less additional gating, but one multiplexer at least is required to implement each function.

If the Boolean functions to be implemented by this method require all the P-terms, then it would be necessary to use a 4-to-16 line decoder which will generate all the 16 P-terms of four Boolean variables.

Decoders are also widely used in microprocessor systems. A very common problem met with in such systems is that of the selection of one out of a number of memory chips by the CPU. For example, the Intel 8205 3-to-8 line decoder can be used for this purpose. A block diagram of the device is shown in figure 10.13(*a*) and its truth table is tabulated in figure 10.13(*b*).

(*a*)

I_0 — O_0
I_1 — O_1
I_2 — O_2
3-to-8 — O_3
line decoder — O_4
E_1 — O_5
E_2 — O_6
E_3 — O_7

(*b*)

Address			Enable			Output							
I_2	I_1	I_0	E_1	E_2	E_3	O_0	O_1	O_2	O_3	O_4	O_5	O_6	O_7
0	0	0	0	0	1	0	1	1	1	1	1	1	1
0	0	1	0	0	1	1	0	1	1	1	1	1	1
0	1	0	0	0	1	1	1	0	1	1	1	1	1
0	1	1	0	0	1	1	1	1	0	1	1	1	1
1	0	0	0	0	1	1	1	1	1	0	1	1	1
1	0	1	0	0	1	1	1	1	1	1	0	1	1
1	1	0	0	0	1	1	1	1	1	1	1	0	1
1	1	1	0	0	1	1	1	1	1	1	1	1	0

Figure 10.13. (*a*) The Intel 8205 3-to-8 line decoder. (*b*) Truth table

Since there are eight separate output lines, 1-out-of-8 memory chips can be selected, providing the decoder has been enabled. The Intel 8085 processor has 16 address lines and two of these, A_{15} and A_{14}, can be kept permanently in the low state and used as the enable signals E_1 and E_2, whilst the other enable line can be connected to the +5 volt line. The next three address lines A_{13}, A_{12} and A_{11} can be connected to the three input lines I_0, I_1 and I_2; respectively. As the values of A_{13}, A_{12} and A_{11} change relative to one another under the control of the microprocessor, different output lines and hence different memory chips are selected. All the remaining address lines A_0-A_{10} would be 'bussed' to the memory chips, as indicated in the block schematic diagram of figure 10.14.

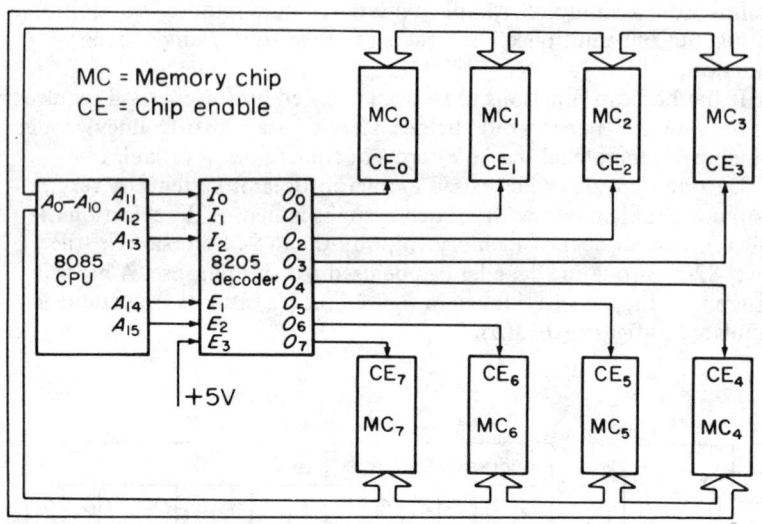

Figure 10.14. *Memory selection system for an Intel 8085 processor*

The lowest memory address that can be accessed by the decoder is

$A_{15}A_{14}A_{13}A_{12}$	$A_{11}A_{10}A_9A_8$	$A_7A_6A_5A_4$	$A_3A_2A_1A_0$	Address bit
0 0 0 0	0 0 0 0	0 0 0 0	0 0 0 0	Binary
0	0	0	0	Hexadecimal

and the address where $A_{13}A_{12}A_{11}$ is about to change from 000 to 001 is

$2^{15}2^{14}2^{13}2^{12}$	$2^{11}2^{10}2^9 2^8$	$2^7 2^6 2^5 2^4$	$2^3 2^2 2^1 2^0$	
0 0 0 0	0 1 1 1	1 1 1 1	1 1 1 1	Binary
0	7	F	F	Hexadecimal

The decimal representation of the second binary number is $2^{11} = 2048$. Hence the first output line O_0 can be used to enable a memory chip having a maximum storage capacity of 2 k bytes. Similarly, each succeeding output line can be used as the enable line for a further 2 k bytes of memory, hence the maximum total memory capacity for the system shown in figure 10.14 is 16k bytes.

In microprocessor terminology, the output lines of the decoder would be referred to as the chip enable lines. A memory chip associated with a particular chip enable line will be enabled when it goes low.

10.6 Read-only memories (ROMs)

The circuit shown in figure 10.15 is that of a 64-bit ROM organised as eight words of eight bits each. It consists of a three-bit address decoder,

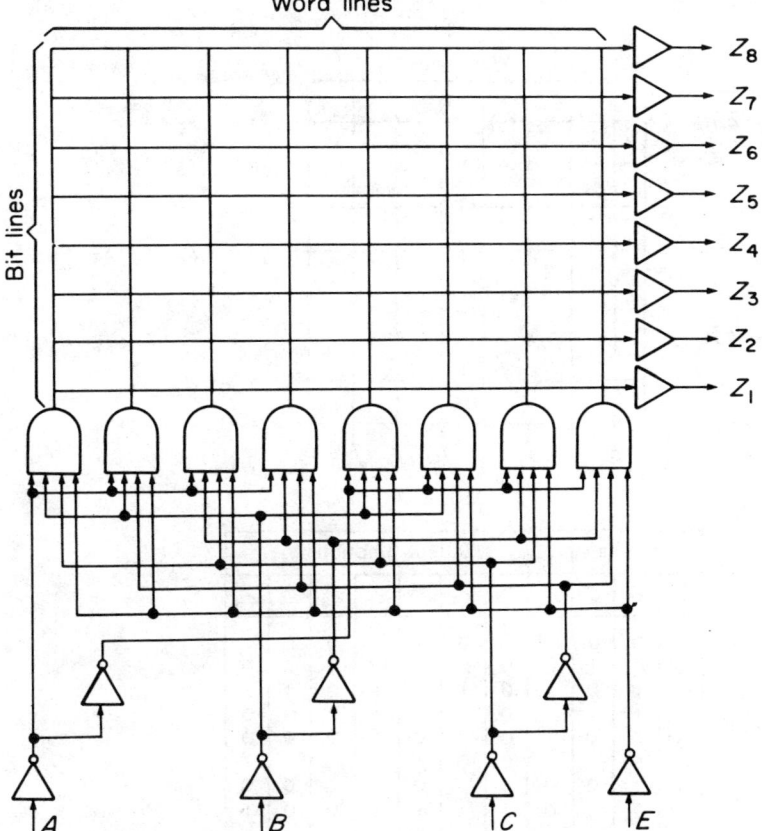

Figure 10.15. A 64-bit ROM with address decoder and enable input

a 64-bit memory matrix and eight non-inverting buffers. An enable input E, when at logical '0', enables all the gates in the address decoder. The vertical lines in the memory matrix are referred to as the word lines, and the horizontal ones as bit lines.

Words are programmed into the ROM at each address and the output on the bit line depends on whether it is connected to the addressed word line or not, the connection being made by the presence of an MOS or bipolar transistor, depending upon which technology is used. If connected to the word line, via a transistor, the bit line is raised to a logical '1' when the word line is addressed, and if not, it remains at logical '0'. In this way the word programmed at the selected address is transferred to the output.

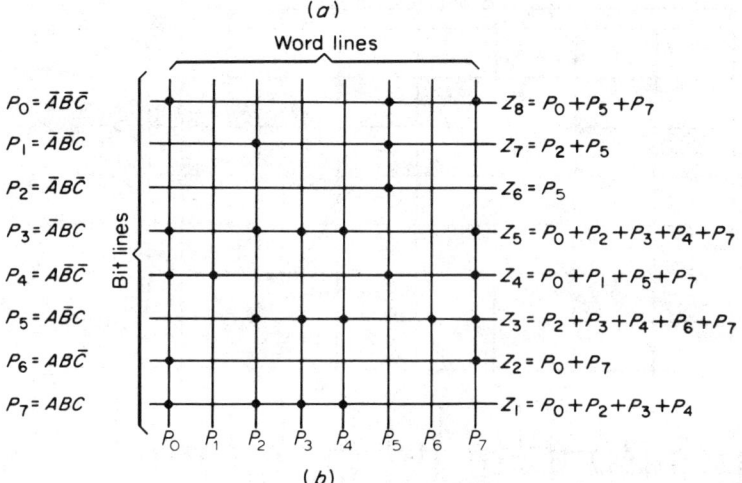

Figure 10.16. (a) Connection matrix for a 64-bit ROM. (b) Truth table

A schematic way of representing a programmed 64-bit ROM is shown in figure 10.16(a), where at those intersections of bit lines and word lines marked by a dot there is an OR input for the output function. For example, the output Z_8 is given by

$$Z_8 = P_0 + P_5 + P_7$$

Hence the ROM shown in figure 10.16(a) is being used to generate eight three-variable functions, each of which is expressed in canonical form.

If a customer wishes to realise the functions shown in figure 10.16(a), he must supply the ROM manufacturer with either a connection matrix, such as the one shown in the diagram, or a truth table, such as the one shown in figure 10.16(b). Alternatively, the customer may have his own programming facilities, in which case he would purchase a programmable read-only memory (PROM) which he would programme himself.

10.7 Addressing techniques for ROMs

The connection matrix shown in figure 10.17(a) is for a 2 × 8 bit word ROM addressed in one dimension only. The total capacity of this ROM is 16 bits, and the Boolean functions generated by it are

$$Z_1 = P_2 + P_5 + P_7$$

and

$$Z_2 = P_0 + P_4 + P_5 + P_7$$

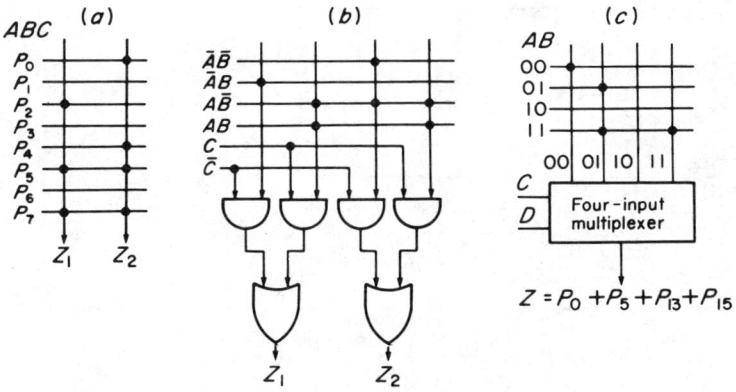

Figure 10.17. (a) A 2 × 8 bit word ROM addressed in one dimension, and (b) in two dimensions. (c) Generation of one four-variable Boolean function using a two-dimensional addressing scheme

The total number of connections into and out of the ROM matrix is ten for this method of addressing.

An alternative two-dimensional method of addressing a 2 × 8 bit word ROM is illustrated in figure 10.17(b), where examination of the connection matrix shows that the same Boolean functions are generated as in the previous example. In effect, each bit line has now been split into two sections and the selection of the appropriate section is done by the AND/OR circuits which are controlled by the Boolean variable C.

Using this two-dimensional addressing technique there is a reduction in the number of connections to the ROM matrix. The total number of connections in and out of the matrix is now eight, a reduction of two when compared with the one-dimensional addressing techniques. The larger the number of input variables, the more dramatic this reduction in matrix connections becomes.

The same ROM matrix in conjunction with one four-input multiplexer can be used to generate one four-variable function:

$$Z = P_0 + P_5 + P_{13} + P_{15}$$

as illustrated in figure 10.17(c). Eight matrix connections are required in this implementation. However, if the same function had been generated by a ROM addressed in one dimension, 17 matrix connections would have been required.

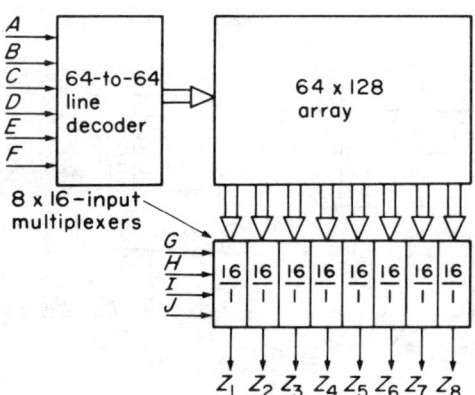

Figure 10.18. *Reduction in number of lines to and from* ROM *by two-dimensional addressing*

Clearly, large capacity ROMs can also be addressed in two dimensions. For example, a 1024-word by eight-bit ROM, using one-dimensional addressing would require a 10-to-1024 line address decoder and eight output lines, giving a total of 1032 connections to the ROM matrix. On the other hand, a two-dimensional addressing scheme such as the one shown in figure 10.18 can be used. Six of the input variables A, B, C, D, E and F are used to drive a 6-to-64 line address decoder whilst the other four variables G, H, I and J are used to provide the control signals to eight 16-input multiplexers. For this scheme a total of 192 connections needs to be made to the ROM matrix, a considerable saving when compared to one-dimensional addressing.

10.8 Design of sequential circuits using ROMs

ROMs are suitable devices for the implementation of clock-driven sequential circuits and, as an example, the NBCD invalid code detector designed in Chapter 8 using JK flip-flops and NAND gates will be implemented here using a ROM.

In this problem serial NBCD data arrives on line X, most significant digit first. Each data bit is synchronised with a clock pulse. It is required to design a circuit using a ROM that generates a fault signal $Z = 1$ each time an invalid code is received.

The block diagram and the internal state diagram are shown in figures 10.19(a, b). The state table shown in figure 10.19(c) is displayed in a suitable form for programming a ROM. For example, in the first row of the table, the present input to the ROM is $A = 0, B = 0, C = 0$ and $X = 0$, and the ROM output word is $A = 1, B = 0, C = 0$ and $Z = 0$.

With the aid of this state table, the connection matrix for a two-dimensionally addressed ROM can be developed, as illustrated in figure 10.19(d). It will be noticed that the implementation requires a 64-bit ROM.

Besides the ROM, additional logic is required to produce the output signal $Z = ABC$ and three D-type flip-flops are required, one in each feedback line to synchronise the operation of the circuit to the clock. These additional requirements with their connections to the ROM are illustrated in the block schematic shown in figure 10.19(e). The outputs from the ROM on the lines A, B and C will be transferred back to the input of the ROM on the trailing edge of the clock pulse.

ROMs can also be used for the design of asynchronous sequential circuits. As a further example of their use, a ROM-based design of a

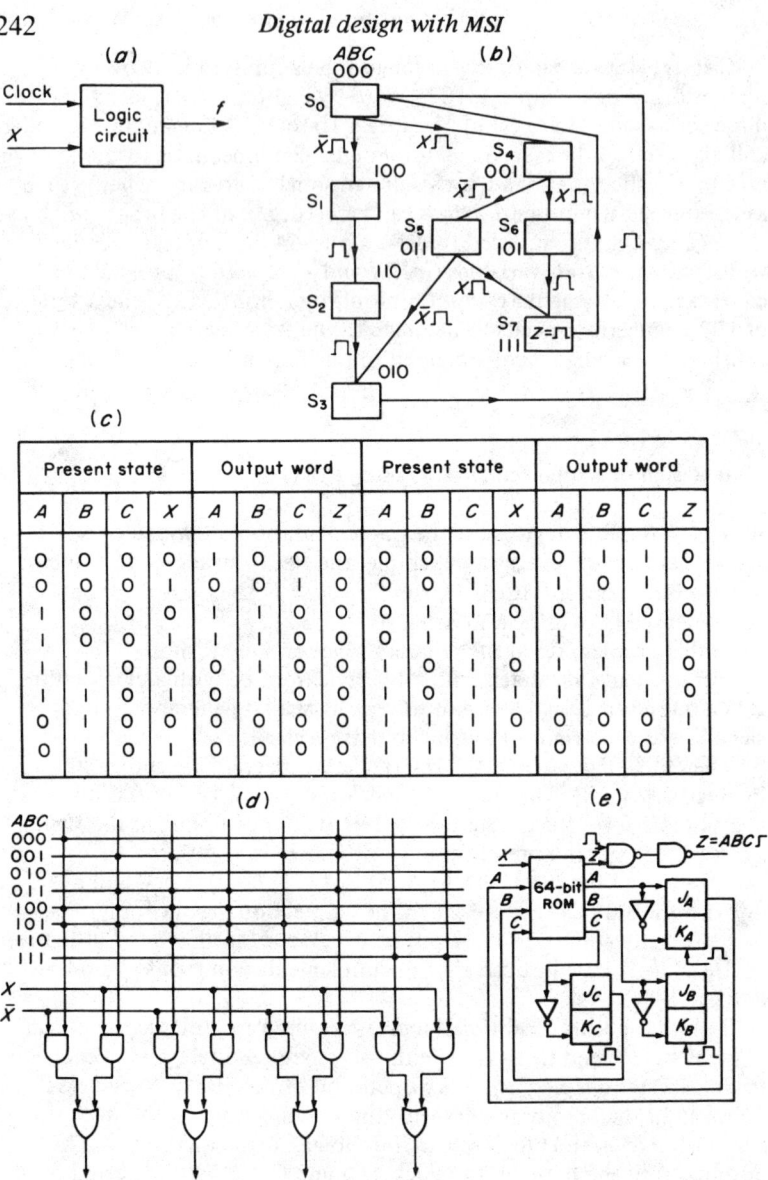

Figure 10.19. The invalid code detector. (a) Block diagram. (b) Internal state diagram. (c) State table. (d) Connection matrix. (e) Block schematic for ROM implementation

sequential network for controlling the traffic lights at a road junction will be developed.

A block diagram for the traffic light controller is shown in figure 10.20(a). The inputs to the controller are the timing signal X and the synchronisation signal P, whilst the output signals are used for driving the green, red and amber lamps.

The timing and synchronisation signals, together with the light sequence, are shown in figure 10.20(b). A suitable internal state diagram is shown in figure 10.20(c). When the controller is initially switched on, it may be in any one of its eight internal states, but a transition from S_2 to S_3 can only be made when both P and $\bar{X}$ are present. Examination of the timing diagram shows that this combination of signals only occurs during the second two-minute period, and hence state S_3 is associated with this time period and

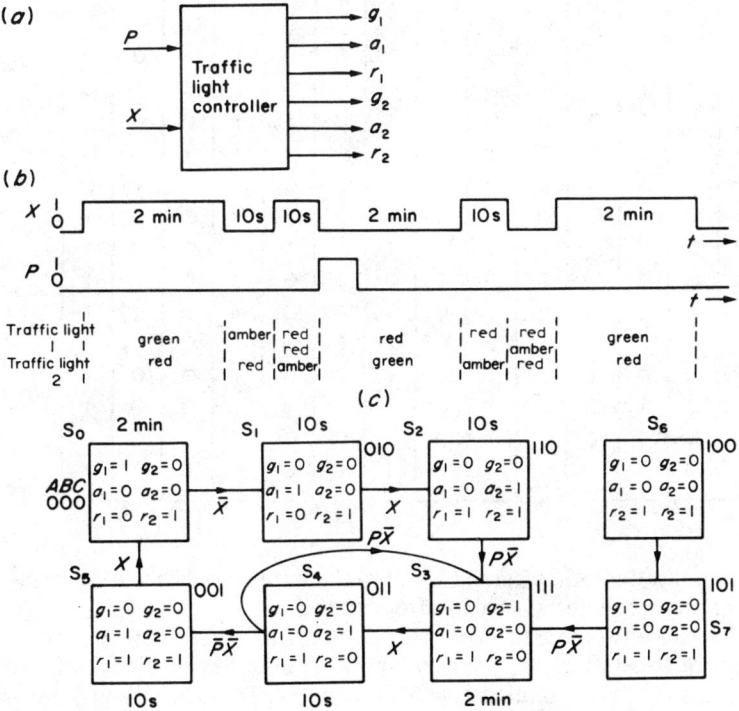

Figure 10.20. (a) Block diagram for a traffic controller. (b) Timing and synchronisation waveforms for a traffic controller. (c) Internal state diagram. (d) State table. (e) ROM matrix. (f) Block schematic showing ROM connections

244 Digital design with MSI

(d)

Present state					Output word						
A	B	C	X	P	A	B	C	g_1	a_1	r_1	g_2	a_2	r_2
0	0	0	0	0	0	1	0	0	1	0	0	0	1
0	0	0	0	1	0	1	0	0	1	0	0	0	1
0	0	0	1	0	0	0	0	1	0	0	0	0	1
0	1	0	0	0	0	1	0	0	1	0	0	0	1
0	1	0	0	1	0	1	0	0	1	0	0	0	1
0	1	0	1	0	1	1	0	0	0	1	0	1	1
1	1	0	0	0	1	1	0	0	0	1	0	1	1
1	1	0	0	1	1	1	1	0	0	1	1	0	0
1	1	0	1	0	1	1	0	0	0	1	0	1	1
1	1	1	0	0	1	1	1	0	0	1	1	0	0
1	1	1	0	1	1	1	1	0	0	1	1	0	0
1	1	1	1	0	0	1	1	0	0	1	0	1	0
0	1	1	0	0	0	0	1	0	1	1	0	0	1
0	1	1	0	1	1	1	1	0	0	1	1	0	0
0	1	1	1	0	0	1	1	0	0	1	0	1	0
0	0	1	0	0	0	0	1	0	1	1	0	0	1
0	0	1	0	1	0	0	1	0	1	1	0	0	1
0	0	1	1	0	0	0	0	1	0	0	0	0	1
1	0	0	0	0	1	0	1	0	0	1	0	0	1
1	0	0	0	1	1	0	1	0	0	1	0	0	1
1	0	0	1	0	1	0	1	0	0	1	0	0	1
1	0	1	0	0	1	0	1	0	0	1	0	0	1
1	0	1	0	1	1	1	1	0	0	1	1	0	0
1	0	1	1	0	1	0	1	0	0	1	0	0	1

Figure 10.20. Cont.

synchronisation of the timing signal to the state sequence will occur when the circuit enters state S_3 from state S_2.

There are two unused states, S_6 and S_7. If the circuit should enter either of these states at power on, or due to some circuit malfunction, both red lamps are simultaneously illuminated. Transition from S_6 to S_7 is unconditional and the circuit will then remain in S_7 until the signal $P\bar{X} = 1$ occurs, when it will make a transition to S_3 and synchronisation will have been achieved.

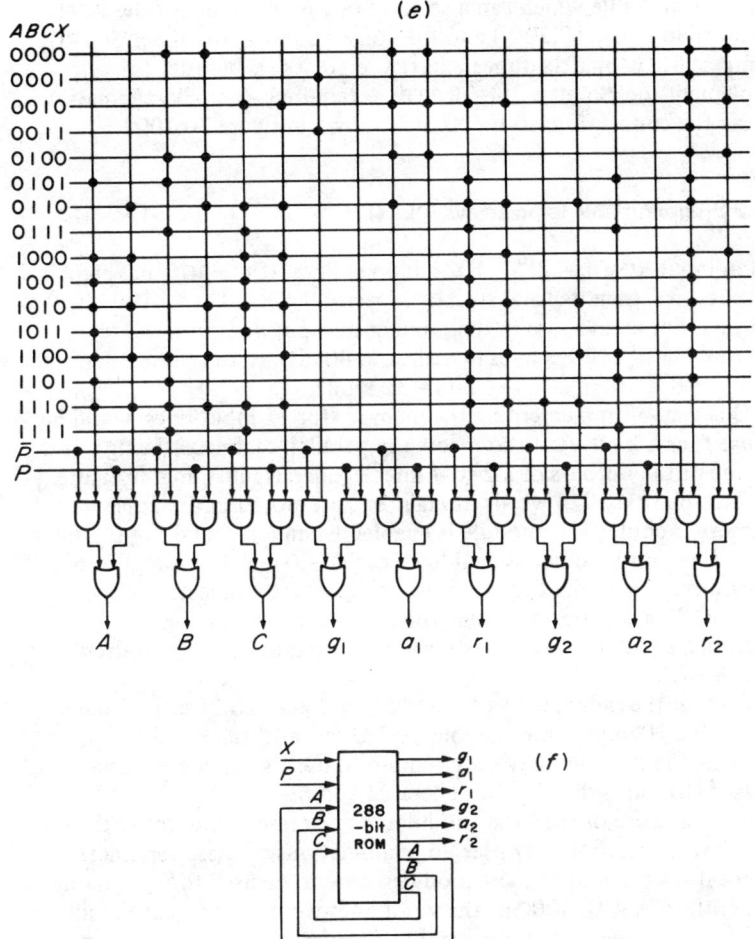

Figure 10.20. Cont.

Synchronisation can also be achieved by a transition from S_4 to S_3 if the next signal combination received by the circuit is $P\bar{X}$. However, synchronisation direct from states S_0, S_1 and S_5 by transition to S_3 is not possible since this would require a change of more than one secondary variable thus setting up the conditions for a race.

The state table which can also serve as a programme for the ROM is shown in figure 10.20(*d*) and the corresponding two-dimensionally addressed ROM matrix appears in figure 10.20(*e*). In order to implement the circuit a 288-bit ROM is required. A block schematic giving the connections to the ROM is shown in figure 10.20(*f*).

10.9 Programmable logic arrays (PLAs)

A ten-input ROM has 2^{10} = 1024 address lines. If the ROM has eight output lines, then its total storage capacity is 8 × 1024 = 8192 bits = 8 k bits. For an increase in the number of input lines from 10 to 12, the number of address lines increases to 4096, and the storage capacity becomes 8 × 4096 = 32 768 bits = 32 k bits.

One way of implementing the memory for 12 input lines would be to use four 8 k bit ROMs connected in parallel, each ROM being selected by one of the outputs of a 2-to-4 line decoder as illustrated in figure 10.21. The first 1024 words of eight bits are stored in the upper memory module. This module is enabled when $K = 0$ and $L = 0$. The second memory module is enabled when $K = 0$ and $L = 1$, and so on. The outputs from the four memory modules are connected in the 'wired-OR' configuration. Since the four memory modules are addressed simultaneously, buffers are required to drive the address lines A to J.

Each of the individual ROM modules in figure 10.21 can produce any of the 1024 possible canonical AND terms of ten variables. In practice, the designer may only require to use a small percentage of the 1024 AND terms that can be generated and this may lead to an uneconomic use of the ROM modules. For example, the Hollerith code has 12 variables, but only uses 96 combinations of these variables. Hence if four ten-input ROM modules were to be used for converting Hollerith to ASCII, 4000 of the word lines would not be used at all. Under these circumstances it would be preferable to use an MSI chip known as a programmable logic array (PLA) rather than a ROM for implementing this code converter.

In essence, a PLA can be regarded as being made up of two separate ROMs: one is referred to as the AND ROM (or logical product generator), and the other is referred to as the OR ROM (or logical sum generator). A block schematic of a very simple PLA is shown in figure 10.22(*a*).

In the block schematic there are two input lines A and B to the AND ROM, and the outputs from this ROM are the two *P*-terms, $P_0 = \overline{A}\,\overline{B}$ and $P_3 = AB$, which are generated on separate lines. The two *P*-terms,

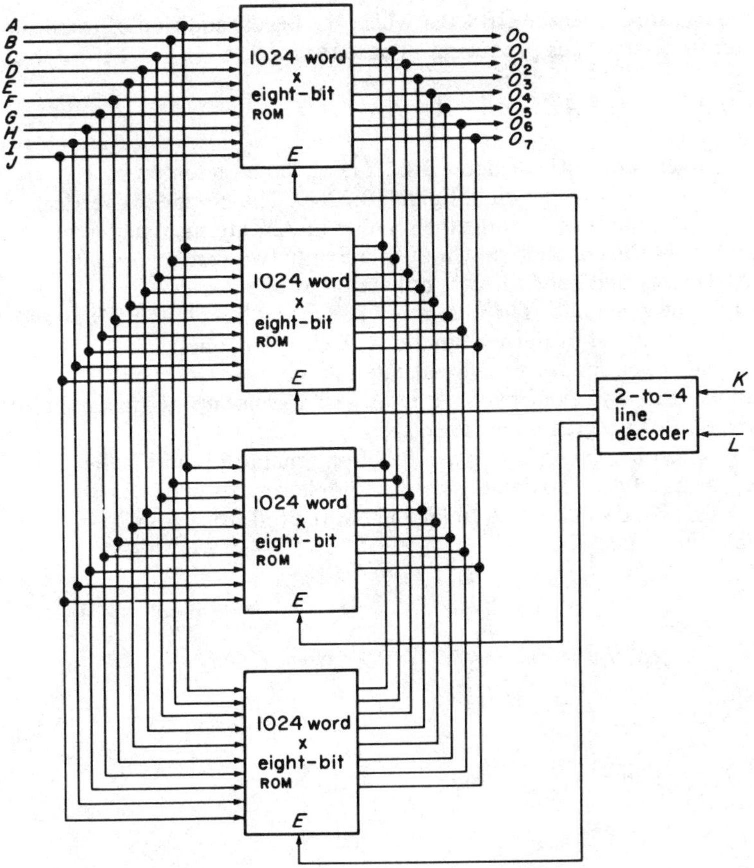

Digital design with MSI

Figure 10.21. Implementation of a 4096-word × eight-bit ROM

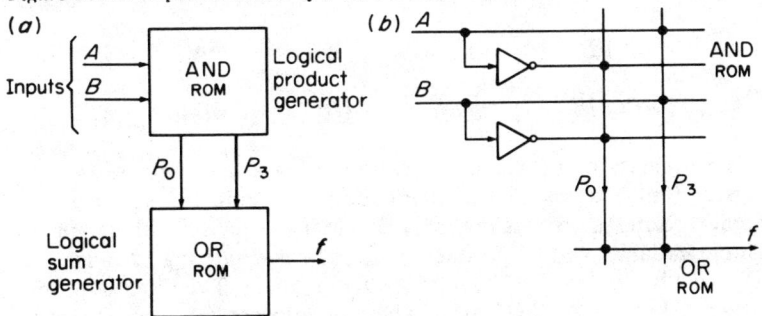

Figure 10.22. (a) Block schematic of a simple PLA. (b) Connection matrix of the PLA

P_0 and P_3, are fed to the OR ROM where the logical addition of these two terms occurs, thus producing an output function

$$f = P_0 + P_3 = \overline{A}\overline{B} + AB = A \odot B$$

The block schematic of figure 10.22(a) can be represented by a connection matrix as shown in figure 10.22(b). The number of vertical lines in the AND array governs the number of AND terms that can be generated. In the example shown there are only two vertical lines in the AND array and consequently only two out of a possible four P-terms can be generated. The number of horizontal lines in the OR-array defines the number of output functions, in this case, one.

Output functions are always generated in the two-level sum-of-products form, and elementary terms as well as canonical terms can be generated by a programmable logic array.

A PLA can be used as a Boolean function generator in much the same way as a ROM. To demonstrate the implementation of Boolean functions, a PLA will be used to implement the following four six-variable functions.

$$f_1 = \underset{1}{ABC\overline{D}E\overline{F}} + \underset{2}{AE\overline{F}} + \underset{3}{ACD} + \underset{4}{\overline{B}\overline{D}}$$

$$f_2 = \underset{5}{AB\overline{E}\overline{F}} + \underset{6}{A\overline{C}\overline{D}} + \underset{4}{\overline{B}\overline{D}} + \underset{7}{EF}$$

$$f_3 = \underset{8}{ACDE\overline{F}} + \underset{9}{AB\overline{C}\overline{D}} + \underset{10}{\overline{A}\,\overline{C}D} + \underset{11}{ABD}$$

$$f_4 = \underset{12}{AB\overline{C}\overline{D}\overline{E}} + \underset{13}{\overline{A}EF} + \underset{4}{\overline{B}\overline{D}} + \underset{14}{C\overline{D}} + \underset{7}{EF}$$

An examination of the above equations reveals that there are 14 separate product terms, each of them numbered, some of them in canonical form, and others in more elementary form. Since there are six input variables and 14 product terms to be generated, a 12 × 14 AND array is required, as illustrated in figure 10.23. The product terms are then fed to a 4 × 12 OR array where the process of logical addition

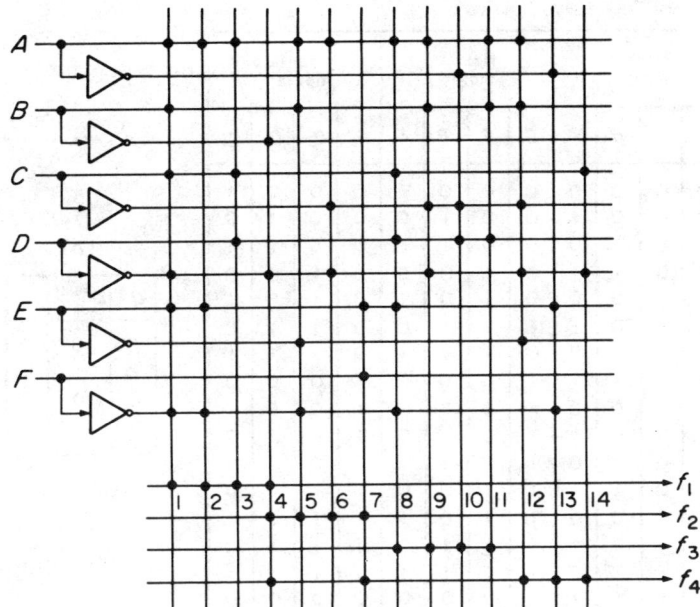

Figure 10.23. Boolean function implementation using a PLA

takes place, thus generating the four required Boolean functions. In all, a PLA with a storage capacity of 224 bits will be required for the implementation of these four functions, whereas a ROM implementation would require a 256-bit storage capacity. This represents approximately a 12½% saving in bit capacity.

10.10 Design of sequential circuits using PLAs

It has been demonstrated above that ROMs can be used for implementing sequential circuits, and it follows that PLAs can be used in exactly the same way. For example, it is possible to fabricate a master PLA chip consisting of an AND array for the product matrix, an OR array for the summing matrix, and clocked flip-flops to give synchronous circuit operation. The particular unique function that the PLA has to perform can be programmed into this LSI master chip by simply changing the gate mask.

In practice, there would be a series of master PLA chips available, each of which has a particular size product matrix and a particular size

250 Digital design with MSI

(a)

Present state				Next state				Flip-flop inputs			
D	C	B	A	D	C	B	A	D_D	D_C	D_B	D_A
0	0	0	0	0	0	0	1	0	0	0	1
0	0	0	1	0	0	1	0	0	0	1	0
0	0	1	0	0	0	1	1	0	0	1	1
0	0	1	1	0	1	0	0	0	1	0	0
0	1	0	0	0	1	0	1	0	1	0	1
0	1	0	1	0	1	1	0	0	1	1	0
0	1	1	0	0	1	1	1	0	1	1	1
0	1	1	1	1	0	0	0	1	0	0	0
1	0	0	0	1	0	0	1	1	0	0	1
1	0	0	1	1	0	1	0	1	0	1	0
1	0	1	0	1	0	1	1	1	0	1	1
1	0	1	1	1	1	0	0	1	1	0	0
1	1	0	0	1	1	0	1	1	1	0	1
1	1	0	1	1	1	1	0	1	1	1	0
1	1	1	0	1	1	1	1	1	1	1	1
1	1	1	1	0	0	0	0	0	0	0	0

(b)

Q^t	$Q^{t+\delta t}$	D^t
0	0	0
0	1	1
1	0	0
1	1	1

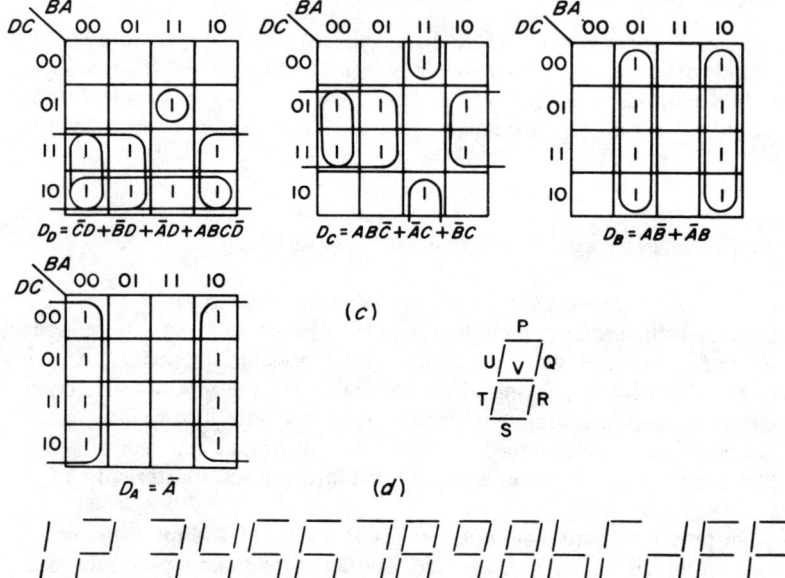

$D_D = \bar{C}D + \bar{B}D + \bar{A}D + ABC\bar{D}$

$D_C = AB\bar{C} + \bar{A}C + \bar{B}C$

$D_B = A\bar{B} + \bar{A}B$

$D_A = \bar{A}$

(c)

(d)

Digital design with MSI 251

(e)

State of counter				Segment inputs for a seven-segment decoder						
D	C	B	A	P	Q	R	S	T	U	V
0	0	0	0	1	1	1	1	1	1	0
0	0	0	1	0	1	1	0	0	0	0
0	0	1	0	1	1	0	1	1	0	1
0	0	1	1	1	1	1	1	0	0	1
0	1	0	0	0	1	1	0	0	1	1
0	1	0	1	1	0	1	1	0	1	1
0	1	1	0	1	0	1	1	1	1	1
0	1	1	1	1	1	1	0	0	0	0
1	0	0	0	1	1	1	1	1	1	1
1	0	0	1	1	1	1	0	0	1	1
1	0	1	0	1	1	1	0	1	1	1
1	0	1	1	0	0	1	1	1	1	1
1	1	0	0	1	0	0	1	1	1	0
1	1	0	1	0	1	1	1	1	0	1
1	1	1	0	1	0	0	1	1	1	1
1	1	1	1	1	0	0	0	1	1	1

(f)

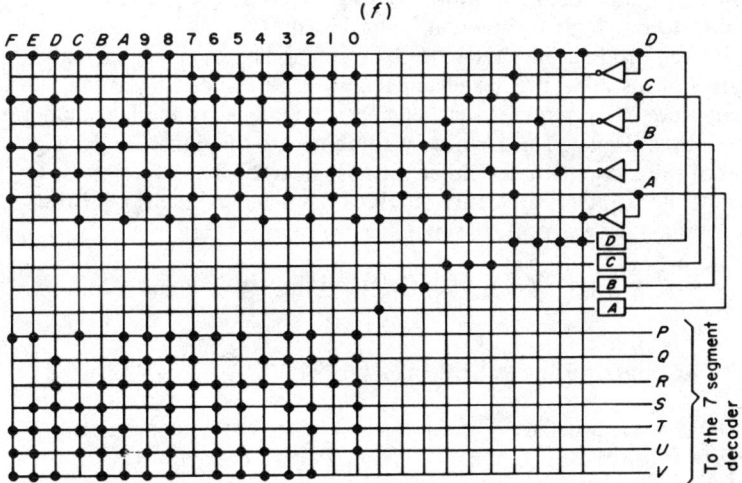

Figure 10.24. (a) State table for the hexadecimal counter. (b) Steering table for the D-type flip-flop. (c) K-maps for the hexadecimal counter. (d) Segmental representation of hexadecimal digits. (e) Truth table for a seven-segment decoder. (f) Implementation of the hexadecimal counter using a PLA

summing matrix, together with a specific number of flip-flops of a particular type. The designer then selects a suitably sized PLA master chip for his circuit and arranges for it to be programmed to implement the required circuit. As an example of the use of PLAs in sequential circuits, a hexadecimal counter will be designed using D-type flip-flops, and the output of the counter will be decoded to give a visual seven-segment display.

The state table for the hexadecimal counter is shown in figure 10.24(a). The inputs to the four flip-flops required for each state change are shown on the right of the state table, and have been obtained with the aid of the steering table for the D-type flip-flop shown in figure 10.24(b). K-maps have been plotted and simplified for each flip-flop input in figure 10.24(c), and the minimum form of the input equations obtained from them are

$$D_D = \bar{C}D + \bar{B}D + \bar{A}D + ABC\bar{D} \qquad D_B = A\bar{B} + \bar{A}B$$

$$D_C = AB\bar{C} + \bar{A}C + \bar{B}C \qquad D_A = \bar{A}$$

The segment allocation for the seven-segment display is defined in figure 10.24(d), as well as the segmental representation of each of the 16 hexadecimal digits. A truth table for the seven-segment decoder is shown in figure 10.24(e), and the implementation of counter and display decode logic is shown in figure 10.24(f).

Sixteen product lines are required for decoding the hexadecimal digits. For instance, the hexadecimal digit A corresponding to the binary code 1010 requires that segments P, Q, R, T, U and V should be illuminated. Hence, for this binary combination the signal for driving each of these segments should be set to logical '1'. The functions for each of the segments are easily obtained from an examination of the truth table. For example, the function for segment P is

$$P = 0 + 2 + 3 + 5 + 6 + 7 + 8 + 9 + A + C + E + F \text{ and so on.}$$

10.11 Arithmetic using MSI chips

In Chapter 4 it was shown how a four-bit adder can be constructed from four full adders (figure 4.4) and how, with the aid of four exclusive-OR gates it can be converted to a four-bit adder/subtractor using 2's complement arithmetic (figure 4.7). With the advent of MSI, four-bit adders are now available on a single chip, a typical example

Digital design with MSI 253

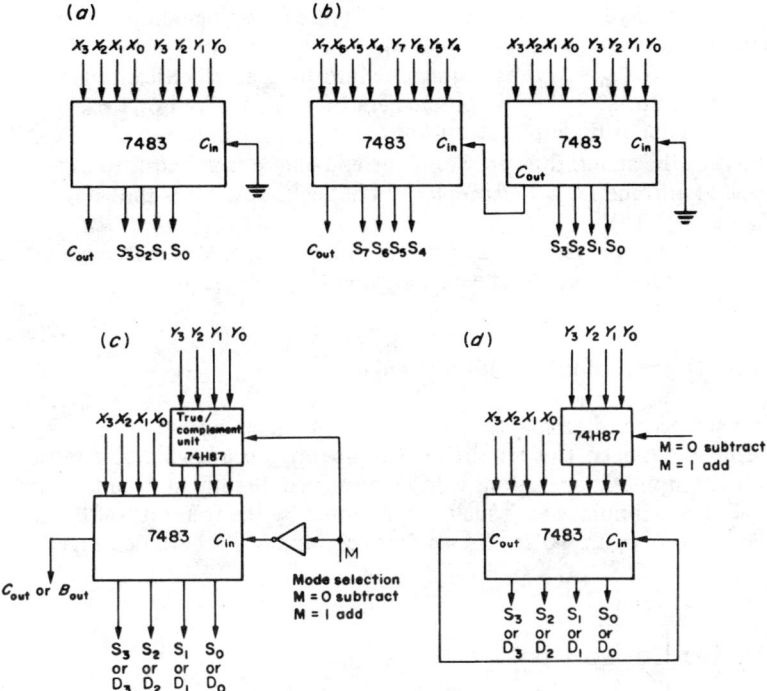

Figure 10.25. (a) The 7483 four-bit adder. (b) The addition of two eight-bit binary numbers. (c) The 2's complement adder/subtractor. (d) The 1's complement adder/subtractor

being the SN 7483. This adder will accept two four-bit numbers and produce a four-bit sum output in conjunction with a carry, as indicated in figure 10.25(a).

To add together two eight-bit numbers, two 7483 chips can be connected in cascade, as shown in figure 10.25(b). The carry-out C_{out} of the least significant chip is connected to the carry-in C_{in} of the second chip. Clearly, the range of addition can be extended further by increasing the number of four-bit adders in the cascade.

MSI adders (such as the 7483) can be used in conjunction with a four-bit true or complement unit (such as the 74H87) to perform complement arithmetic. Examples of 2's and 1's complement four-bit adder/subtractors are shown in figures 10.25(c, d).

The 74H87 forms the 1's complement of the input binary number $Y_3 Y_2 Y_1 Y_0$ if the mode control $M = 0$. If, however, $M = 1$, the true value of $Y_3 Y_2 Y_1 Y_0$ appears at the output of the true/complement unit. In the case of the 2's complement circuit shown in figure 10.25(c), when $M = 0$, $C_{in} = 1$, thus adding one into the least significant place, and hence forming the 2's complement of $Y_3 Y_2 Y_1 Y_0$. For 1's complement arithmetic, the output carry C_{out} is connected to the C_{in} input of the adder, thus providing the end-about carry required in this type of arithmetic. A 1's complement adder/subtractor is shown in figure 10.25(d).

To extend the range of addition for complement arithmetic, adders can be connected in cascade, as shown in figure 10.25(b).

10.12 Decimal addition with MSI adders

It is sometimes desirable to perform arithmetic operations using decimal numbers. This requirement frequently occurs where the result of the arithmetic operations is to be displayed directly in decimal form.

Decimal numbers are usually represented by the four-bit NBCD code tabulated in figure 10.26. If two decimal numbers to be added together

d	A	B	C	D
0	0	0	0	0
1	0	0	0	1
2	0	0	1	0
3	0	0	1	1
4	0	1	0	0
5	0	1	0	1
6	0	1	1	0
7	0	1	1	1
8	1	0	0	0
9	1	0	0	1

Figure 10.26. NBCD code

produce a sum $S \leqslant 9$, then the sum of the two NBCD codes representing the decimal numbers gives the correct answer. For example,

```
      3    0011
   +
      4    0100
      ─    ────
      7    0111
```

On the other hand, if the sum of the two numbers $S > 9$, then the addition of the two NBCD codes representing the numbers will give an incorrect answer. This is illustrated by the following example:

$$
\begin{array}{r}
7 \quad 0111 \\
+7 \quad 0111 \\
\hline
14 \quad 1110
\end{array}
$$

The four-digit combination obtained by the addition of the two NBCD codes in this case does not appear in the code listing, and hence it is a forbidden combination of digits. The true representation of $(14)_{10}$ in the NBCD code requires two groups of four digits:

$$(14)_{10} = 0001, 0100$$

Hence, if the sum $S > 9$, a correction has to be made to the resulting forbidden combination. This correction can be made by subtracting $(10)_{10} = (1010)_2$ from the forbidden combination, or, alternatively, the 2's complement of $(1010)_2$ can be added to the incorrect result. The 2's complement of $(1010)_2$ is $(0110)_2$, and hence in decimal terms $(6)_{10}$ has to be added in when the sum is performed as is illustrated in the following example:

$$
\begin{array}{r}
7 \quad 0111 \\
+7 \quad 0111 \\
\hline
1110 \\
0110 \\
\hline
14 \quad 1,0100
\end{array}
$$

When $S > 9$ a carry is required for the next most significant stage of the addition. Consequently, a logic function must be developed which will detect the six forbidden code combinations, 1010, 1011, 1100, 1101, 1110 and 1111. These forbidden combinations are shown plotted on the K-map in figure 10.27(*a*) and are simplified in the usual way. The carry function obtained from this map for representing these combinations is

$$f_c = AB + AC$$

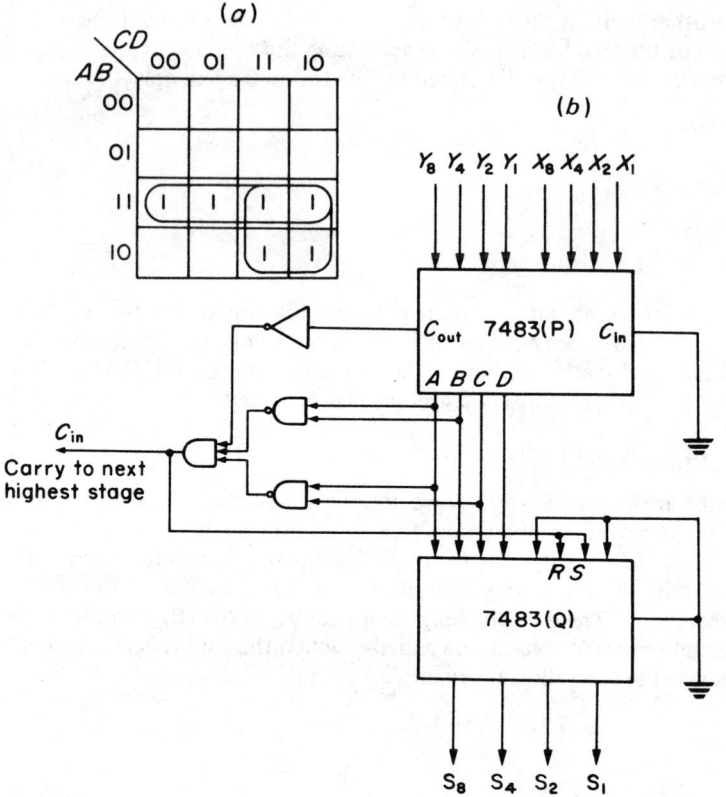

Figure 10.27. (a) The forbidden combinations plotted on a K-map. (b) A single-stage NBCD adder

In the event of the addition of two decimal digits producing a sum $S > 15$, a carry C_{out} will be generated by the four-bit adder. For example,

```
    9    1001
+   9    1001
        ──────
        1,0010
         0110
        ──────
        1,1000
```

Hence for a decimal adder the equation for the carry into the next stage is given by

$$C_{in} = C_0 + AB + AC$$

The implementation of one stage of an NBCD adder is shown in figure 10.27(b). It requires two 7483 four-bit adders, three NAND gates and an inverter. The adder marked P is used for the addition of two four-bit NBCD codes $Y_8 Y_4 Y_2 Y_1$ and $X_8 X_4 X_2 X_1$. Outputs from this adder are fed to a NAND gate circuit which generates part of the C_{in} function for the next stage. They are also fed to the adder marked Q, whose only function is to add in $(0110)_2$ when the sum of the two decimal digits $S > 9$. When this condition exists a carry-in, C_{in}, to the next stage is generated and this signal is also fed to the inputs marked R and S on adder Q, thus generating the number $(0110)_2$ at its four right-hand input terminals.

If n such single-stage decimal adders are connected in cascade then an n-decade adder is the result.

Problems

10.1 Implement the following three-variable Boolean functions using four-input multiplexers:

(a) $f = \Sigma\ 0, 2, 3, 5, 7$ control variables A and B
(b) $f = \Sigma\ 1, 3, 4, 6, 7$ control variables B and C
(c) $f = \Sigma\ 0, 2, 4, 5, 6, 7$ control variables A and C

10.2 Implement the following four-variable Boolean functions using four-input multiplexers and NAND gates:

(a) $f = \Sigma\ 0, 1, 3, 5, 6, 8, 9, 11, 12, 13$ control variables A and B
(b) $f = \Sigma\ 0, 7, 8, 9, 10, 11, 15$ control variables B and C
(c) $f = \Sigma\ 0, 1, 3, 5, 9, 10, 11, 13, 14, 15$ control variables C and D
(d) $f = \Sigma\ 1, 8, 9, 10, 12, 13, 14, 15$ control variables A and D

10.3 Implement the following five-variable Boolean functions using four-input multiplexers:

(a) $f = \Sigma\ 0, 1, 2, 3, 4, 8, 9, 11, 12, 13, 14, 18, 19, 20, 21, 25, 26, 29, 30$ and 31

(b) $f = \Sigma\ 5, 6, 7, 8, 9, 10, 14, 15, 16, 17, 18, 19, 22, 23, 24, 25,$ 26, 29, 30 and 31.

10.4 Implement the following six-variable Boolean function using

(a) four-input multiplexers and NAND gates, and
(b) eight-input multiplexers and four-input multiplexers.

$f = \Sigma\ 0, 1, 3, 5, 7, 12, 14, 16, 18, 20, 22, 26, 28, 30, 32, 34,$ 37, 39, 41, 43, 45, 50, 51, 53, 60, 61, 62 and 63.

10.5 Use multiplexers for implementing a four-bit full adder.

10.6 Design an NBCD-to-decimal decoder

(a) that will reject all false data, and
(b) in its simplest form, using NAND gates and inverters.

10.7 A combinational circuit is defined by the equations

$f_1 = AB + \bar{A}\bar{B}\bar{C}$

$f_2 = A + B + \bar{C}$

$f_3 = \bar{A}B + A\bar{B}.$

Design a circuit which will implement these three equations using a decoder and NAND gates external to the decoder.

10.8 The tabulation below gives details of four frequently used codes: the 8–4–2–1, the 2–4–2–1, XS3 and XS3 Gray. Using 4-to-10 line decoders and external logic, design three code converters for converting from 8–4–2–1 to each of the other three codes.

Binary number	8–4–2–1	2–4–2–1	XS3	XS3 Gray
0000	0	0		
0001	1	1		
0010	2	2		0
0011	3	3	0	

Binary number	8–4–2–1	2–4–2–1	XS3	XS3 Gray
0100	4	4	1	4
0101	5		2	3
0110	6		3	1
0111	7		4	2
1000	8		5	
1001	9		6	
1010			7	9
1011		5	8	
1100		6	9	5
1101		7		6
1110		8		8
1111		9		7

10.9 A combinational circuit is defined by the equations

$$f_1 = ABC + \overline{A}\overline{B}C$$

$$f_2 = \overline{A} + \overline{B} + C + D$$

$$f_3 = A + B + \overline{C}D + \overline{A}D$$

$$f_4 = ACD + A\overline{C}\overline{D} + B\overline{C}D + BC\overline{D}.$$

Design a circuit which will implement these four equations using a decoder with NAND gates external to the decoder.

10.10 Develop a 3-to-8 line decoder using NOR gates only, and draw its logic diagram.

10.11 Develop the block diagram for a 6-to-64 line decoder using nine 3-to-8 line decoders.

10.12 A ROM is to be used for converting from pure binary to NBCD. determine the ROM capacity required for converting an eight-bit binary number to NBCD, and draw a ROM matrix indicating the locations where 1's are stored with a dot.

10.13 Determine the ROM capacity required to implement

(a) An NBCD to 2-out of-5 code converter.
(b) A four-bit magnitude comparator.
(c) A binary multiplier that multiplies two three-bit numbers.

10.14 Implement the following four Boolean functions using a ROM.

(a) $f_1 = \bar{A}BCD + A\bar{C}\bar{D} + \bar{B}CD$
(b) $f_2 = A + B + C$
(c) $f_3 = \bar{A}B + A\bar{B} + \bar{C}D + C\bar{D}$
(d) $f_4 = (A + B + CD)(\bar{A} + B + \bar{C}D) + \bar{A}\bar{B}\bar{C}$.

10.15 Design a scale-of-seven counter having a seven segment display output and using D-type flip-flops that are incorporated on a PLA. Give a detailed drawing of the memory matrix of the PLA and the system interconnections.

10.16 Use a ROM pair to realise each of the following switching functions.

(a) $f(A, B, C) = \Sigma\ 0, 1, 2, 6, 7$
(b) $f(A, B, C, D) = \Sigma\ 0, 1, 8, 9, 10, 11, 12, 14, 15$
(c) $f(A, B, C, D, E) = \Sigma\ 0, 1, 16, 17, 18, 19, 20, 21, 22, 23, 24, 26, 28, 29, 30, 31$.
(d) $f(A, B, C, D, E, F) = \Sigma\ 0, 1, 2, 3, 4, 5, 10, 11, 12, 13, 14, 15, 19, 20, 21, 22, 23, 24, 27, 28, 29, 30, 31, 47, 48, 49, 60, 61, 62, 63$.

10.17 Use a ROM pair to implement the following code conversions.

(a) NBCD–to–XS3
(b) NBCD–to–XS3 Gray
(c) NBCD–to–2,4,2,1
(d) NBCD–to–6, 3, 1,–1

10.18 Design a four-bit carry look-ahead adder using a PLA.

10.19 Develop the circuit for a three-decade, XS3 decimal adder/subtractor using 7483 adders and any additional discrete logic that may be required. The 9's complement arithmetic is to be used for the subtraction process.

11

Hazards

11.1 Introduction

One the of the major causes of the malfunction of digital circuits can, in practice, be traced to the presence of *race hazards*. It is therefore necessary for the digital designer to have a clear understanding of the mechanism that produces such hazards and to be aware of their effects on circuit performance.

There are three types of hazard which occur in digital circuits:

(*a*) static hazards,
(*b*) dynamic hazards,
(*c*) essential hazards.

Static and dynamic hazards can be present in combinational networks and they are also to be found in gate-implemented, event-driven circuits. Essential hazards are peculiar to fundamental-mode asynchronous circuits.

The causes of hazards are identified in this chapter, and methods for detecting their presence in a digital circuit are described. Design methods to eliminate these hazards are also introduced.

11.2 Gate delays

If a two-input NAND gate is used as an inverter in a combinational network, as illustrated in figure 11.1, there will be a finite time delay t_g before any change appearing at the input of the gate produces the required change at the output. This is demonstrated in the time diagrams, where the change in A from 0 to 1 is followed by a change in

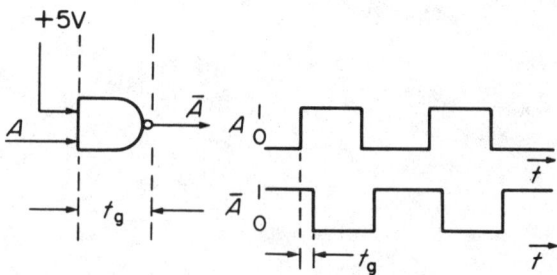

Figure 11.1. *The effect of gate delays when inverting a signal A*

$\bar{A}$ from 1 to 0 t_g seconds later. Similarly, when A changes from 1 to 0 the corresponding change in $\bar{A}$ from 0 to 1 also occurs t_g seconds later.

11.3 The generation of spikes

If the signal A and its complement $\bar{A}$, generated by the NAND gate shown in figure 11.1, are both fed to the inputs of a two-input AND gate, as shown in figure 11.2, then according to the laws of Boolean algebra, the output of the gate should be $A \cdot \bar{A} = 0$ at all instants. However, it will be observed from an examination of the time diagrams that in the time periods that have been shaded, A and $\bar{A}$ are simultaneously equal to 1, so that during these periods the gate output is $A \cdot \bar{A} = 1$. The output of the gate $A\bar{A}$ consists of a series of positive-going spikes which are initiated when A is changing from 0 to 1, each of time duration t_g seconds. This is, of course, the gate delay of the inverter shown in figure 11.1.

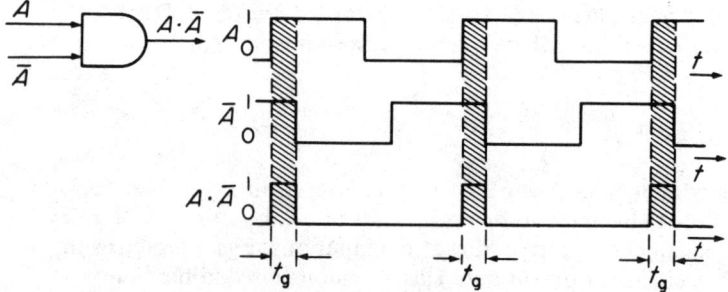

Figure 11.2. *Generation of spikes by an AND gate*

The circuit used to generate the signal $A \cdot \bar{A}$ is said to exhibit a static 0-hazard because the output signal (which should be permanently 0) goes to 1 for short transient periods.

If, on the other hand, the signals A and $\bar{A}$ are applied to the inputs of a two-input OR gate, as shown in figure 11.3, then the output of the gate is $f = A + \bar{A}$, which, according to the laws of Boolean algebra, should be logical '1' at all instants. The waveforms of A and $\bar{A}$ (figure 11.3) show that during the shaded time periods, they are both

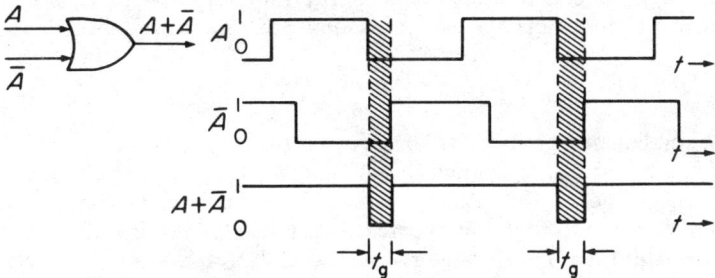

Figure 11.3. Generation of spikes by an OR gate

simultaneously equal to 0. In these shaded periods, which are of short time duration, the output goes to logical '0'. The circuit is said to exhibit a static 1-hazard because its output, which is normally 1, goes to 0 for short time periods. It will be observed, that for the OR gate, the negative-going spikes are initiated at the instant when A is changing from 1 to 0.

The generation of spikes by NAND and NOR gates is illustrated in figure 11.4. Negative-going spikes are generated by a NAND gate at the instant when A is changing from 0 to 1. The circuit exhibits a static 1-hazard. In the NOR circuit, positive-going spikes are generated at the

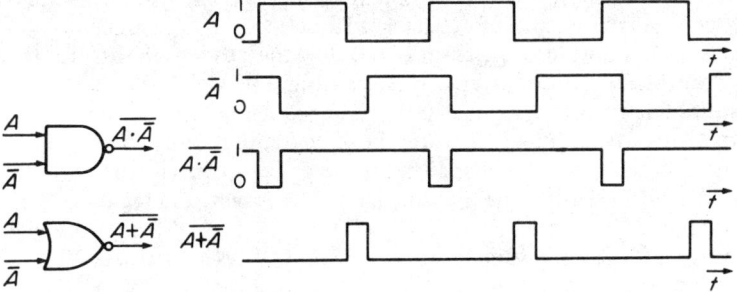

Figure 11.4. Generation of spikes by NAND and NOR gates

instant when A is changing from 1 to 0. This circuit exhibits a static 0-hazard.

11.4 The production of static hazards in combinational networks

When an input to a combinational network is changing, spikes may be generated at the output of the circuit. These spikes occur because different path lengths through the network, from output to input, may introduce different time delays. For example, the Boolean function

$$f = AB + \bar{A}C$$

may be implemented by the NAND circuit shown in figure 11.5. There are two paths through the network, the first via g_1, g_2 and g_3, and the second via g_4 and g_3. If it is assumed that all gates have equal time delays of t_g seconds, then it is apparent that the delay through the first path is greater than that through the second path.

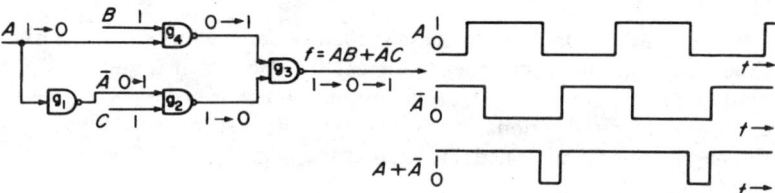

Figure 11.5. *The production of a static 1-hazard in a combinational network*

The changes taking place in the circuit are illustrated in figure 11.5 for the condition of $B = 1$, $C = 1$ and A changing from 1 to 0. For this change in A, the output of g_4 changes from 0 to 1 and produces a change in the output of g_3 from 1 to 0. For the other path through the network, the output of g_1 first changes from 0 to 1, followed by the output of g_2 changing from 1 to 0, thus producing a change in the output of g_3 from 0 to 1. Because the g_4, g_3 path has the shorter time delay, it is clear that the change in the output propagated along this path occurs earlier in time than the change propagated along the g_1, g_2, g_3 path.

Since it has been assumed that $B = 1$ and $C = 1$, the network equation reduces to $f = A + \bar{A}$. When the equation of a circuit can be reduced to this form there is the possibility of the production of a

Hazards

static 1-hazard. In the example chosen here, the time diagrams shown in figure 11.5 indicate that for a short period of time after A changes from 1 to 0, both A and $\bar{A}$ are equal to 0, and hence $A + \bar{A} = 0$. Consequently, at the beginning of this time period $A + \bar{A}$ changes from 1 to 0, returning to 1 again when $\bar{A}$ changes from 0 to 1. The circuit has generated a static 1-hazard. The production of negative-going spikes is illustrated in figure 11.5 and their presence confirms the earlier deduction, made by following the signal changes through on the circuit diagram, that the output changes are $1 \to 0 \to 1$.

The dual function of $f = AB + \bar{A}C$ is

$$f_d = (A + B)(\bar{A} + C).$$

This equation can be implemented using NOR gates as shown in figure 11.6. When $B = 0$ and $C = 0$ in this circuit its equation reduces to $f = A\bar{A}$. If the equation of a given circuit can be reduced to this form there is a possibility of a static 0-hazard being generated when A is changing from 0 to 1.

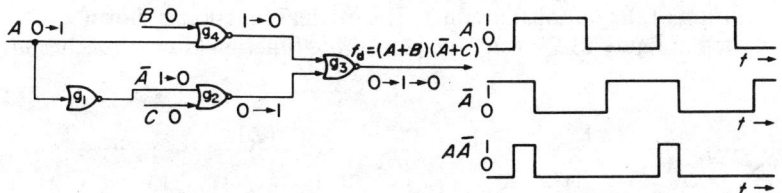

Figure 11.6. The production of a static 0-hazard in a combinational network

The production of a static 0-hazard in the circuit whose equation is $f_d = (A + B)(\bar{A} + C)$ is illustrated in figure 11.6. Immediately after A changes from 0 to 1, both A and $\bar{A}$ are simultaneously 1, and hence $A\bar{A} = 1$. It remains at this value until $\bar{A}$ falls to 0 when $A\bar{A}$ resumes its value of 0 again.

Signal changes in the circuit are also illustrated in figure 11.6, where it has been assumed that $B = C = 0$ and that A changes from 0 to 1. If all the gates in the circuit have the same delay t_g, then path g_4, g_3 has the shortest time delay and the change in the output due to A changing from 0 to 1 will propagate along this path faster than along path g_1, g_2, g_3. This results in the output changing from 0 to 1. When the same change arrives at the output along the alternative path, the output changes back to 0 again.

11.5 The elimination of static hazards

The equation of the NAND circuit shown in figure 11.5 is

$$f = AB + \bar{A}C$$

The optional product for this equation is BC, and this can be added to the original equation without altering its value. Thus

$$f = AB + \bar{A}C + BC$$

For the condition $B = C = 1$ the equation reduces to

$$f = A + \bar{A} + 1$$

and even if A and $\bar{A}$ are, for a short period of time, simultaneously equal to 0, the value of f remains at 1.

The effect of adding the optional product can be studied by examining the K-map plot of the function before and after the inclusion of the optional product. The original function is shown plotted in figure 11.7(a) and the plot of the function after the inclusion

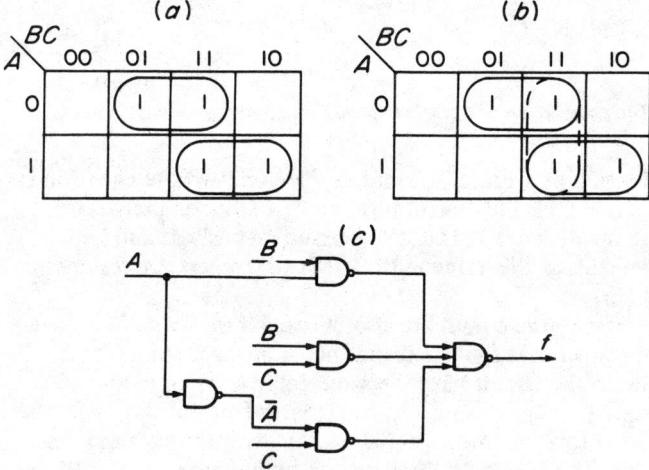

Figure 11.7. (a) Plot of $f = AB + \bar{A}C$. (b) Plot of $f = AB + \bar{A}C + BC$. (c) Implementation of the hazard-free function $f = AB + \bar{A}C + BC$

of the optional product is shown in figure 11.7(b). Comparison of the two plots shows that before the addition of the optional product there are two 1's in adjacent cells not covered by the same prime implicant. When these two adjacent 1's are covered by the same prime implicant, by the inclusion of the optional product, as in figure 11.7(b), the hazard is removed from the circuit.

Static 1-hazards can therefore be detected by looking for adjacent 1's, on a K-map plot of the function, that are not covered by the same prime implicant. They can then be removed at the design stage by including additional prime implicants which cover pairs of adjacent 1's not covered by the same prime implicant.

The hazard-free circuit for the Boolean function $f = AB + \bar{A}C$ is shown in figure 11.7(c), and it will be observed that an additional gate is required for generating the inverse of optional product BC.

For the NOR circuit of figure 11.6,

$$f_d = (A + B)(\bar{A} + C)$$

The optional sum for this equation is $B + C$, and this can be included in the above equation without altering its value, so that

$$f_d = (A + B)(\bar{A} + C)(B + C)$$

If $B = C = 0$, then

$$f_d = A \cdot \bar{A} \cdot 0 = 0$$

With the inclusion of the optional sum, the value of the function is always 0, irrespective of whether A and $\bar{A}$ are simultaneously equal to 1.

The static 0-hazard is removed from the circuit by the inclusion of the optional sum $B + C$ in the circuit equation, and the resulting hazard-free circuit is shown in figure 11.8. Removal of the hazard requires an additional gate which generates the inverse of the optional sum $B + C$.

A K-map is normally plotted such that cells marked with a 1 indicate the combinations of the variable that cause the value of the plotted function to become 1. The static 1-hazard is generated in the NAND circuit of figure 11.5 when $A = 1, B = 1, C = 1$ and A is changing to 0. On the K-map of figure 11.7(a) a transition is being made from cell 111 to cell 011 when the hazard occurs, and is revealed by the adjacent 1's in the two cells not covered by the same prime implicant.

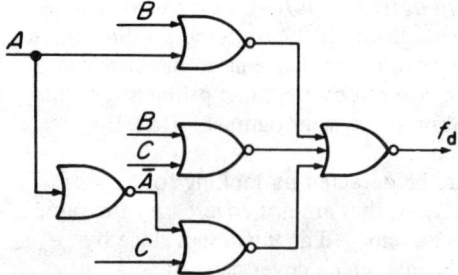

Figure 11.8. Implementation of the hazard-free function $f_d = (A + B)(\bar{A} + C)(B + C)$

When looking for a static 0-hazard, a K-map plot of the function which indicates those combinations of the variable that cause the function value to be 0 is required. To obtain a plot of the 0-terms, the inverse of the function f_d must be plotted. The circuit equation is

$$f_d = (A + B)(\bar{A} + C)$$

Dualising,

$$f = AB + \bar{A}C$$

and inverting,

$$\bar{f}_d = \bar{A}\bar{B} + A\bar{C}$$

The inverse function is shown plotted in figure 11.9(a) and it will be noticed that the two 0's in the adjacent cells 000 and 100 are not covered by the same prime implicant. The inclusion of the additional prime implicant $\bar{B}\bar{C}$ in figure 11.9(b) ensures that the two adjacent 0's are now covered by the same prime implicant.

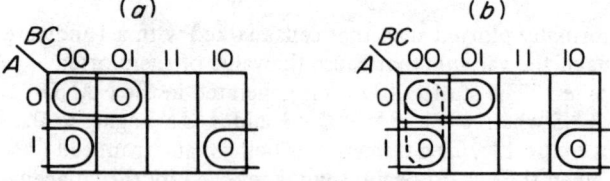

Figure 11.9. (a) Plot of $\bar{f}_d = \bar{A}\bar{B}\ A\ A\bar{C}$. (b) Plot of $\bar{f}_d = \bar{A}\bar{B} + A\bar{C} + \bar{B}\bar{C}$

Hazards

The function including the additional prime implicant becomes

$$\bar{f}_d = \bar{A}\bar{B} + A\bar{C} + \bar{B}\bar{C}$$

dualising

$$f = (\bar{A} + \bar{B})(A + \bar{C})(\bar{B} + \bar{C})$$

and inverting

$$f_d = (A + B)(\bar{A} + C)(B + C),$$

which is the hazard-free function obtained by introducing the optional sum obtained above.

The rules for finding static 0-hazards can be summarised as follows.
(1) Plot the inverse function.
(2) Look for adjacent 0's not covered by the same prime implicant.
(3) Insert additional prime implicants to cover all adjacent 0's that are not covered by the same prime implicant.
(4) Modify the inverse equation by the inclusion of the additional prime implicants.
(5) Re-invert the equation to its hazard-free form.

11.6 Design of hazard-free combinational networks

In this section the function represented by the equation

$$f = \Sigma\, 2, 5, 6, 7, 10, 13, 15$$

will be implemented in hazard-free form using (*a*) NAND gates, and (*b*) NOR gates. It will be assumed that gates having a maximum number of inputs (fan-in) of three are available.

Since the circuit will first be implemented using NAND gates, it is necessary to obtain a function which is free of static 1-hazards. The first step in the design is to plot all those combinations of the variables which make the value of the function logical '1' on a K-map as shown in figure 11.10(*a*) and to simplify in the normal way. The plot is now examined to see if there are any 1's in adjacent cells not covered by the same prime implicant. In this case a pair of such cells are 0111 and 0110, and an additional prime implicant is added to the plot to remove

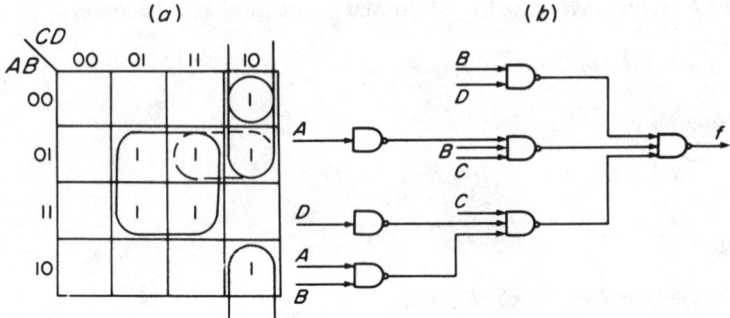

Figure 11.10. (a) Plot of $f = \Sigma\ 2, 5, 6, 7, 10, 13, 15$. (b) NAND hazard-free implementation

the uncovered adjacency. The 1's that constitute the additional prime implicant are enclosed by dotted lines on the K-map plot.

Reading from the map, the hazard-free function is

$$f = BD + \bar{A}C\bar{D} + \bar{B}C\bar{D} + \bar{A}BC$$

To meet the fan-in restriction, the equation can be factorised and then appears in the following form

$$f = C\bar{D}\,(\bar{A} + \bar{B}) + BD + \bar{A}BC$$

The factorisation of an equation in this way does not re-introduce hazards. In this problem the hazard would have occurred when $A = 0$, $B = 1$ and $C = 1$, with D changing from 1 to 0. Insertion of these conditions in the factorised equation gives

$$f = \bar{D}\,(1 + 0) + D + 1$$

$$= \bar{D} + D + 1$$

which is the required condition for the removal of the hazard.

The NAND implementation of the hazard-free function is shown in figure 11.10(b).

To obtain the hazard-free NOR realisation of the given function, it is first necessary to plot those combinations of the variables that make the value of the function logical '0' on a K-map, and then simplify in the normal way. The 0-plot, is, of course, a plot of the inverse function

Hazards

and is derived from figure 11.10(a) simply by marking the vacant cells on that map with 0's, as shown in figure 11.11(a).

The presence of 0's in adjacent cells not covered by the same prime implicant reveals that the simplified function will produce a static 0-hazard under certain prescribed conditions. In this example there are two such pairs of adjacent cells: (a) 0000 and 0001, and (b) 1000 and 1001. The introduction of an additional prime implicant $\overline{B}\overline{C}$ covers

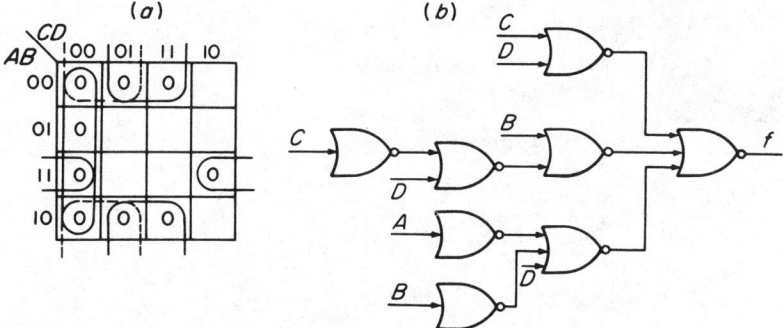

Figure 11.11. (a) The 0-plot of $f = \Sigma\ 2, 5, 6, 7, 10, 13, 15$. (b) NOR hazard-free implementation

these uncovered adjacencies and removes the static 0-hazard. The 0's constituting the additional prime implicant are enclosed by dotted lines in figure 11.11(a). Reading the inverse function from the map

$$\bar{f} = \overline{C}\overline{D} + \overline{B}D + \overline{B}\overline{C} + AB\overline{D}$$

and factorising, so that the fan-in restriction can be met, gives

$$\bar{f} = \overline{C}\overline{D} + \overline{B}(\overline{C} + D) + AB\overline{D}$$

Dualising,

$$f_d = (\overline{C} + \overline{D})(\overline{B} + \overline{C}D)(A + B + \overline{D})$$

and inverting,

$$f = (C + D)(B + C\overline{D})(\overline{A} + \overline{B} + D)$$

The implementation of this hazard-free function with NOR gates is shown in figure 11.11(b).

11.7 Detection of hazards in an existing network

The existing AND/OR circuit shown in figure 11.12(a) is to be analysed to see if it has any static hazards. Since the network consists of both AND and OR gates it may generate both static 0- and static 1-hazards. The equation of the network is

$$f = AB\bar{C} + (A+B)(\bar{A}+\bar{D})$$

and it may be expanded into the form

$$f = AB\bar{C} + A\bar{A} + A\bar{D} + \bar{A}B + B\bar{D}$$

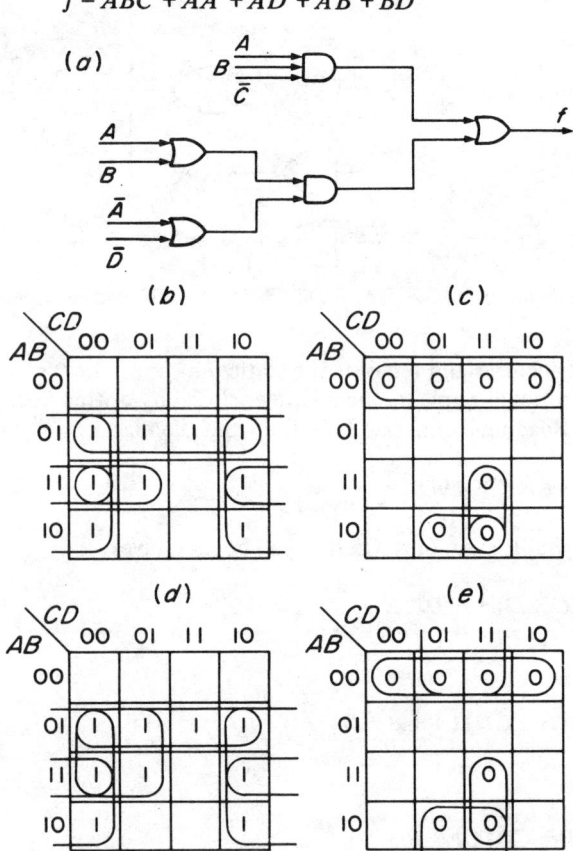

Figure 11.12. (a) AND/OR circuit for the function $f = AB\bar{C} + (A+B)(\bar{A}+\bar{D})$. (b) K-map plot of the function. (c) Plot of the inverse function. (d) The resimplification of the function. (e) The inverse plot of the resimplified function

Hazards

This expression contains the term $A\bar{A}$ which, under normal circumstances, would be removed since by the laws of Boolean algebra its value is 0. However, since A and $\bar{A}$, in combinational networks can be simultaneously logical '1', they are treated as independent variables in this equation which may be regarded as the equation which holds for transient conditions.

The hazards can be detected by examining the expanded equation to see whether it reduces to either $X\bar{X}$ or $X + \bar{X}$ under defined input conditions. For example, if $B = 0, C = 0$ or 1 and $D = 1$, the equation reduces to $f = A\bar{A}$. Hence for these input conditions, a static 0-hazard occurs when A is changing from 0 to 1.

When deriving the transient equation of a circuit, some of the theorems of Boolean algebra may not be used. Those which make use of the identities $A\bar{A} = 0$ and $A + \bar{A} = 1$ to reduce an equation may not be used to manipulate the circuit equation into its transient form. For example, the Boolean theorem $A + \bar{A}B = (A + \bar{A})(A + B) = A + B$ cannot be used since it relies on the identity $A + \bar{A} = 1$ to reduce the expression $A + \bar{A}B$ to the form $A + B$. Earlier in this chapter, it has been shown that A and $\bar{A}$ may be simultaneously equal to zero in which case $A + \bar{A} \neq 1$ and the theorem given above is no longer valid for all instants of time.

If $B = 1, C = 0$ and $D = 1$ the equation reduces to $f = A + A\bar{A} + \bar{A}$. For these conditions a static 1-hazard occurs when A is changing from 1 to 0. It should be noted that since A is changing from 1 to 0, $A\bar{A} = 0$, and it can only have a value of logical '1' when A is changing from 0 to 1.

If $B = 1, C = 0$ and $D = 0$ the equation reduces to $f = A + A\bar{A} + A + \bar{A} + 1$. In this case, irrespective of the instantaneous values of A and $\bar{A}$, $f = 1$, hence no hazard is generated.

Alternatively, the static 1-hazard can be detected by plotting those combinations of the variables that make the value of the function $f = 1$, as shown in figure 11.12(b). Examination of this map shows that the two 1's in the adjacent cells 1101 and 0101, respectively, are not covered by the same prime implicant. The introduction of an additonal prime implicant $B\bar{C}$ will ensure the coverage of these two cells by the same prime implicant and will remove the static 1-hazard.

To detect the static 0-hazard, the function has first to be inverted and then plotted on the K-map.

$$f = AB\bar{C} + (A + B)(\bar{A} + \bar{D})$$

Dualising,

$$f_d = AB + A\bar{A}D + \bar{A}B\bar{D} + AB\bar{C} + \bar{A}\bar{C}D$$

and inverting,

$$\bar{f} = \bar{A}\bar{B} + \bar{A}A\bar{D} + A\bar{B}D + \bar{A}\bar{B}C + ACD$$

and using the redundancy theorem this reduces to

$$\bar{f} = \bar{A}\bar{B} + A\bar{B}D + ACD + \bar{A}AD$$

The first three terms of the inverse function are plotted on the K-map shown in figure 11.2(c), while the fourth term (the transient term) cannot be represented on the map.

The map of the inverse function reveals that the 0's in cells 1001 and 1011 are adjacent to the 0's in the cells 0001 and 0011, respectively, and are not covered by the same prime implicant. By introducing the additional prime implicant $\bar{B}D$ to cover these four cells, the static 0-hazard is removed. This term is now added to the inverse equation which is re-inverted to form the hazard-free function.

The analysis of the circuit given in figure 11.12(a) reveals that it can generate both kinds of static hazard. In practice, it would be a more satisfactory solution to redesign the circuit using the K-map plot of figure 11.12(b) which, for convenience, is repeated in figure 11.12(d). On this map the function has been simplified in such a way that the resulting function is free of static 1-hazards. The hazard-free function is

$$f = \bar{A}B + B\bar{D} + A\bar{D} + B\bar{C}$$

All that remains to be done now is to examine the plot of the inverse function for static 0-hazards. The inverted function is

$$\bar{f} = \bar{A}\bar{B} + \bar{B}D + ACD$$

and is shown plotted in figure 11.12(e). Since there are no adjacent 0's not under the same prime implicant there are no static 0-hazards present, and the function f is also free of this type of hazard.

11.8 Dynamic hazards

A second type of hazard that can occur in combinational networks is called a dynamic hazard, which occurs when the network output is supposed to change from 0 to 1 or, alternatively, from 1 to 0. If the output has been designed to change from 1 to 0, but in practice changes from 1→0→1→0, then a dynamic hazard is present in the circuit. Similarly, an output required to change from 0→1 would, in the presence of a dynamic hazard, change from 0→1→0→1. In either case there is a minimum of three changes appearing at the output, as illustrated in figure 11.13.

Figure 11.13. Dynamic hazards

Dynamic hazards may occur in those circuits where there are three or more signal paths for the same variable, each path having different gate delay times. For example, the function

$$f = (AC + B\overline{C})(\overline{A} + C)$$

may be implemented with AND and OR gates, as shown in figure 11.14(a). Inspection of the circuit reveals that there are three different paths through the network for the variable C and, consequently, the possibility of the occurrence of a dynamic hazard exists. The three paths through the network are

(a) via gates g_1 and g_2,
(b) via gates g_3, g_5 and g_2, and
(c) via gates g_4, g_5 and g_2.

The eight possible combinations of the variables A, B and C are tabulated in figure 11.14(b). For each of these combinations it is assumed that the next change to occur will be in the variable C, as indicated in the fourth column of the table. The presence of a dynamic hazard is recorded in the last column of the table.

For the first four combinations in the table, $\overline{A} = 1$, and this signal is one of the inputs to the OR gate g_1. The output of this gate will therefore remain at 1, irrespective of any change that may take place in C, and consequently the output of the gate g_2 will also remain unchanged. Since for these four combinations a change in C at the input of g_1 does not produce a change in the output of g_2, a dynamic hazard cannot be generated.

However, in the case of $A = 1$, $B = 1$ and $C = 0$, with C changing from 0 to 1, the upper input of g_2 changes from 0 to 1. The other input to g_2 is $AC + B\overline{C} = 1$, and consequently the output of g_2 also changes

(a)

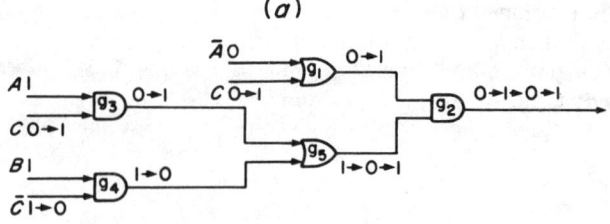

(b)

A	B	C	Change in C	Dynamic hazard
0	0	0	0→1	No
0	0	1	1→0	No
0	1	0	0→1	No
0	1	1	1→0	No
1	0	0	0→1	No
1	0	1	1→0	No
1	1	0	0→1	Yes if $t_{g4} < t_{g3}$
1	1	1	1→0	No

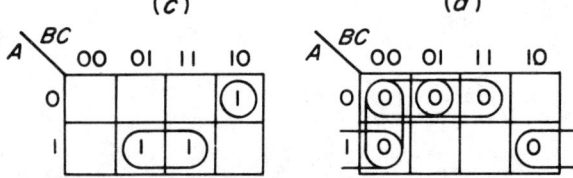

Figure 11.14. (a) Network with a dynamic hazard. (b) Occurrence of a dynamic hazard in the network. (c) and (d) K-map plots for determining the presence of static 1- and static 0-hazards

from 0 to 1. Assuming that $t_{g4} < t_{g3}$, the effect of the change of $\bar{C}$ from 1 to 0 at the input of g_4 must next be traced. The effect of this change is that the output of g_4 changes from 1 to 0, followed by the output of g_5 changing from 1 to 0, and the output of g_2 now makes a second change from 1 to 0. Finally, the effect of a change of C from 0 to 1 at the input g_3 must be traced, and it will be found that this leads to a third change in the output g_2 from 0 to 1. These changes are illustrated on the network diagram in figure 11.14(a), and it is clear that under the prescribed input conditions a dynamic hazard will occur at the output when C changes from 0 to 1.

A similar analysis can be made for all the other cases in the table of figure 11.14(b), and it will be found that there are no other dynamic hazards present in this network.

If the function implemented by this network is plotted on a K-map as shown in figure 11.14(c), it will be observed that there are no static 1-hazards. Similarly, a plot of the inverse function (figure 11.14d) reveals that there are no static 0-hazards either.

11.9 Essential hazards

This type of hazard is peculiar to event-driven circuits and is caused by a race between a primary and a secondary signal. The state diagram for an event-driven digital machine having a race-free secondary assignment is shown in figure 11.15. Assuming that the machine is in state S_0 and a change in the value of X from 0 to 1 occurs, a transition from S_0 to S_1 should take place and, on arriving in S_1, the machine should remain in that state. However, correct operation of the machine as described above will depend upon the relative values of the inversion time t_i for the primary signal X and the turn-on time t_t for the secondary signal B.

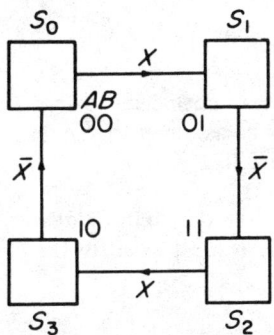

Figure 11.5. State diagram for a machine which can have an essential hazard

If the machine arrives in state S_1 before the value of $\bar{X}$ has changed from 1 to 0, a further transition to S_2 will be made. Since $X = 1$ when the machine arrives in state S_2, it follows that a further transition will take place to state S_3, where the machine will now remain, provided that the change in $\bar{X}$ has now occurred. Hence, if $t_i > t_t$, incorrect circuit operation will occur as a consequence of the race between the inversion of the primary signal X and the turn-on of the secondary signal B.

An examination of the equation for the secondary signal A reveals more clearly the origin of the hazard. The turn-on set of $A = B\bar{X}$, the turn-off set of $A = \bar{B}\bar{X}$, and

$$A = B\bar{X} + (\overline{\bar{B}\bar{X}})A$$

$$= B\bar{X} + (B + X)A$$

The first term of this equation provides the turn-on signal for A when the machine is in state S_1. If B changes to 1 before $\bar{X}$ has changed to 0 the value of $B\bar{X} = 1$ and the secondary variable A is turned on.

The method of dealing with this type of hazard is to insert a delay in the output line of the circuit generating the secondary variable B. This will ensure that the change in B does not arrive at the input to the circuit generating the secondary variable A until the value of $\bar{X}$ has changed.

Problems

11.1 Plot the K-map of the functions

(a) $f(A, B, C, D) = \Sigma\ 0, 2, 4, 5, 6, 8, 9, 11, 12, 14, 15$, and
(b) $f(A, B, C, D) = \Sigma\ 3, 4, 5, 6, 11, 12, 13, 14, 15$

and determine hazard-free implementations in both cases, using (i) NAND gates, and (ii) NOR gates. It can be assumed that gates having a maximum fan-in of three are available.

11.2 Find all the static hazards in the two networks shown in figures P11.2(a, b). Specify the input conditions that must exist for the

(a)

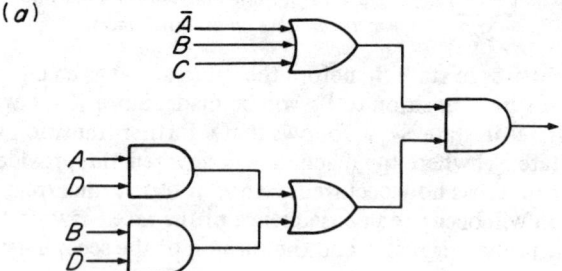

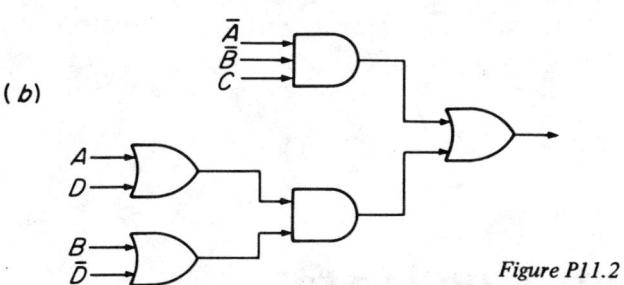

(b)

Figure P11.2

hazards to occur and draw the logic diagram for modified networks that are hazard-free.

11.3 With the aid of a K-map, explain why the function

$$f = (\bar{A} + \bar{B})(B + C)$$

has a static 0-hazard for an input change of $ABC = 100$ to $ABC = 110$, but not for an input change of $ABC = 100$ to $ABC = 000$.

11.4 Examine the circuits shown in figure P11.4 for dynamic hazards. What are the values of the input variables before and after the occurrence of the hazards? Specify the order in which the circuits must switch in order for the hazards to occur.

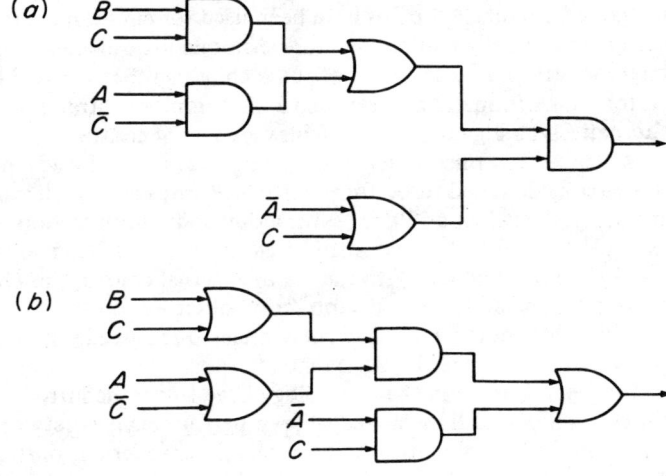

Figure P11.4

12

An introduction to microprocessors

12.1 Introduction

The preceding chapters have dealt with the methods that are used by digital engineers for designing digital circuits. The design techniques used require a knowledge of Boolean algebra and its application to combinational and sequential logic problems.

In the early days of digital design, engineers used discrete gates and flip-flops to implement their circuits. However, with the advent of integrated circuits, SSI chips have been used for circuit implementation, each chip generally containing more than one logic element, and an interconnection of a selection of these chips can then be used to perform an arithmetic process such as addition or subtraction, or alternatively, a logical process such as sequential control.

Advances in semiconductor technology have seen the advent of MSI circuits designed to perform particular processes which might be arithmetical, such as addition, subtraction and multiplication. A typical example of an MSI chip was introduced in Chapter 10, namely the SN7483 four-bit adder. With the aid of external control this chip can be used for the addition and subtraction of either four-bit binary or four-bit NBCD numbers, and when connected in cascade the addition and subtraction range can be extended.

Further advances in the technology have led to the introduction of LSI chips. Some of these are capable of performing a variety of functions, both arithmetical and logical, under external control. Such a chip contains an arithmetic/logic unit which implements the selected

An introduction to microprocessors

function. Control of the arithmetic/logic unit is by a programme stored in memory which allows it to perform a desired sequence of arithmetical and logical processes under programme control. This type of LSI chip is called a microprocessor and it usually forms the central processing unit (CPU) of a microcomputer system.

In this chapter the methods developed earlier are used to design the control logic for a sequentially operated multiplier circuit. The synchronous sequential machine thus obtained is converted to a dedicated microprocessor by the introduction of memory and additional registers. Finally, some of the basic features of contemporary microprocessors are introduced and incorporated into the basic machine to form a more sophisticated device.

12.2 Binary multiplication

The 'pencil and paper' method for multiplying together two binary numbers is illustrated in the example shown below.

Multiplicand	1110	14
Multiplier	1010	10
Partial product 1	0000	
Partial product 2	1110	
Partial product 3	0000	
Partial product 4	1110	
Product	10001100	140

There are three main features to the process:

(1) If the multiplier bit is 1, then a partial product is formed by writing down the multiplicand. Alternatively, if the multiplier bit is 0, then the partial product is formed by writing down a row of 0's.

(2) Four partial products are formed, one for each bit of the multiplier, and they are all added together to form the final product.

(3) As the multiplication progresses from the least to the most significant bit of the multiplier, each succeeding partial product is shifted one place to the left.

To implement binary multiplication using a digital machine, two processes introduced in previous chapters have to be performed, namely, addition and shifting. Addition can be carried out using an adder circuit, whilst the shifting process can be achieved by clocking a shift register. A multiplier designed on the basis of these two processes is called a shift and add multiplier.

There would, however, be one change to the pencil and paper method in the machine implementation. The above example shows that all the partial products are formed before their addition takes place to produce the final product. In a machine this would require four registers, one for each partial product, and clearly this would increase as the number of multiplier bits increases. From the hardware point of view this would be extremely uneconomic, and in practice the addition takes place each time the multiplicand appears as a partial product, that is, every time the multiplier bit is 1.

12.3 The hardware requirements of a binary multiplier

An examination of the multiplication of two four-bit numbers carried out in § 12.2 indicates the following preliminary list of hardware requirements:

(1) a four-bit register for the multiplicand;
(2) a four-bit register for the multiplier;
(3) a double-length eight-bit register for the product;
(4) an arithmetic unit that will add, and
(5) control logic for controlling the shift and add operations.

In practice, this preliminary list is more than is required. A little thought by the reader will reveal that initially, the double-length register for the product contains no data at all, and consequently it would seem reasonable to use a portion of this register for holding the multiplicand on a temporary basis. As the multiplication progresses and successive bits of the multiplicand are used, they are moved out of the product register one bit at a time, thus leaving space available for the accumulation of the partial products. For obvious reasons the portion of the double-length register used for the accumulation of the partial products is called the accumulator.

12.4 The binary multiplier

A basic diagram for the machine is shown in figure 12.1(a). It consists of the four-bit multiplicand register, a four-bit adder, a nine-bit product register, and a box labelled control logic. The additional bit is required in the product register to hold any carry that may be generated when two four-bit numbers are summed in the four-bit adder.

The control logic box has three functions: (i) it must examine the multiplier bit to determine whether it is 0 or 1; (ii) and (iii) it must generate the shift and add signals. It will also be noticed that the control logic is supplied with clock and 'start' signals for synchronising and starting the multiplier operation.

If the multiplier bit is 1, an add signal A is generated by the control logic. The contents of the multiplicand register and the least four significant bits of the accumulator are then fed to the four-bit adder where they are added and returned to the accumulator. A shift pulse S is now generated and the data stored in the product register are shifted one place to the right, thus moving the least significant bit of the multiplier out of the product register and replacing it with the next most significant bit. In the event of the multiplier bit being zero, only a shift pulse is generated, and no addition takes place.

Comparing the machine operation with the paper and pencil multiplication, it will be noticed that on paper, the multiplicand shifts left and the multiplier remains in a fixed position relative to the product. On the other hand, in the machine implementation, the multiplier and the accumulating product move right relative to the multiplicand, which remains in a fixed position.

A suitable internal state diagram for the machine is shown in figure 12.1(b). Once a start signal X is received, an add or shift pulse is generated in state S_0, depending upon the value of the multiplier bit M. Assuming that $M = 1$, an add pulse is generated and on the trailing edge of the next clock pulse to be received, a transition will be made to S_1. Alternatively, if $M = 0$, a shift pulse is generated and a transition is made to S_2 on the trailing edge of the next clock pulse. The outer square of the state diagram may be regarded as the shift and add sequence and a transition path round this square will be traversed if all four multiplier bits are 1's. Alternatively, if all four multiplier bits are 0's, then a transition path will be traced round the inner diamond of the state diagram.

Since the machine has eight internal states, three flip-flops (P, Q and R) are required for its implementation. The input equations for the JK flip-flops are derived in the way described in Chapter 8:

$$J_P = QR + Q\bar{M} \qquad K_P = QR + Q\bar{M}$$
$$J_Q = R + \bar{M}X + \bar{M}P \qquad K_Q = R + \bar{M}$$
$$J_R = M(X + P + Q) \qquad K_R = 1$$

(a)

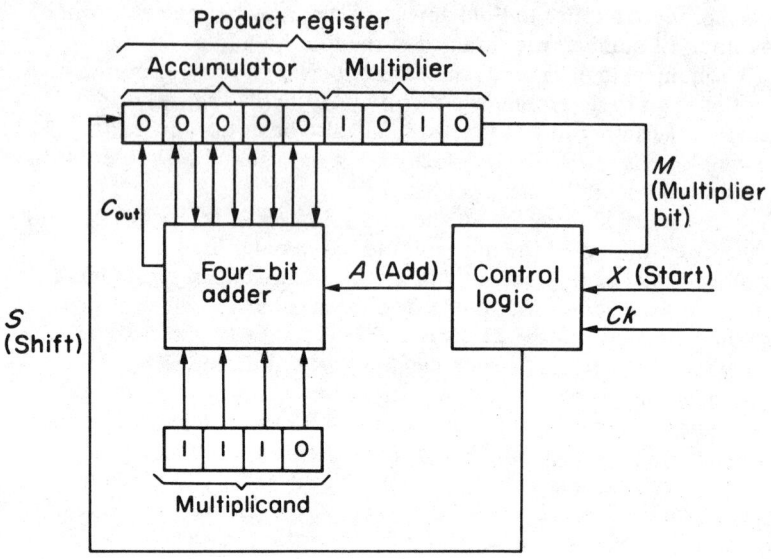

(b)

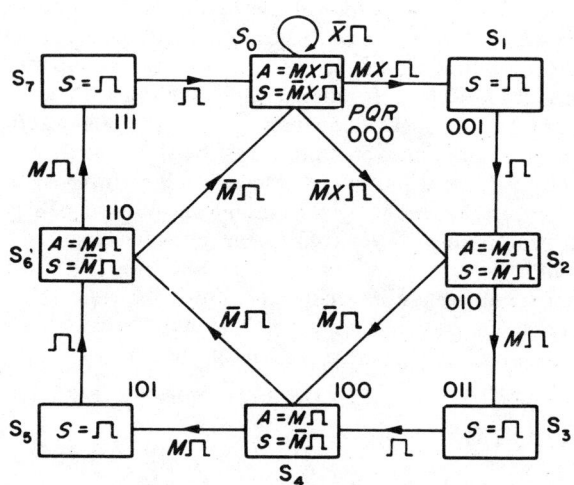

(c)

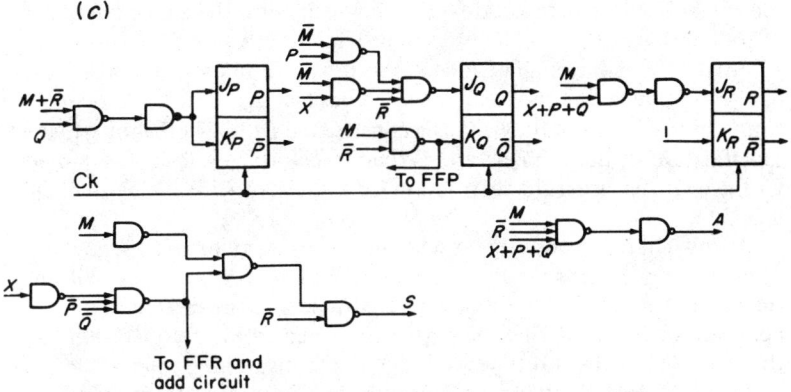

Figure 12.1. (a) The binary multiplier. (b) Internal state diagram. (c) Implementation

and the output equations are

$$S = \bar{M}(X + P + Q) + R$$

$$A = M\bar{R}(X + P + Q)$$

The implementation of the machine is shown in figure 12.1(c).

The binary multiplier designed in this section consists of three basic items.

(1) working registers,
(2) an arithmetic unit,
(3) a control unit.

In practice, the above items constitute the three main elements of a microprocessor. To convert the binary multiplier into a dedicated microprocessor, some way must be found to bring data to the product and multiplicand registers and to supply shift and add instructions. The method used is to store instructions in consecutive memory locations in the correct order, and to access the memory sequentially by means of a device called a programme counter.

12.5 Flow charts

The basic design tool for the engineer who designs digital circuits is the state diagram. Given a verbal or written specification of the problem, the engineer constructs a state diagram which, with the aid of a number of well defined rules, can be used to provide a hardware solution. Alternatively, if the engineer decides to provide a microprocessor-based solution to the problem, then the basic design tool is the flow chart. In this case the flow chart is constructed from a verbal or written specification of the problem and is then transformed into a programme by the engineer using the instruction set associated with the chosen machine.

A flow chart is a convenient way of specifying an ordered sequence of operations. It consists of blocks connected by directed lines. Within the blocks specified operations are performed and the directed lines between the blocks define the path to be taken from one operation to the next. There are two types of block in the diagram: (i) the rectangular block in which operations to be performed are specified, and (ii) diamond-shaped boxes where decisions are made based on conditions which are specified within the box.

In effect, the flow chart is analogous to the state diagram, where each state is analogous to a rectangular block on a flow chart. It is the place where a specified operation is performed. For example, an examination of the state diagram in figure 12.1(b) indicates that the shift and add operations are allocated to the states. In S_1 a shift signal is generated, whereas in S_2 either an add signal or a shift signal is generated, depending upon the value of M. The logic signals allocated to the transition paths in the state diagram are analogous to the contents of the decision boxes on the flow chart. For example, when in S_2, the transition path selected depends upon the value of M, that is, whether the multiplier bit is 0 or 1. For $M = 1$, a transition is made to S_3 and for $M = 0$ a transition is made to S_4.

A typical example of a flow chart is shown in figure 12.2. The flow chart chosen is one which describes the shift and add multiplication process. The first rectangular block A specifies the initial conditions that must be set up before the programme can be executed. In this block the contents of the product register are set to zero, and a count of 4, which equals the number of shifts to be performed, is set up in another register. Additionally, the multiplicand and the multiplier are stored in memory locations 50 and 51, respectively.

Block B in figure 12.2 is a decision box from which there are two exits. If the multiplier bit is 1 the decision is 'yes' and the next operation appears in the rectangular box C, where the add operation takes place, the multiplicand being added to the contents of the product register to form the next partial product. Alternatively, if the multiplier bit is 0, the decision taken is 'no', and a jump is inserted

An introduction to microprocessors

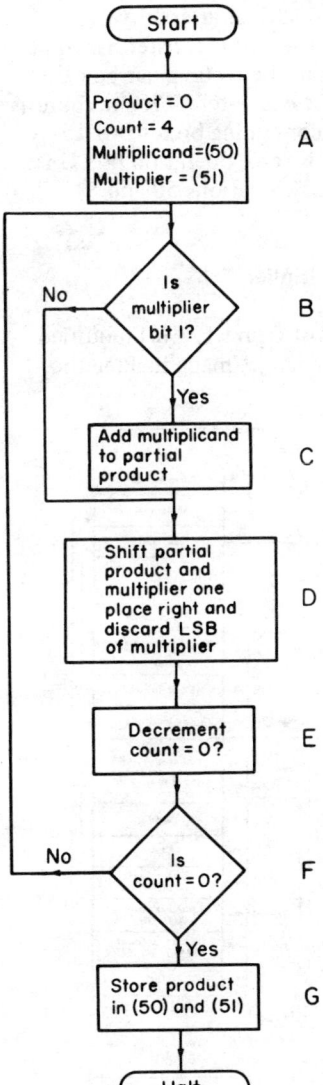

Figure 12.2. Flow chart for multiplication

which leads to the omission of the addition box C. The two paths rejoin one another and enter the operation box D, where the shift process takes place.

Having now performed either a shift and add, or alternatively, a shift-only operation, the number stored in the count register has to be decremented. This operation is carried out in the rectangular box E. After decrementing the count, decision box F is entered. If the count is not zero, a jump is executed and the loop containing boxes BCDEF is traversed again. Alternatively, if the count is zero, operation box G is entered, and the product is stored in memory locations 50 and 51.

12.6 The programme-controlled binary multiplier

Figure 12.3(a) shows the binary multiplier of figure 12.1(a) modified to operate under programme control. The changes made include the

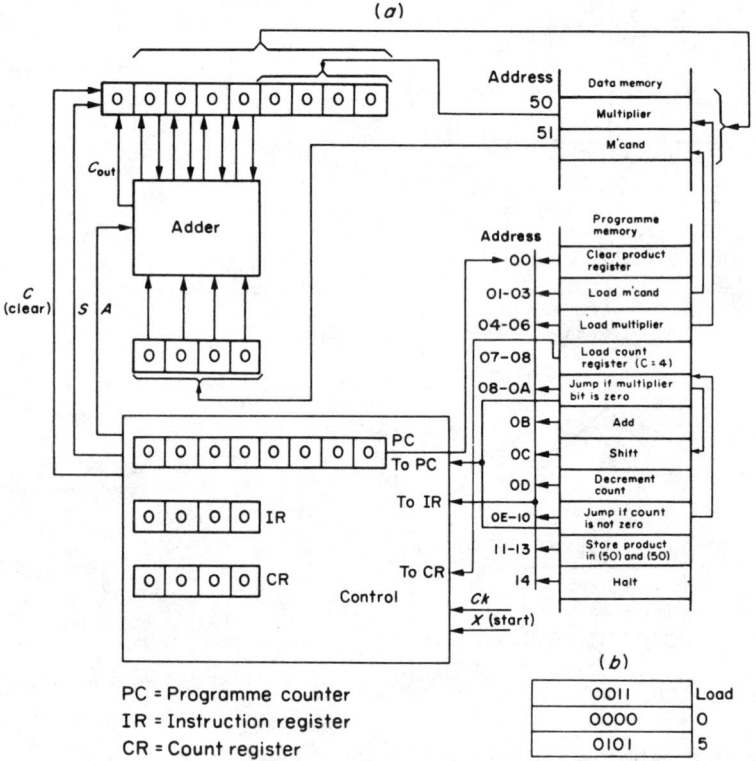

Figure 12.3. (a) The programme-controlled multiplier. (b) The 'load' instruction. (PC = programme counter, IR = instruction register, CR = count register.)

introduction of memory, a programme counter, an instruction register and a count register. The last two items, operated in conjunction with additional logic and directed by the programme, constitute the control of the multiplier.

The memory is divided into two distinct sections. First, there is the data section which, in this example, holds the multiplier and multiplicand in memory locations 50 and 51, respectively. This type of information is stored in a random access memory (RAM), which is used in microprocessor systems to store data which may change during the implementation of a programme. In this case, for example, the multiplier and the multiplicand are displaced from locations 50 and 51, at the end of the programme, by the final product. The second memory section stores the programme steps and also data constants. Generally speaking, this type of information is stored in ROM. In practice, ROM is used where the information stored is not changing and this infers a dedicated system.

The programme written into the memory follows closely the flow chart of figure 12.2. Memory locations 00—08 represent the initialisation box A. The first decision is taken in memory locations 08—0A and this corresponds to decision box B on the flow chart. The next three instructions (add (0B), shift (0C), and decrement (0D)), correspond to boxes C, D and E on the flow chart. Memory locations 0E—10 represent the decision box F, and the operation of storing the product, specified in box G, corresponds to the store instruction in memory locations 11—13. Finally, the machine is halted by the instruction appearing in memory location 14.

12.7 Word length

The number of binary digits which are used to handle both instructions and data is called the word length of the machine. For the multiplier shown in figure 12.2, the word length is four bits, selected in this case because the machine is designed to multiply two four-bit words. Consequently, each memory location specified in figure 12.3 stores a four-bit word; for example, location 50 stores the multiplier (1010) and location 51, the multiplicand (1110). Since the final product contains a maximum of eight bits, two memory locations, 50 and 51, are needed to store it.

The choice of word length also governs the number of instructions available. If the machine instructions consist of four bits only, then the total number available is $2^4 = 16$, and this imposes a very severe limitation to the instruction set of the machine. However, it should be recognised that the instruction set can be extended by forming each instruction from two consecutive words in memory. With this modification, the machine can have a maximum number of instructions equal to 256. In practice, microprocessors have been designed with word lengths of four, eight and twelve bits, and recently machines having a 16-bit word length have been introduced.

12.8 The programme counter

The programme counter is an eight-bit register which is situated in the control section of the machine. Its function is to access sequentially the memory locations of the programme unless it is instructed to do otherwise. The register is eight bits wide, and hence it follows that all memory addresses also consist of eight bits. In practice, it is usual to express memory addresses in hexadecimal notation rather than in terms of binary digits. For this reason the addresses appearing in figure 12.3 are all specified by two hexadecimal digits.

On switching on, it must be arranged that the programme counter holds the starting address of the programme (i.e. 00H, where H indicates that the address is expressed in terms of hexadecimal digits). When the instruction has been fetched from memory to the control section of the machine, the programme counter is incremented by one.

Some of the instructions, such as 'clear product register' are represented by one four-bit word in memory; however, it will be noticed that the next instruction 'load multiplicand' consists of three consecutive four-bit words. In such cases the instruction is dealt with a word at a time, and the programme counter is incremented by one as each successive word is handled by the machine. To execute the programme it is only necessary to set the programme counter to the starting address 00H and inject an execute signal.

In certain circumstances, the programme counter may not be incremented by one, but will be directed to a non-sequential memory location for the next instruction. For example, if on arriving at memory location 08H in figure 12.3 the multiplier bit is found to be zero, there will be a jump to the location 0CH which contains the 'shift' instruction. In this case the 'jump' instruction will contain the non-sequential address 0CH which will be loaded into the programme counter as the 'jump' instruction is executed.

12.9 Instructions and the instruction register

The instruction register for the simple machine shown in figure 12.3 is situated in the section labelled control and is four-bits wide. The first four-bit word of every instruction in the programme memory is taken to the instruction register, where it is decoded and where it stays until the instruction has been executed. The instruction 'load', for example, may be represented by the four-bit word (0011B) (3H); when the control section receives the word it decodes it and recognises it as the instruction 'load'. It will be observed in figure 12.3 that every instruction is brought out from the programme memory and is then transferred to the instruction register.

A typical example of an instruction is shown in figure 12.3(b). The 'load multipler' instruction consists of three four-bit words. The first four-bit word is the actual instruction which is transferred from memory to the instruction register, while the second two words give the address in the data memory where the multiplier is to be found.

The instructions in the programme memory of this machine can be placed in three separate categories:

(1) transfer,
(2) control,
(3) operational.

For this machine there are four instructions concerned with the transfer of data:

(1) load multiplier,
(2) load multiplicand,
(3) store product,
(4) load count register.

The first three of these instructions are three-word instructions, whilst the last one in the group only requires two words.

In the case of the 'load multiplicand' instruction, the first of the three words, stored in memory location 01H, represents the load instruction and this is fetched from memory to the instruction register in the control section of the machine where it is decoded and executed. The other two words stored in memory locations 02H and 03H, are the address in data memory where the multiplicand is to be found. An interpretation of the instruction, in this case is 'load the data stored in memory location 51H into the multiplicand register'.

The 'load multiplier' instruction can be described in the same way as the 'load multiplicand' instruction; both of them are concerned with the transfer of data from memory to a register.

Similarly, the 'store' instruction is concerned with a transfer of data within the confines of the machine. In this case, the instruction is used to store the contents of the product register in locations 50H and 51H. When the instruction is accessed at location 11H, the second and third words provide the address of the memory location 50H into which the four least significant bits of the product are stored. The four most significant bits are then automatically written into the next sequential memory location, namely 51H.

One other instruction is concerned with the transfer of data and that is the 'load count' instruction. This is a two-word instruction, the first word being the instruction that is decoded in the control section, whilst the second word is the numerical value of the count which has to be loaded into the count register, and in this example has a value of four (0100B).

Two of the instructions in the programme memory can be allocated to the category-designated control:

(1) jump,
(2) halt.

For the 'jump' instruction three four-bit words are required, the first being the actual instruction which is transferred to the control section for decoding (08H). The second and third words of the instruction (09H and 0AH) give the address in programme memory to which the jump has to be made (0CH). These two words are also taken to the control section where they are loaded into the programme counter to implement the address jump.

The 'halt' instruction is a single-word instruction used in this case to stop the machine when the programme has been executed. External action is then required to restart the machine.

The four remaining instructions are concerned with operation:

(1) decrement,
(2) clear,
(3) shift,
(4) add.

In the case of the first three, the operations are on a particular register and are not concerned with the transfer of data. For example, the

An introduction to microprocessors

'clear' instruction is used to insert 0's into the product register, the 'decrement' instruction simply decrements the count stored in the count register by one, and the 'shift' instruction moves the contents of the product register one place to the right. The last of this group of instructions 'add' is associated with operations on data. Such operations are performed by the arithmetic/logic unit of the machine. For the machine shown in figure 12.3 the arithmetic/logic unit can only provide the single facility of addition. More sophisticated machines have ALUs that provide additional facilities such as, subtract, compare, AND, OR, and exclusive-OR, etc.

It will have been observed by the reader that the instructions add and shift represent the core of the multiplication process. These two instructions each require a single line of programme. However, an examination of figure 12.3 reveals that the whole programme consists of 22 lines. Consequently 20 lines of the programme are used to organise the machine so that the two crucial instructions, shift and add, can be implemented.

12.10 The hexadecimal number system

The addresses of the memory locations in figure 12.3 have been expressed in terms of hexadecimal digits. It is also normal practice to represent the words stored in memory locations in the same way. Although the microprocessor, as a working machine, receives its instructions as binary digits and also processes data expressed in binary form, it happens that hexadecimal notation is a very convenient shorthand method of representing binary numbers.

HD	Decimal	HD	Decimal
0	0	8	8
1	1	9	9
2	2	A	10
3	3	B	11
4	4	C	12
5	5	D	13
6	6	E	14
7	7	F	15

Figure 12.4. The hexadecimal number system with its corresponding decimal representation

The hexadecimal number system has 16 digits. This, of course, raises problems since there are only ten numerical digits and consequently six additional digits have to be invented to represent the decimal numbers from 10 to 15. It is normal practice to use the first six letters of the alphabet, A, B, C, D, E and F, to represent these numbers.

A tabulation of the hexadecimal digits with their corresponding decimal representations is shown in figure 12.4. A number system which has both numerical and alphabetical representation of digits is called an alphanumeric system.

The base of the hexadecimal system is 16, and a hexadecimal number such as 8AC2 is converted to its decimal representation using the formula given below:

$$(N)_{10} = a_n b^n + a_{n-1} b^{n-1} \ldots a_1 b^1 + a_0 b^0$$

where b represents the base of the number system, a represents the hexadecimal digit, and $n + 1$ represents the number of integral digits. Hence

$$(N)_{10} = 8 \times 16^3 + A \times 16^2 + C \times 16^1 + 2 \times 16^0$$
$$= 8 \times 16^3 + 10 \times 16^2 + 12 \times 16^1 + 2 \times 16^0$$
$$(N)_{10} = 35\,522$$

There are 16 combinations of four binary digits, and these combinations can be used to represent the 16 hexadecimal digits. Each one of the four-bit combinations is allocated to a hexadecimal digit as indicated in the tabulation shown in figure 12.5. With the aid of this conversion table a hexadecimal number can be readily transformed to its binary equivalent, as illustrated below.

2	A	C	8	Hexadecimal
becomes				
0010	1010	1100	1000	Binary

Binary	HD	Binary	HD	Binary	HD	Binary	HD
0000	0	0100	4	1000	8	1100	C
0001	1	0101	5	1001	9	1101	D
0010	2	0110	6	1010	A	1110	E
0011	3	0111	7	1011	B	1111	F

Figure 12.5. *Binary/hexadecimal equivalences*

An introduction to microprocessors

It is apparent that a four-digit hexadecimal number transforms to a 16-bit binary number.

In the elementary machine of figure 12.3 memory addresses are represented by two hexadecimal digits which, when converted, become eight binary digits. For example, the multiplicand is stored in memory location 51H = 01010001B. This binary representation is referred to as the machine code for the address. It also follows that the four-bit word stored in a particular memory location can be represented by one hexadecimal digit.

Converse transformations are also readily performed. For example,

1011	0010	1001	1010	Binary

becomes

| B | 2 | 9 | A | Hexadecimal |

12.11 Comparison of the simple machine with a practical microprocessor

In practice, a microprocessor is a much more sophisticated device than the programme-controlled multiplier discussed in § 12.6. The two machines differ in the following important respects.

(1) A microprocessor has more working registers. For example, the Intel 8085 has five 16-bit registers and two eight-bit registers.

(2) The adder would be replaced by an arithmetic/logic unit in the microprocessor and this would provide more arithmetical and logical facilities such as AND, OR and exclusive-OR for logical operations, and complement, subtract, increment and decimal addition for arithmetical operations. Facilities are also often provided for binary to BCD conversion.

(3) The word length of the most popular microprocessors in common use today is either 8 or 16 bits.

(4) In machines such as the 8085, the registers, ALU, control and timing, instruction decoder and clock generator are all situated on one 40-pin chip.

(5) Memory will be either ROM, RAM or a mixture of both. In the latter case, separate ROM and RAM chips are required.

(6) A microprocessor will normally have input/output (I/O) ports so that access to and from the processor is possible. For example, in

the SDK85 system, the *I/O* ports are an integral part of the associated ROM and RAM chips.

(7) Interconnection between elements in a microprocessor system is normally via a common bus system.

12.12 A general block diagram for a microprocessor system

A typical microprocessor system is identified by the four blocks A, B, C and D shown in figure 12.6.

(1) Block A represents the CPU. This contains the ALU, working registers, instruction decoder and control and timing.

(2) Block B represents the RAM. This stores variable data and results.

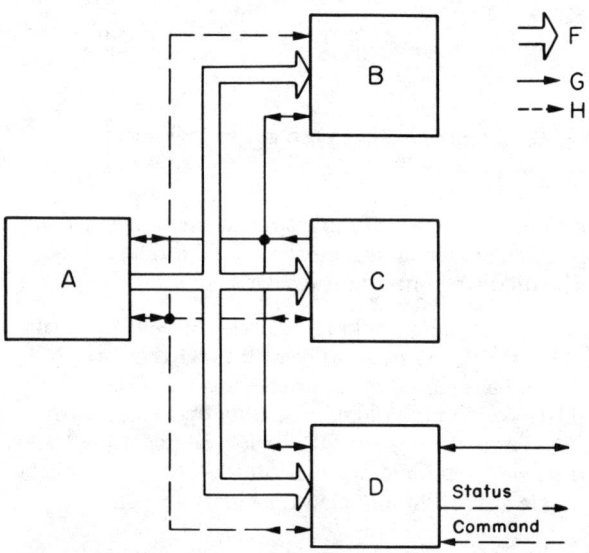

Figure 12.6. Microprocessor block diagram

(3) Block C represents the ROM. This stores the programme of instructions and data constants.

(4) Block D represents the *I/O* section. This may be a simple *I/O* port or a more complex *I/O* controller, and it represents the interface between the processor and the outside world. This block allows peripherals such as *A/D* and *D/A* converter, teletypes, printers, floppy disc controllers and DMA (direct memory access) controllers to

communicate with the microprocessor system. The function of the interface is to coordinate the activities of peripheral and microprocessor, and this it does by monitoring its status signals and, in response, generating command signals in their correct sequence.

The interconnecting lines between the blocks represent the bus system which is identified by three separate parts:

(1) F — the unidirectional address bus, commonly 16 lines wide,

(2) G — the bi-directional data bus, commonly eight lines wide. In some machines such as the Intel 8085 the eight data lines are time-multiplexed with the eight least significant address lines.

(3) H — the control bus, which carries control signals generated by the CPU and distributes them throughout the system. However, in certain cases, control signals may be generated by a peripheral, for example the 'ready' signal generated by the 8355 ROM in the Intel 8085 system.

To complete this section a functional block diagram of the 8085 CPU is shown in figure 12.7. It contains all of the features described above, such as the ALU and programme counter. Additionally, it will be noticed that there is a register called the accumulator. This is the most important of all the working registers. The results of arithmetical and logical operations in the ALU are stored in this register, and it also serves as one input to the ALU.

The timing and control section is the source of control for the whole of the system. However, in two specific cases control signals are received by this section from external sources. For example, the 'hold' signal is generated by a DMA controller and indicates to the CPU that a peripheral such as a floppy disc wishes to transfer a block of data to memory or, alternatively, receive a block of data from memory. The other signal received by the control section is the 'ready' signal referred to above.

One of the working registers is called a stack pointer. This is associated with the stack, which is either an array of registers or a section of memory set aside by the programmer to allow data or addresses to be accessed from the top of the stack on a last-in, first-out (LIFO) basis. Stack operation in the 8085 is controlled by two instructions: 'push' and 'pop'. The 'push' instruction stores data or addresses on the top of the stack, and the 'pop' instruction implements the reverse process. The stack pointer is a register which holds the address of the top element of the stack. Typically, the stack is used in microprocessor systems to store the machine status during sub-routines and interrupts.

298 An introduction to microprocessors

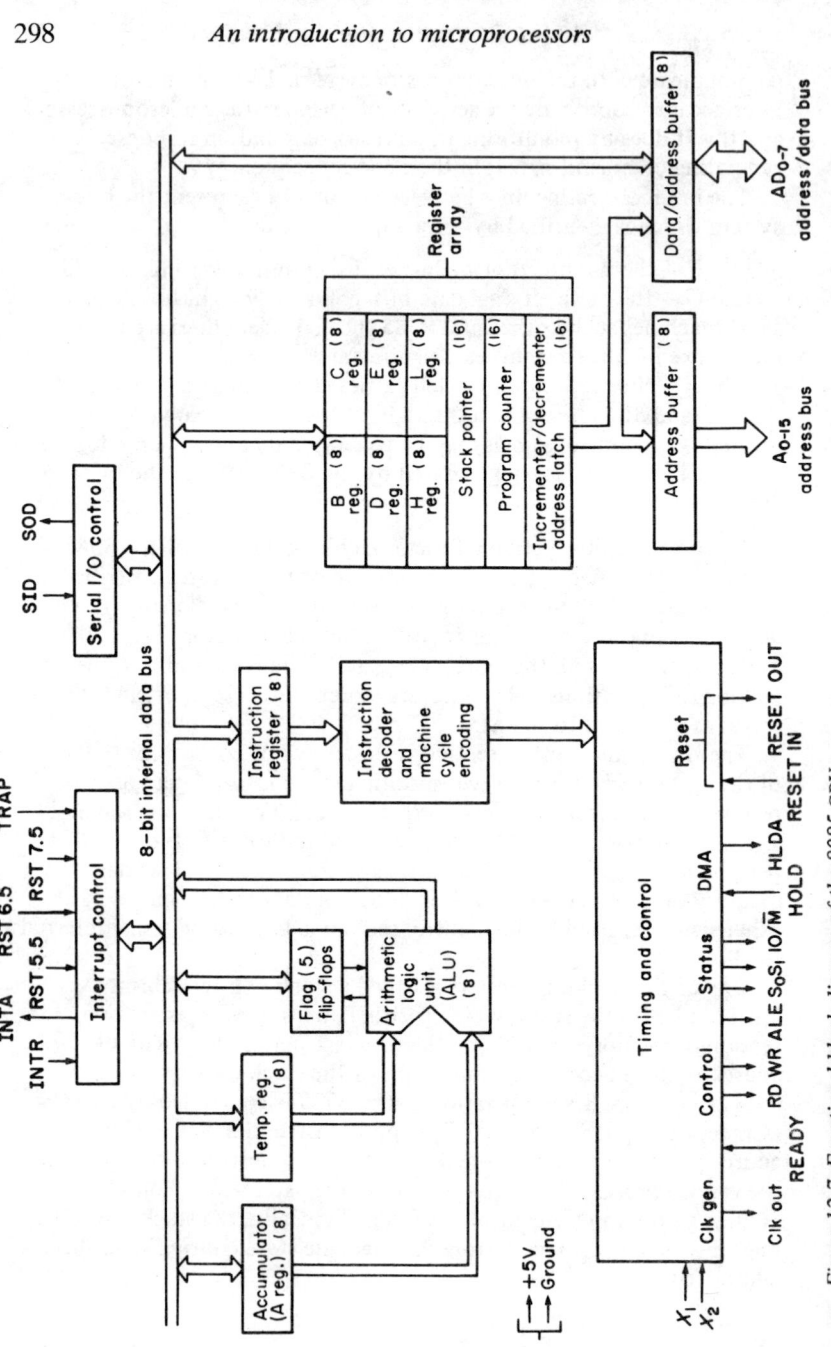

Figure 12.7. Functional block diagram of the 8085 CPU

An introduction to microprocessors 299

There is one other important register on the 8085 CPU block diagram which is designated flag flip-flops. In essence, this is a five-bit register which holds information concerning the accumulator contents. The five flags are:

(1) *Carry*. When this flag is set it indicates that there has been a carry-out of the most significant bit of the accumulator after an arithmetic operation has been executed.

(2) *Zero*. The flag is set when all eight bits in the accumulator are zero.

(3) *Parity*. If the accumulator contains an even number of 1's the flag is set, otherwise it is zero.

(4) *Sign*. This flag is set to the condition of the most significant bit in the accumulator. If this is 1 the sign flag is set and the accumulator holds a negative number.

(5) *Auxiliary carry*. If there is a carry from bit 3 to bit 4 in the accumulator the flag is set. It is commonly used in NBCD arithmetic processes.

12.13 Programming a microprocessor

A microprocessor is a machine that processes binary data with the aid of instructions that are supplied to the machine in binary form. For an eight-bit machine, each instruction is supplied as a string of eight binary digits, and the maximum number of instructions available is $2^8 = 256$. An instruction such as 'complement accumulator' could, for example, be written as 00101111.

In the case of the binary multiplier shown in figure 12.3, all the instructions stored in the programme memory could have been expressed as four-bit binary words. Since the programme is only 22 lines long it would be a simple matter to write it in binary form, that is, in machine code language. However, writing programmes of any length in machine code is both time-consuming and tedious. Additionally, it is prone to errors which are difficult to eliminate.

An alternative way of writing the programme is to use a hexadecimal notation for the instructions. For example, the 'complement accumulator' instruction expressed in hexadecimal form is simply 2F. It is now clear that much less labour is involved in writing a programme, and because 16 digits are used instead of two, a hexadecimal listing is easier to check for errors than its corresponding binary listing.

In practice it is perfectly possible to hand-programme a machine. The programmer writes down the object programme as a hexadecimal listing, having compiled it with the aid of the instruction set supplied by the manufacturer. This normally gives the binary and hexadecimal forms of the instruction and how many eight-bit words (or bytes) it contains. For example, the instruction set for the Intel 8085 informs the designer that the 'complement accumulator' instruction is designated in mnemonic form as CMA, that its hexadecimal code is 2F, and that it is a one-word (or byte) instruction. Alternatively, the instruction 'load accumulator from memory' is designated in mnemonic form as LDA, its hexadecimal code is 3A, and it is a three-byte instruction. Having completed the programme it can now be loaded into the memory using, for example, a hexadecimal keyboard or a teletype, providing the peripheral has been interfaced to the machine.

Alternatively, the programmer can compile the programme in mnemonic form. This would be termed an assembly language programme. Since mnemonic representations of instructions appear in the instruction set, it is clear that the programme can be hand-assembled into machine code. This involves such tasks as converting mnemonic codes to their hexadecimal equivalents, counting the number of words to calculate addresses, and inserting data and addresses correctly in instructions. In practice, a computer programme called an assembler may be written, and this allows the mechanisation of the assembly process. The assembler will accept a programme in assembly language, usually called a source code, as input data, and produce a machine code programme as output data.

Answers to problems

Chapter 1

1.1 (a) $\overline{f}_1 = \overline{A}(B + \overline{C})$ (b) $\overline{f}_2 = (A + \overline{B}\overline{C})(\overline{B} + D + A\overline{C})$
(c) $\overline{f}_3 = (\overline{A} + B)[\overline{C} + A(\overline{D} + \overline{E})] + B[\overline{A} + \overline{C} + E(B + D)]$

1.2 (a) $f_1 = A\overline{C} + A\overline{B} + AD + B\overline{C}D$ (b) $f_2 = \overline{A} + B + \overline{C} + D$
(c) $f_3 = \overline{A}B\overline{D} + AB\overline{C} + A\overline{B}D$

1.3 (a) $\overline{f}_1 = \overline{A}(\overline{B} + C)$ (b) $\overline{f}_2 = (\overline{A} + \overline{C})(\overline{B} + \overline{C})(\overline{A} + \overline{B})$
(c) $\overline{f}_3 = \overline{A}B + AB\overline{C}$ (d) $\overline{f}_4 = (\overline{B} + \overline{D})(A + \overline{C})(B + D)$

1.4 (a) $f_1 = \overline{A}\overline{B}\overline{C}$ (b) $f_2 = \overline{B}\overline{C}$ (c) $f_3 = \overline{A} + \overline{C}$

1.5

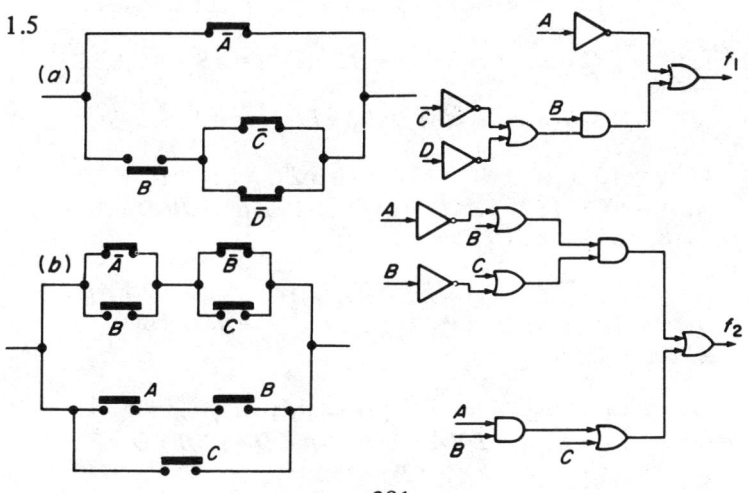

301

(c)

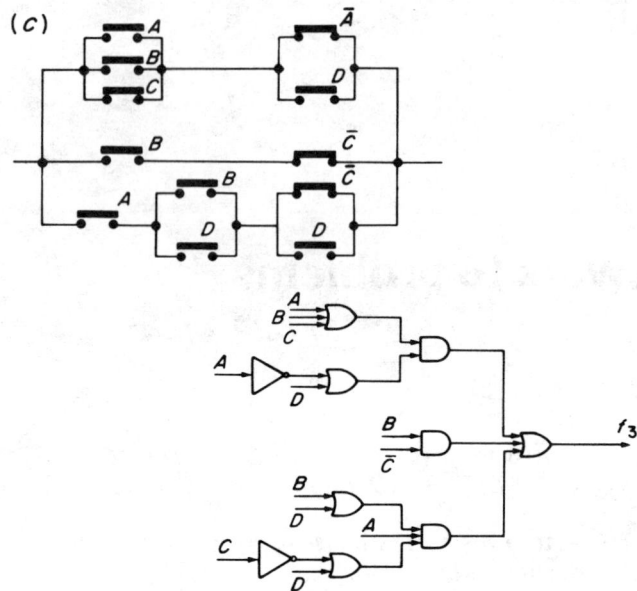

Chapter 2

2.1 (a) $f_1 = \bar{A}\bar{B}C + \bar{A}B\bar{C} + \bar{A}BC + A\bar{B}C + ABC$ (b) $f_2 = \bar{A}\bar{B}C + \bar{A}B\bar{C} + A\bar{B}\bar{C} + A\bar{B}C + AB\bar{C} + ABC$ (c) $f_3 = \bar{A}\bar{B}\bar{C}D + \bar{A}\bar{B}CD + \bar{A}BC\bar{D} + \bar{A}BCD + A\bar{B}CD + A\bar{B}CD + AB\bar{C}\bar{D} + AB\bar{C}D + ABC\bar{D} + ABCD$

2.2 (a) $f_1 = \bar{B}C + B\bar{C}$ (b) $f_2 = \bar{A} + BC$ (c) $f_3 = AB + AC + BC$

2.3 (a) $f = S_2 S_3 S_4 S_5 S_6 S_7$ (b) $f = P_3 + P_4$ (c) $\bar{f} = S_3 S_5 S_6 S_7$

2.4 (a) $f = \bar{A}\bar{B} + A\bar{C} + BC$ (b) $f = \bar{A}\bar{B} + \bar{B}\bar{C} + CD + A\bar{C}\bar{D}$ (c) $f = AB + \bar{A}\bar{B}C + \bar{A}\bar{B}D + ACD$ (d) $f = AC + \bar{A}\bar{B}E + \bar{B}CD + BCD + \bar{C}\bar{D}\bar{E} + CDE + \bar{A}B\bar{C}\bar{E}$

2.5 (a) $f = AC + BC$ (b) $f = \bar{A}C + \bar{A}D + \bar{B}C$ (c) $f = A\bar{C} + \bar{C}D + \bar{A}\bar{B}C + ABD + A\bar{B}\bar{D}$ (d) $f = ABD + \bar{A}C\bar{E} + \bar{B}C\bar{D} + \bar{A}\bar{B}C + AB\bar{C}\bar{E} + A\bar{B}\bar{D}\bar{E}$.

2.6 (a) $f = (\bar{A} + C)(A + \bar{B} + \bar{C})$ (b) $f = \bar{B}(A + \bar{C})(\bar{A} + C + D)$ (c) $f = (\bar{A} + \bar{B})(\bar{A} + \bar{D})(A + \bar{B} + D)(A + D + E)(B + \bar{D} + \bar{E})$ (d) $f = (A + B)(\bar{A} + \bar{B})(C + D)(\bar{B} + D)$

2.7 (a) $F = \bar{A}CD$ (b) $F = \bar{B}CD + B\bar{C}D + A\bar{B}\bar{C}\bar{D}$
(c) $F = A\bar{B} + \bar{A}B\bar{C}$

2.8 $0 = \bar{A}\bar{B}, 1 = \bar{A}\bar{C}\bar{D}, 2 = \bar{A}\bar{C}D, 3 = \bar{A}C\bar{D}$ or $BC\bar{D}, 4 = BCD,$
$5 = \bar{B}\bar{C}\bar{D}, 6 = \bar{B}\bar{C}D, 7 = \bar{B}CD$ or $AC\bar{D}, 8 = ACD, 9 = AB$

Chapter 3

3.1 (a)

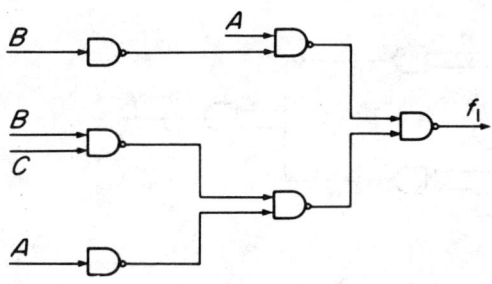

(b)

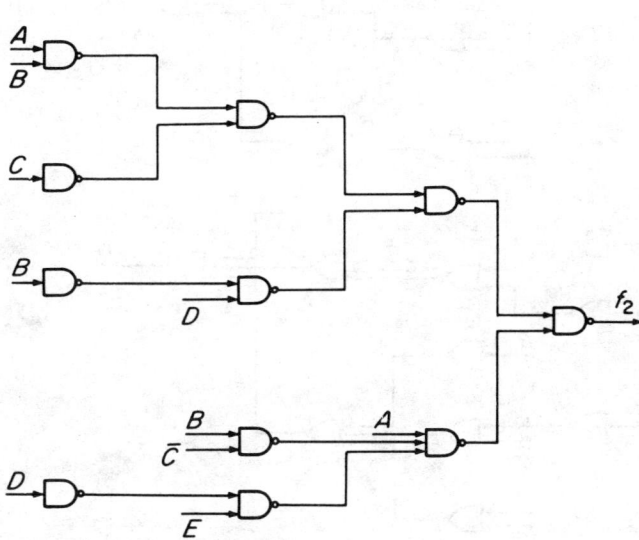

Answers to problems

3.2

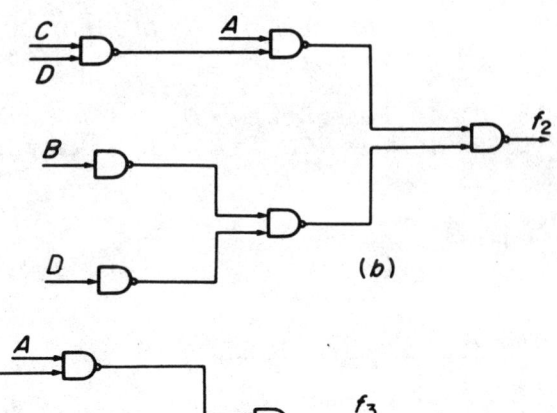

(a)   (b)

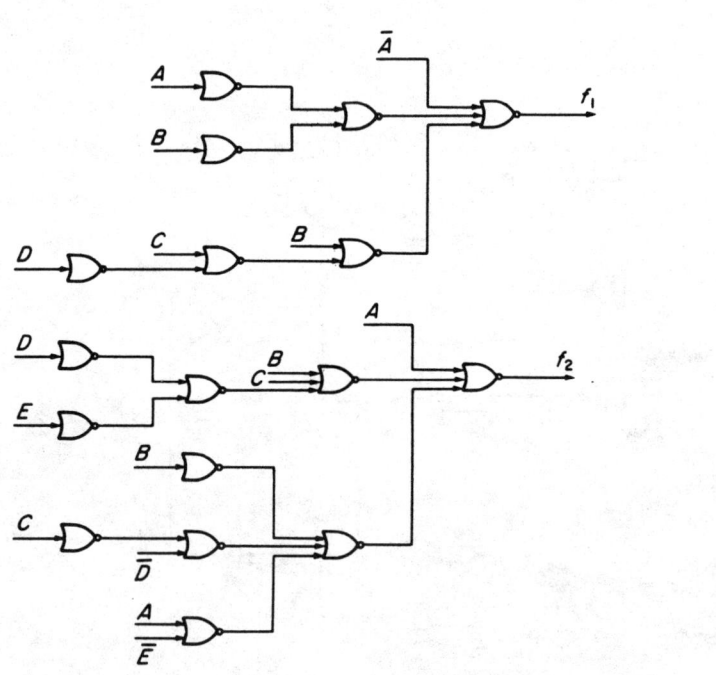

3.3

Answers to problems

3.4

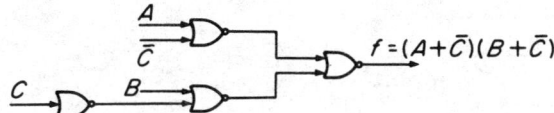

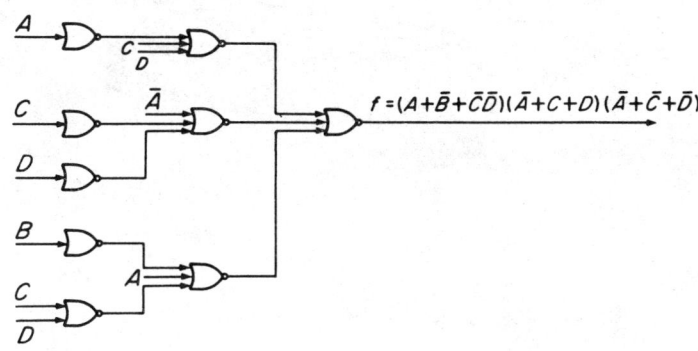

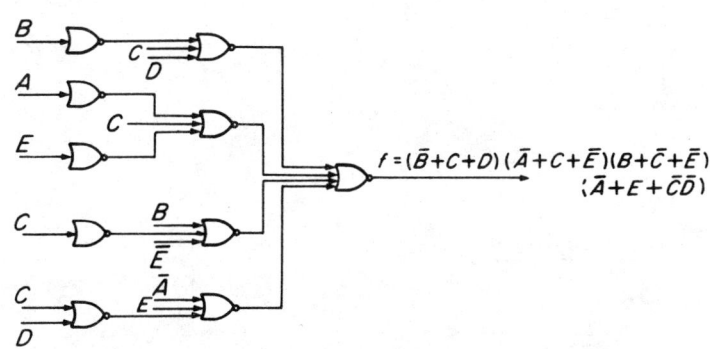

3.5 (a)

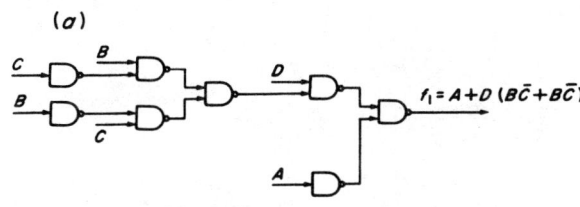

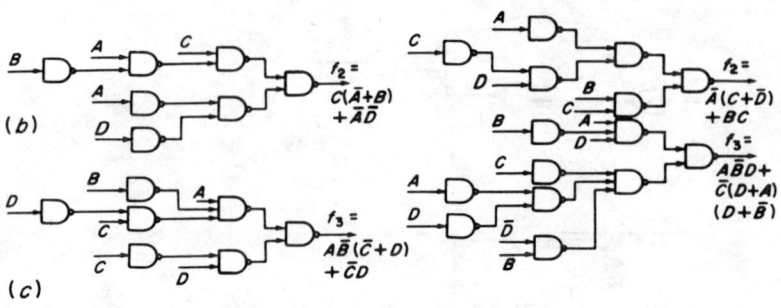

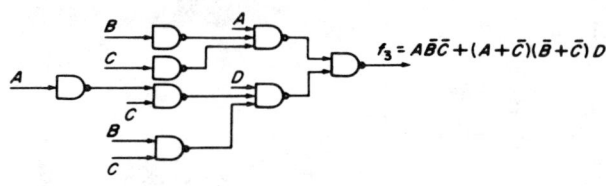

3.6

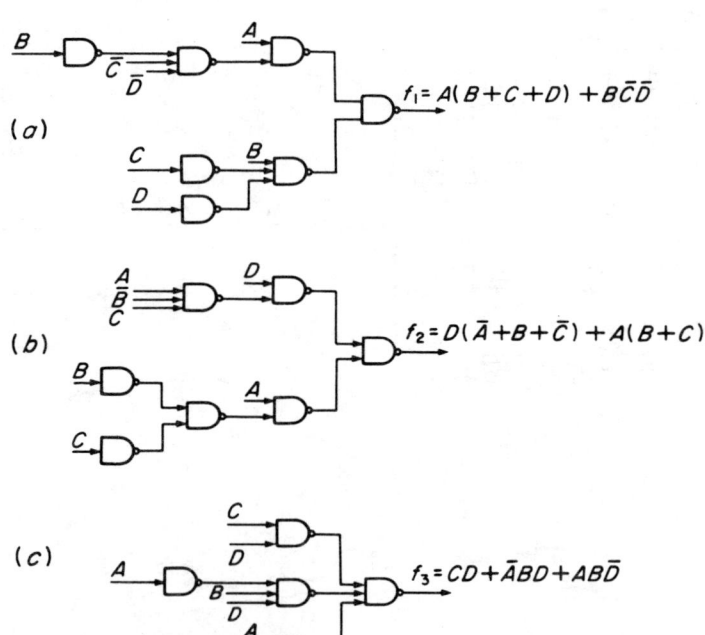

3.7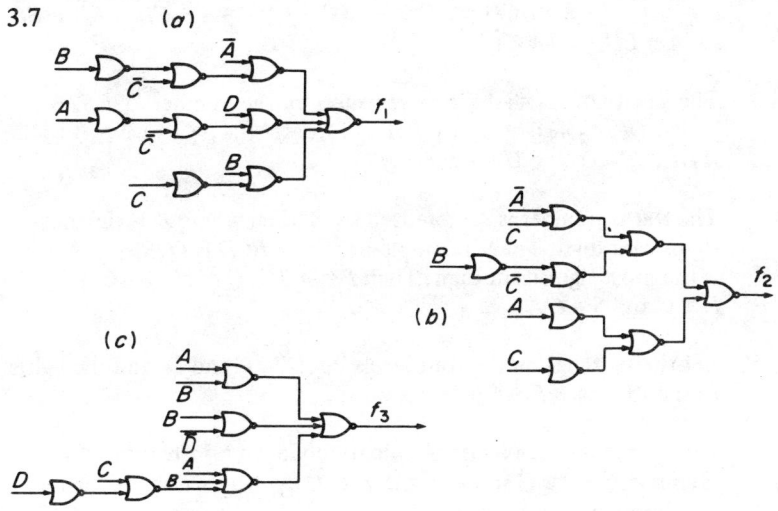

3.8 (a) $ABC + A\bar{B}\bar{C}$ (b) $ABC + A\bar{B}\bar{C}$ (c) $A + \bar{B}C + B\bar{C}$
(d) $A + \bar{B}C + B\bar{C}$

Chapter 4

4.1 The four NBCD digits are A, B, C and D, digit A being the most significant.
(i) $f = AD + BC\bar{D} + \bar{B}CD$ (ii) $f = A + BCD$ (iii) $f = \bar{A}\bar{B}$

4.2 For single-digit numbers: $A > B$, $f_1 = A\bar{B}$; $A < B$, $f_2 = \bar{A}B$; $A = B$, $f_3 = \bar{A}\bar{B} + AB$.
For four-digit numbers: $f_{A=B} = f_{30}f_{31}f_{32}f_{33}$, where $f_{30} = \bar{A}_0\bar{B}_0 + A_0B_0$; $f_{A>B} = f_{13} + f_{33}f_{12} + f_{33}f_{32}f_{11} + f_{33}f_{32}f_{31}f_{10}$, where $f_{10} = A_0\bar{B}_0$ etc; $f_{A<B} = f_{23} + f_{33}f_{22} + f_{33}f_{32}f_{21} + f_{33}f_{32}f_{31}f_{20}$, where $f_{20} = \bar{A}_0B_0$ etc.

4.3 Binary number $= ABCD$, 2's complement form $= PQRS$.
$P = A \oplus (B + C + D)$, $Q = B \oplus (C + D)$, $R = C \oplus D$, $S = D$.

4.4 The four NBCD digits are A, B, C and D, digit A being the most significant. Segment equations: $a = A + C + \overline{B \oplus D}$; $b = \bar{B} + \overline{C \oplus D}$;

$c = \overline{\overline{B}C\overline{D}}; d = e + B\overline{C}D + \overline{B}C; e = \overline{D}(\overline{B} + C); f = A + \overline{C}\overline{D} + B\overline{D} + B\overline{C};$
$g = A + C\overline{D} + (B \oplus C).$

4.5 The square of $A_2 A_1 A_0$ is represented by the number $PQRSTU$, where $P = A_2 A, Q = A_2 \overline{A}_1 + A_2 A_0, R = A_0(A_1 \oplus A_2), S = A_1 \overline{A}_0, T = 0$, and $U = A_0$.

4.6 The NBCD number is represented by $ABCD$, where A is the most significant digit. The 9's complement of $ABCD$ is $PQRS$, where P is the most significant digit. Then $P = \overline{A}\overline{B}\overline{C}, Q = B\overline{C} + \overline{B}C, R = C$ and $S = \overline{C}$.

4.7 Let the switches on the four levels be A, B, C and D, and the lights L, then $L = A \oplus B \oplus C \oplus D$.

4.8 Let A, B and C be bits of the binary code, and p the parity bit, then $p = A \oplus B \odot C$ and $F = A \odot B \odot C \odot p$ where $F = 1$ when an invalid code is received.

4.9 If the five-bit number is $ABCDE$, where A is the most significant digit, then $M = A(B + E)(C + D) + BE(A + C + D) + CD(A + B + E)$ where M is the majority logic function.

4.10

(a)

2-4-2-1 Gray code			
0	0	0	0
0	0	0	1
0	0	1	1
0	0	1	0
0	1	1	0
1	1	1	0
1	0	1	0
1	0	1	1
1	0	0	1
1	0	0	0

(b)

XS3 Gray code			
0	0	1	0
0	1	1	0
0	1	1	1
0	1	0	1
0	1	0	0
1	1	0	0
1	1	0	1
1	1	1	1
1	1	1	0
1	0	1	0

4.11 If the XS3 code is represented by A, B, C and D where A is the most significant digit and the XS3 Gray code is represented by

PQR and *S*, where *P* is the most significant digit then $P = A$, $Q = A \oplus B$, $R = B \oplus C$ and $S = C \oplus D$.

4.12 (i) +11, −53 (ii) +11, −75 (iii) +11, −74

4.13

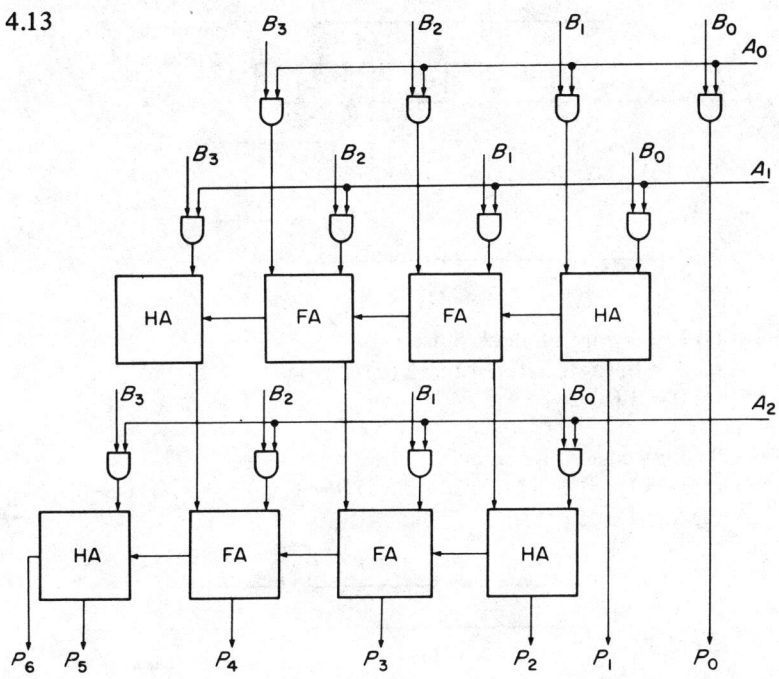

4.14 0,0010101 1,1100011

Chapter 5

5.1 (i) $a = 1, b = 0, c = 1$ (ii) $a = 0, b = 1, c = 1$ (iii) $a = 1, b = 0, c = 0$

310 *Answers to problems*

5.2

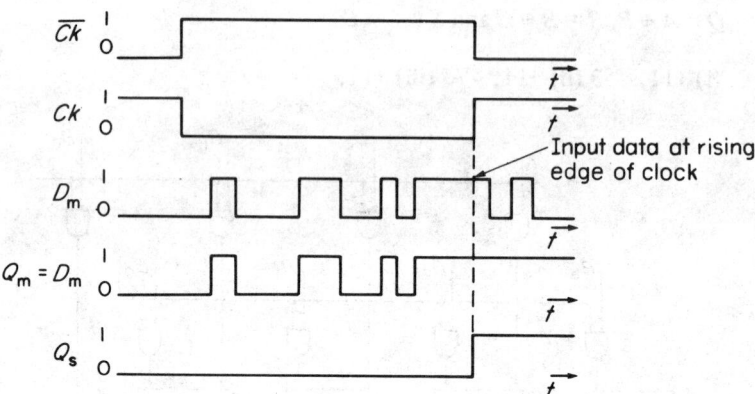

5.3 (i) Rising edge of clock pulse 1
(a) $1 \to 0$, (b) $0 \to 1$, (c) $1 \to 1$ (d) $0 \to 0$ (e) $1 \to 1$, (f) $1 \to 0$,
(g) $0 \to 1$ (h) $1 \to 1$

Trailing edge of clock pulse 1
(a) $0 \to 1$, (b) $1 \to 1$, (c) $1 \to 0$, (d) $0 \to 1$, (e) $1 \to 1$, (f) $0 \to 0$,
(g) $1 \to 1$, (h) $1 \to 0$.

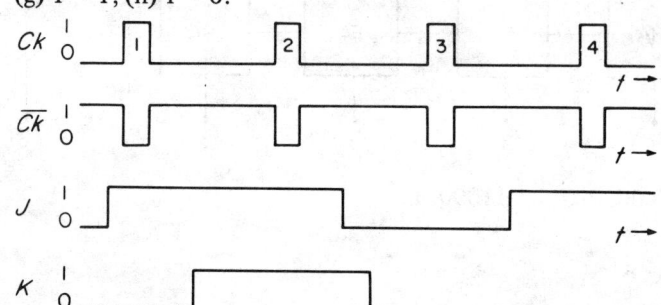

(v) $Q_s = 1$
(vi) $Q_s = 1$

5.4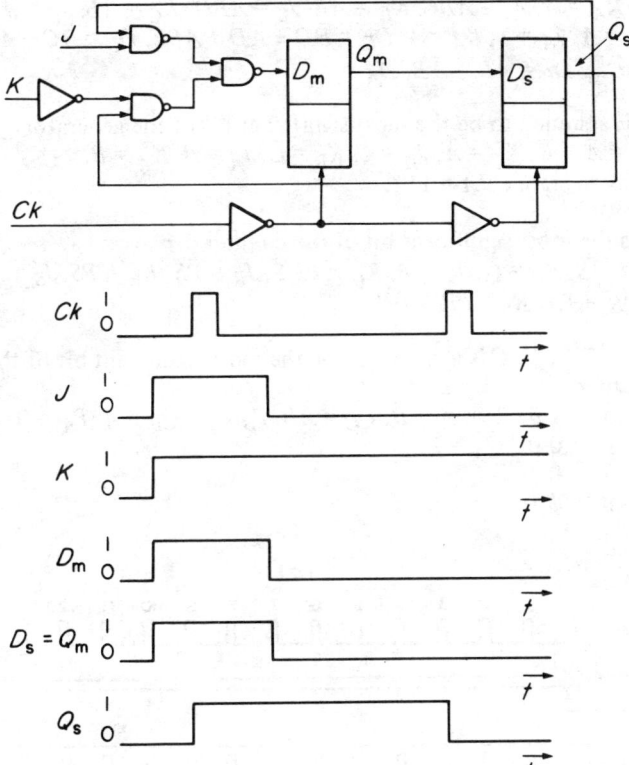

5.5 $Q^{t+\delta t} = (\bar{J}'\bar{Q} + \bar{K}Q)^t$

5.6 For both cases

```
         (X̄Y + XȲ)⎍
    ┌─────────────────→┌─────┐
    │ Q=0 │            │ Q=1 │
    └─────┘←─────────────────┘
         (X̄Y + XȲ)⎍
```

Chapter 6

6.1 D is assumed to be the most significant bit of the counter.
$T_D = ABC + ABD$, $T_C = AB\bar{D}$, $T_B = A$, $T_A = 1$, $S_D = ABC$, $R_D = ABD$ or $AB\bar{C}$, $S_C = AB\bar{C}\bar{D}$, $R_C = ABC$, $S_B = A\bar{B}$, $R_B = AB$, $S_A =$

$\bar{A}$, $R_A = A$, $J_D = ABC$, $K_D = AB$, $J_C = AB\bar{D}$, $K_C = AB$, $J_B = A$, $K_B = A$, $J_A = 1$, $K_A = 1$, $D_D = \bar{B}D + \bar{A}D + ABC$, $D_C = \bar{B}C + \bar{A}C + AB\bar{C}\bar{D}$, $D_B = A\bar{B} + \bar{A}B$, $D_A = \bar{A}$.

6.2 C is assumed to be the most significant bit of the generator.
$J_C = \bar{A} + \bar{B}$, $K_C = \bar{A}$, $J_B = C$, $K_B = \bar{C}$, $J_A = B$, $K_A = \bar{B}$.
Lock in state $CBA = 111$.

6.3 P is the most significant bit of the counter
$J_P = R\bar{S}$, $K_P = \bar{Q}$, $J_Q = \bar{P}$, $K_Q = PR\bar{S}$, $J_R = \bar{P}S$, $K_R = PS$, $J_S = \bar{P}Q\bar{R} + PR$, $K_S = P\bar{R} + \bar{P}R$

6.4 Z^t $Z^{t+\delta t}$ PQ. C is assumed to be the most significant bit of the counter.

0	0	1	∅
0	1	0	∅
1	0	∅	1
1	1	∅	0

$P_C = \bar{A} + \bar{B}$, $Q_C = AB$, $P_B = \bar{A}$, $Q_B = A$, $P_A = 0$, $Q_A = 1$.

6.5

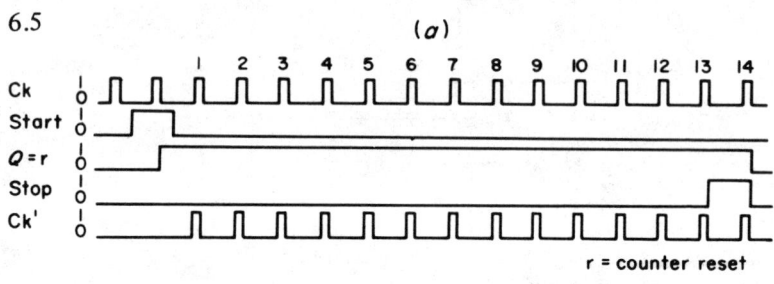

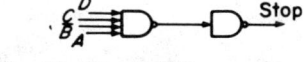

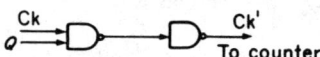

r = counter reset

6.6 (a) $f_u = 5$ MHz

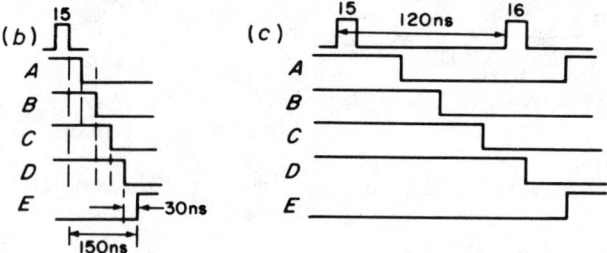

6.7

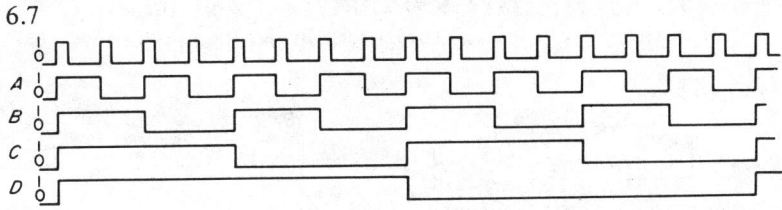

6.8

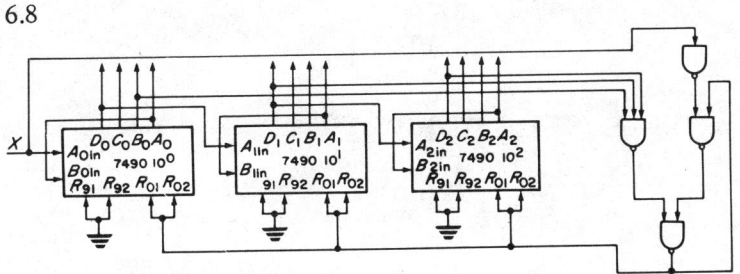

Chapter 7

7.1

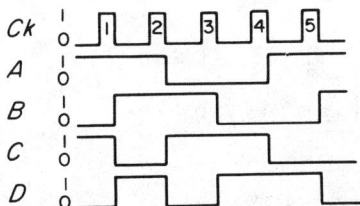

7.2 Chosen sequence: S_0-S_1-S_2-S_5-S_{11}-S_7-S_{15}-S_{14}-S_{13}-S_{10}-S_4-S_8-S_0.
Feedback function assuming A is least significant flip-flop in register $f = B\bar{D} + AC\bar{D} + \bar{A}BC + A\bar{C}D + \bar{A}\bar{C}D$, $0 = \bar{A}\bar{B}\bar{C}D$,
$1 = A\bar{C}\bar{D}$, $2 = B\bar{C}\bar{D}$, $3 = A\bar{B}C\bar{D}$, $4 = A\bar{C}D$, $5 = AB\bar{D}$, $6 = ABCD$,
$7 = \bar{A}BC$, $8 = \bar{B}CD$, $9 = \bar{A}B\bar{C}D$, $10 = \bar{A}\bar{B}C$, $11 = \bar{A}BD$

7.3 Because of ambiguities in the shift register sequence developed from the given binary sequence it is necessary to design a modulo-11 SR counter. The modulo-11 sequence chosen from the de Bruijn diagram is S_0-S_1-S_2-S_5-S_{11}-S_7-S_{14}-S_{13}-S_{10}-S_4-S_8-S_0. Feedback logic $f = \bar{A}\bar{C}D + BCD + A\bar{C}D + A\bar{B}C\bar{D}$ where A is the least significant stage of the shift register. Output logic $g = D + A\bar{C}$

7.4 (a)

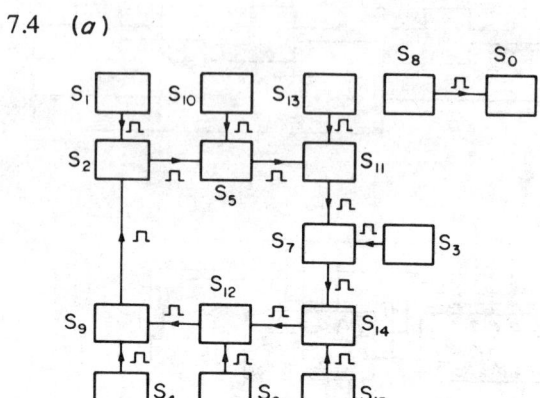

7.5 Feedback logic $f = B \oplus C + \bar{A}\bar{B} + m\bar{B}$ (A is the least significant stage of the shift register)

7.6 $P = \bar{A}BC + AB\bar{C}$, $Q = \bar{A}\bar{C}$, $R = \bar{A}\bar{B} + \bar{B}\bar{C} + \bar{A}\bar{C} + ABC$, $S = A\bar{B}C + \bar{A}\bar{C}$

7.7 (a)

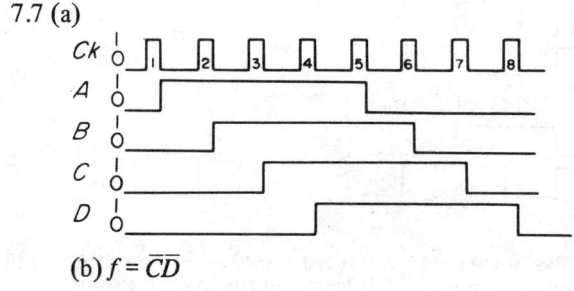

(b) $f = \overline{CD}$

Chapter 8

8.1

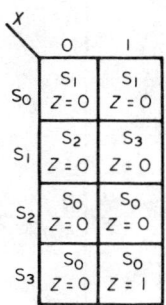
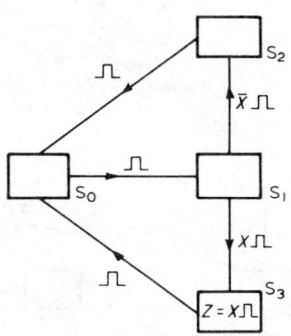

Examines 3 digit words to give an output $Z = 1$ if the last two bits of the word are 1s.

8.2

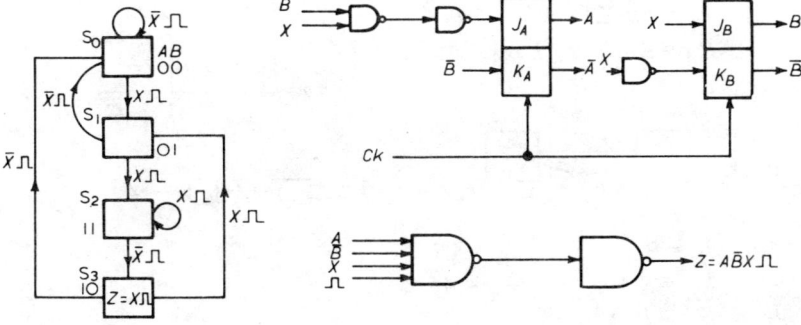

316   Answers to problems

8.3 Circuit solution depends upon state assignment.

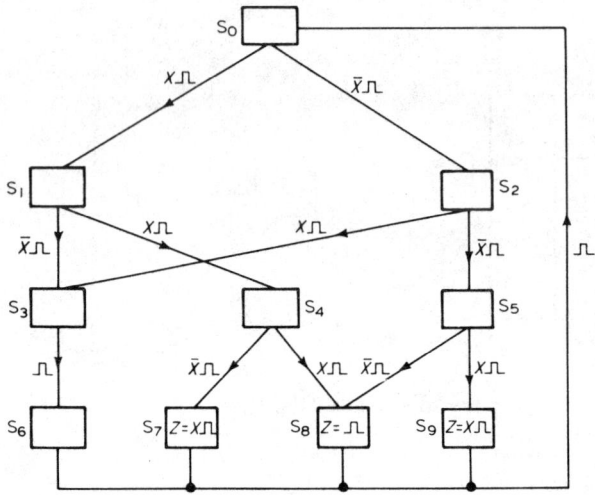

8.4 $S_A = \overline{A}mX$ $J_A = \overline{m}X$
      $R_A = AmX$ $K_A = mX$
      $Z = AX$

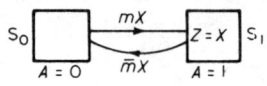

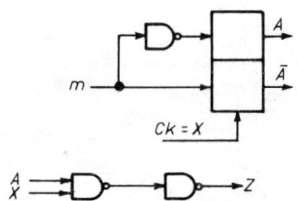

8.5 $J_A = B$ $J_B = \overline{A}S$
      $K_A = \overline{S} + \overline{B}$ $K_B = A\overline{S}$
      $Z = \overline{A}B\sqcap$

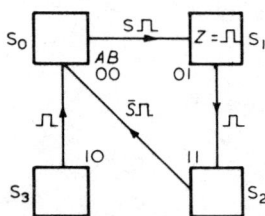

Answers to problems

8.6

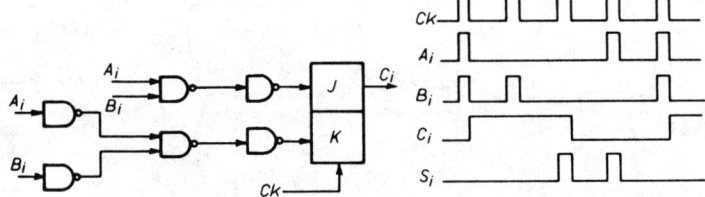

8.7

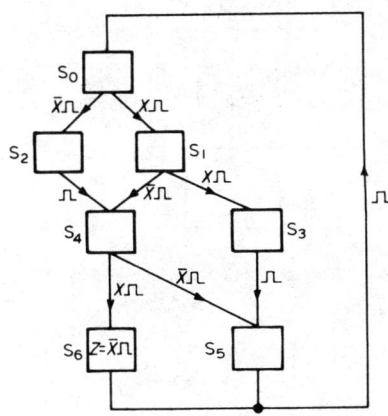

8.8

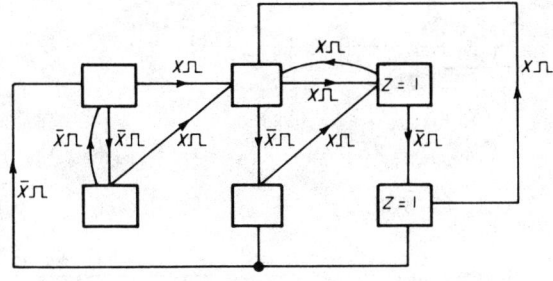

318 Answers to problems

8.9

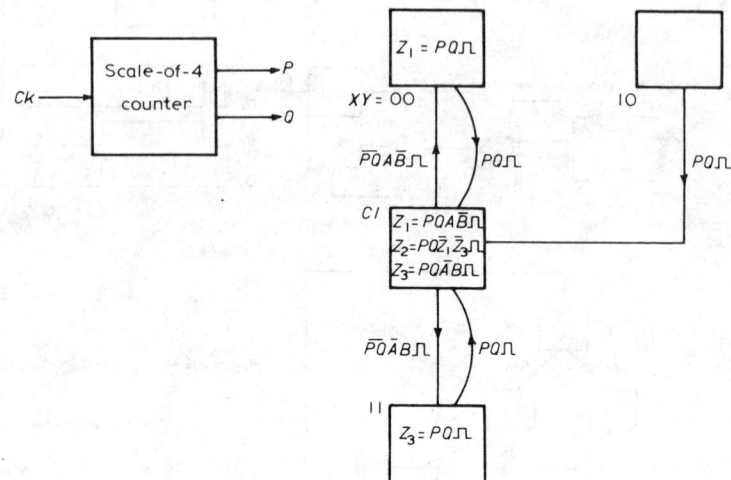

8.10

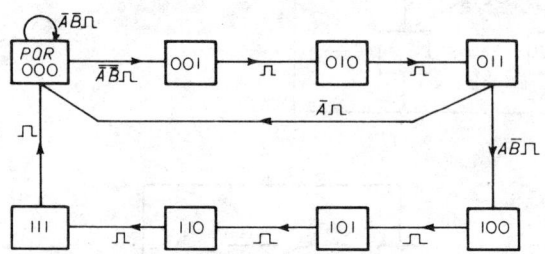

8.11 (a) $P_f = (S_4 S_5)(S_0)(S_1)(S_2)(S_3)(S_6)$
(b) $P_f = (S_3 S_6 S_{10} S_{11} S_{13})(S_2 S_5 S_{12})(S_4 S_7 S_{14})(S_0)(S_1)(S_8)(S_9)$

8.12 $P_f = (S_0 S_1)(S_4 S_5)(S_2 S_3)(S_6)$

8.13

(a)
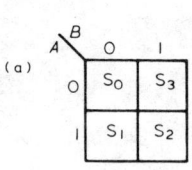

(b)

DC\BA	00	01	11	10
00	S_6	S_7	S_8	S_9
01	S_5	S_3	S_4	S_2
11			S_0	S_1
10				

Answers to problems 319

8.14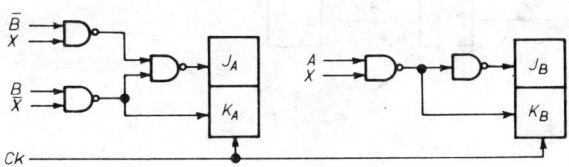

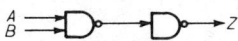

Chapter 9

9.1 The basic state diagram for the problem.

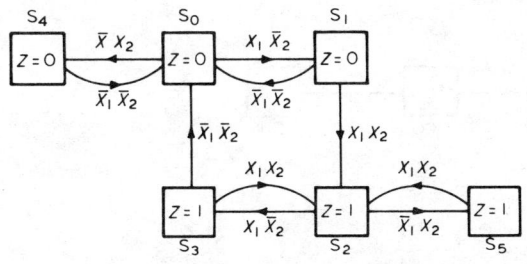

9.2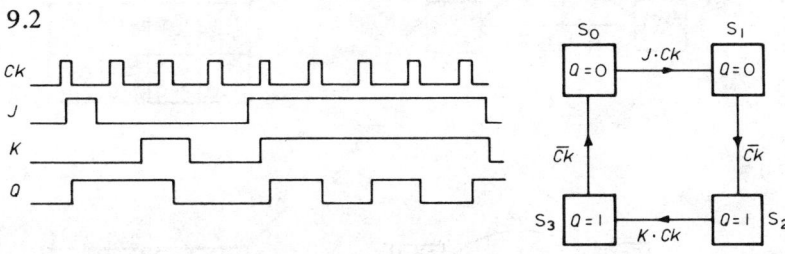

9.3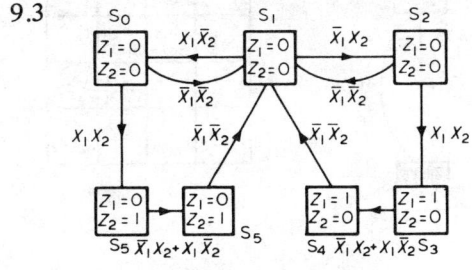

9.4

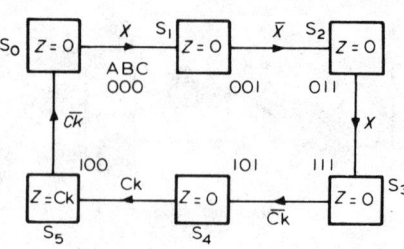

9.5

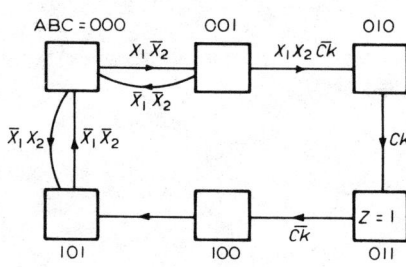

9.6

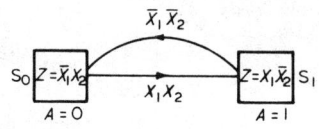

9.7

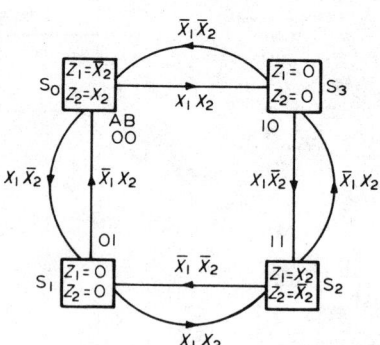

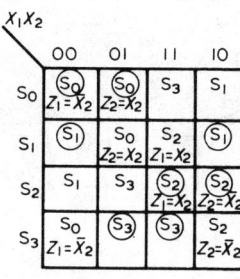

Answers to problems

9.8 Races (1) $S_0 \to S_3$ on signal $X_1 X_2$ – critical
(2) $S_1 \to S_2$ on signal $X_1 X_2$ – critical

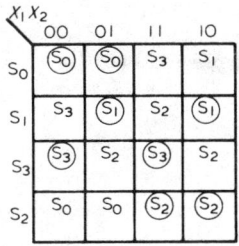

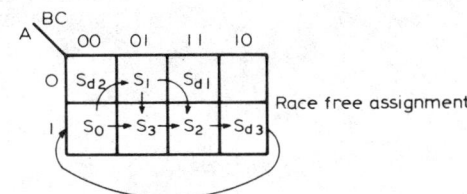

Race free assignment

9.9

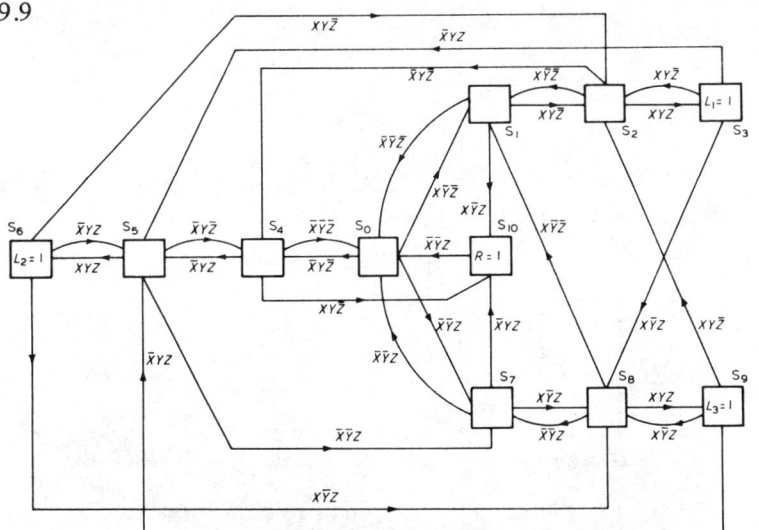

9.10 For state assignment $S_0 = AB = 0, S_1 = 01, S_2 = 11, S_3 = 10.$

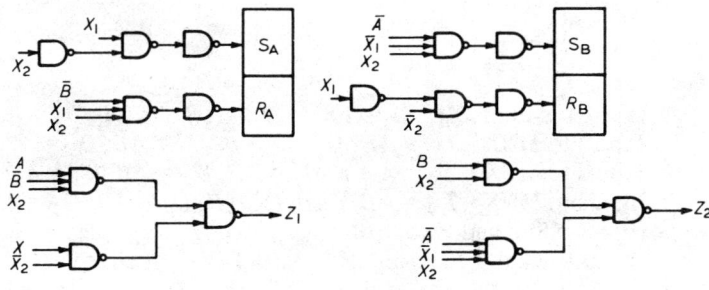

Chapter 10

10.1 (a) $D_0 = \bar{C}, D_1 = 1, D_2 = C, D_3 = C$ (b) $D_0 = A, D_1 = \bar{A}, D_2 = A, D_3 = 1$ (c) $D_0 = 1, D_1 = 0, D_2 = 1, D_3 = 1$

10.2 (a) $D_0 = \bar{C} + D, D_1 = \bar{C}D + C\bar{D}, D_2 = \bar{C} + D, D_3 = \bar{C}$ (b) $D_0 = A + \bar{D}, D_1 = A, D_2 = 0, D_3 = D$ (c) $D_0 = \bar{A}\bar{B}, D_1 = 1, D_2 = A, D_3 = A + \bar{B}$ (d) $D_0 = 0, D_1 = \bar{B}\bar{C}, D_2 = 1, D_3 = B + \bar{C}$.

10.3 (a)

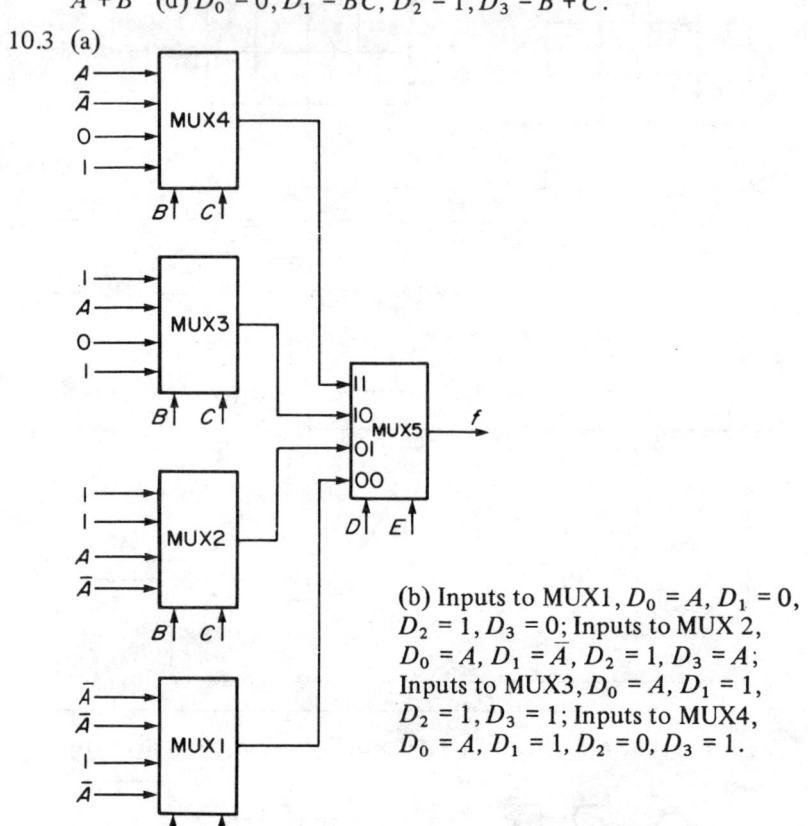

(b) Inputs to MUX1, $D_0 = A, D_1 = 0, D_2 = 1, D_3 = 0$; Inputs to MUX 2, $D_0 = A, D_1 = \bar{A}, D_2 = 1, D_3 = A$; Inputs to MUX3, $D_0 = A, D_1 = 1, D_2 = 1, D_3 = 1$; Inputs to MUX4, $D_0 = A, D_1 = 1, D_2 = 0, D_3 = 1$.

10.4 (a) Inputs to MUX1, $D_0 = \bar{A} + \bar{B}, D_1 = \bar{A}B, D_2 = 0, D_3 = \bar{A} + B$;
Inputs to MUX2, $D_0 = \bar{A}\bar{B}, D_1 = A + \bar{B}, D_2 = A\bar{B}, D_3 = A$;
Inputs to MUX3, $D_0 = A + B, D_1 = \bar{A}B, D_2 = \bar{A}B, D_3 = \bar{A} + B$;
Inputs to MUX4, $D_0 = \bar{A}\bar{B} + AB, D_1 = \bar{B}, D_2 = A\bar{B}, D_3 = AB$.
(b) MUX5, E and F control signals; MUX's 1, 2, 3, and 4, B, C and D control signals. Inputs to MUX1 $D_0 = 1, D_1 = 0, D_2 = 0$,

$D_3 = \overline{A}, D_4 = \overline{A}, D_5 = \overline{A}, D_6 = 0, D_7 = A$. Inputs to MUX2 $D_0 = \overline{A}$, $D_1 = 1, D_2 = A, D_3 = A, D_4 = 0, D_5 = A, D_6 = 0, D_7 = A$. Inputs to MUX 3 $D_0 = A, D_1 = 0, D_2 = 0, D_3 = \overline{A}, D_4 = 1, D_5 = \overline{A}$, $D_6 = \overline{A}, D_7 = 1$. Inputs to MUX4 $D_0 = \overline{A}, D_1 = 1, D_2 = A$, $D_3 = 0, D_4 = A, D_5 = 0, D_6 = 0, D_7 = A$.

10.5 A and B = sum digits, C_{in} = carry in; inputs to sum MUX, B and C_{in} control signals, $D_0 = A, D_1 = \overline{A}, D_2 = \overline{A}, D_3 = A$; inputs to carry MUX, B and C_{in} control signals, $D_0 = 0, D_1 = A, D_2 = A$, $D_3 = 1$.

10.6 (a)

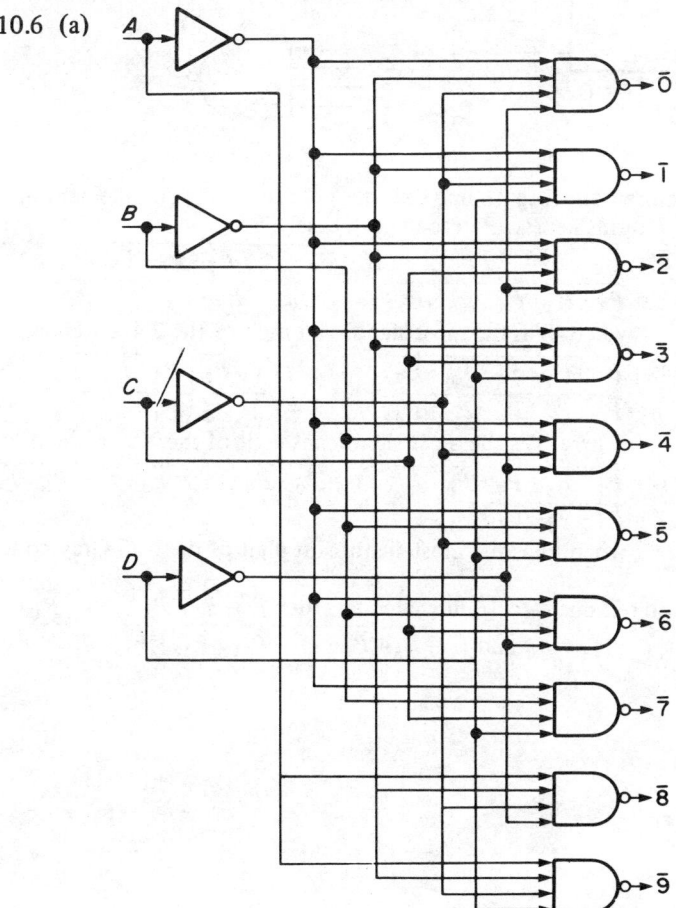

(b) $\bar{0} = \bar{A}B\bar{C}D$, $\bar{1} = \bar{A}BCD$, $\bar{2} = BC\bar{D}$, $\bar{3} = \bar{B}CD$, $\bar{4} = B\bar{C}\bar{D}$,
$\bar{5} = B\bar{C}D$, $\bar{6} = BC\bar{D}$, $\bar{7} = BCD$, $9\bar{8} = A\bar{D}$, $\bar{9} = AD$.

10.7

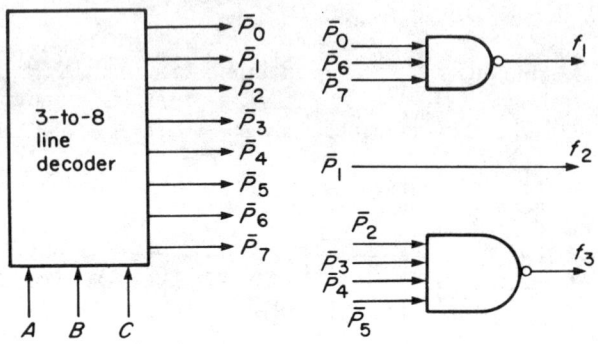

10.8 Required one 4-to-16 line decoder generating the complement of the P terms i.e. $\bar{P}_0$, $\bar{P}_1$ etc in each case. Then

(a) $W = \overline{\bar{P}_5 \cdot \bar{P}_6 \cdot \bar{P}_7 \cdot \bar{P}_8 \cdot \bar{P}_9}$, $\quad X = \overline{\bar{P}_4 \cdot \bar{P}_6 \cdot \bar{P}_7 \cdot \bar{P}_8 \cdot \bar{P}_9}$
$Y = \overline{\bar{P}_2 \cdot \bar{P}_3 \cdot \bar{P}_5 \cdot \bar{P}_8 \cdot \bar{P}_9}$, $\quad Z = \overline{\bar{P}_1 \cdot \bar{P}_3 \cdot \bar{P}_5 \cdot \bar{P}_7 \cdot \bar{P}_9}$,
where W is the most significant digit of the 2-4-2-1 code.

(b) $W = \overline{\bar{P}_5 \cdot \bar{P}_6 \cdot \bar{P}_7 \cdot \bar{P}_8 \cdot \bar{P}_9}$, $\quad X = \overline{\bar{P}_1 \cdot \bar{P}_2 \cdot \bar{P}_3 \cdot \bar{P}_4 \cdot \bar{P}_9}$,
$Y = \overline{\bar{P}_0 \cdot \bar{P}_3 \cdot \bar{P}_4 \cdot \bar{P}_7 \cdot \bar{P}_8}$, $\quad Z = \overline{\bar{P}_0 \cdot \bar{P}_2 \cdot \bar{P}_4 \cdot \bar{P}_6 \cdot \bar{P}_8}$,
where W is the most significant digit of the XS3 code.

(c) $W = \overline{\bar{P}_5 \cdot \bar{P}_6 \cdot \bar{P}_7 \cdot \bar{P}_8 \cdot \bar{P}_9}$, $\bar{X} = \overline{\bar{P}_0 \cdot \bar{P}_9}$, $\bar{Y} = \overline{\bar{P}_3 \cdot \bar{P}_4 \cdot \bar{P}_5 \cdot \bar{P}_6}$
$Z = \overline{\bar{P}_2 \cdot \bar{P}_3 \cdot \bar{P}_6 \cdot \bar{P}_7}$,
where W is the most significant digit of the XS3 Gray code.

10.9 Required one 4-to-16 line decoder, then $f_1 = \overline{\bar{P}_2 \cdot \bar{P}_3 \cdot \bar{P}_{14} \cdot \bar{P}_{15}}$,
$f_2 = \bar{P}_{12}$, $f_3 = \bar{P}_0$ and $f_4 = \overline{\bar{P}_5 \cdot \bar{P}_6 \cdot \bar{P}_8 \cdot \bar{P}_{11} \cdot \bar{P}_{12} \cdot \bar{P}_{13} \cdot \bar{P}_{14} \cdot \bar{P}_{15}}$.

10.10

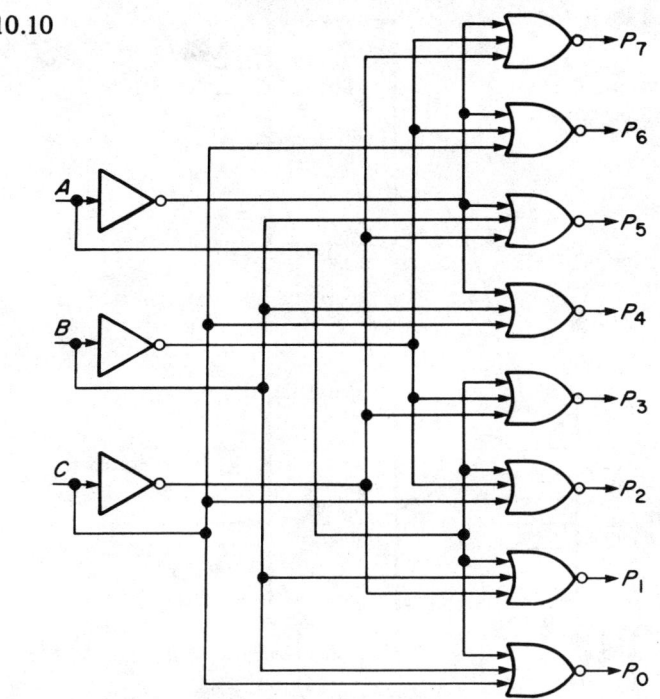

10.11

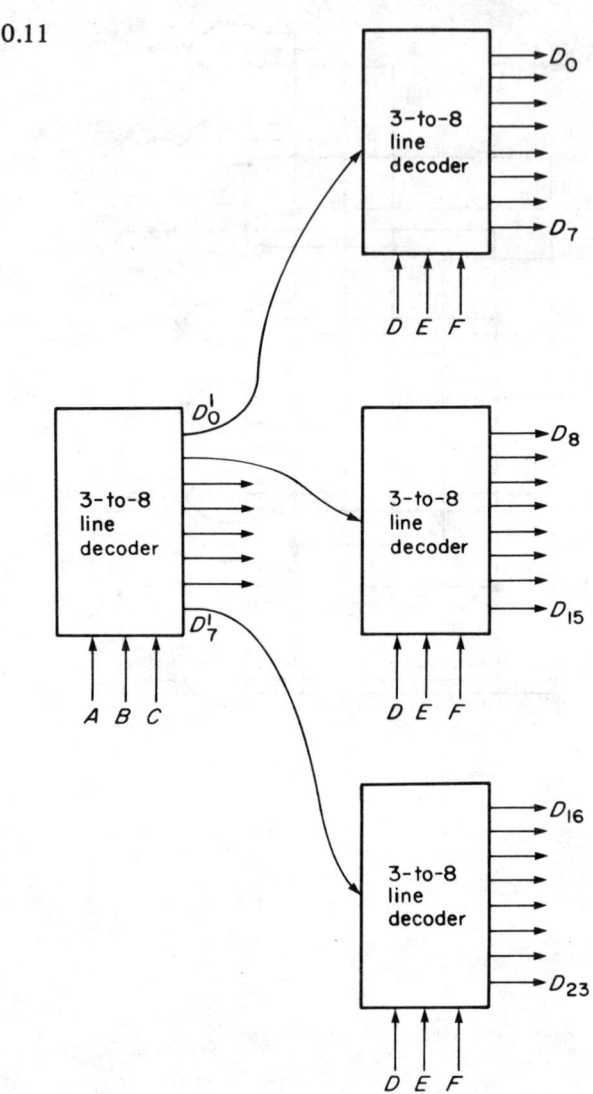

10.12 ROM capacity = 3072 bits

10.13 (a) 80 bits (b) 768 bits (c) 384 bits

10.14

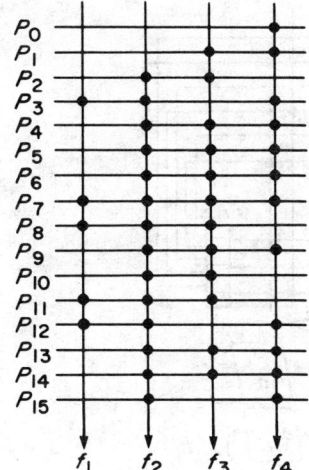

Answers to problems

10.15

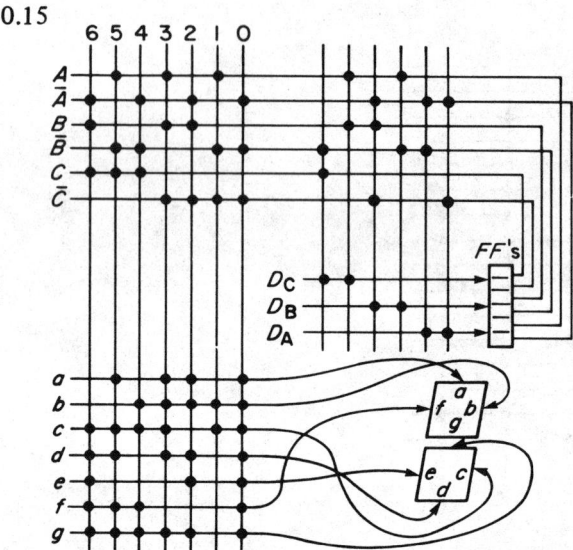

10.16 (and 10.18). A typical solution to this type of problem is shown in the answer to 10.17.

10.17

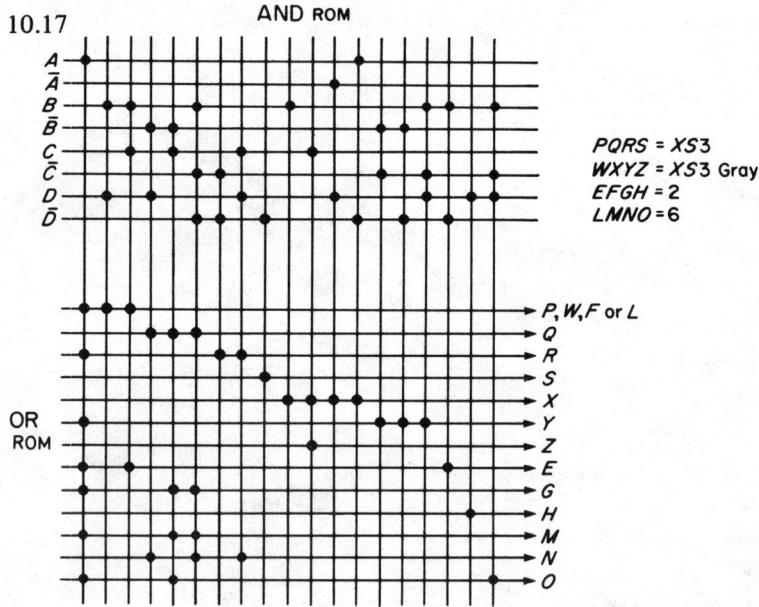

PQRS = XS3
WXYZ = XS3 Gray
EFGH = 2
LMNO = 6

10.19 One decade of the three-decade XS3 subtractor/adder.

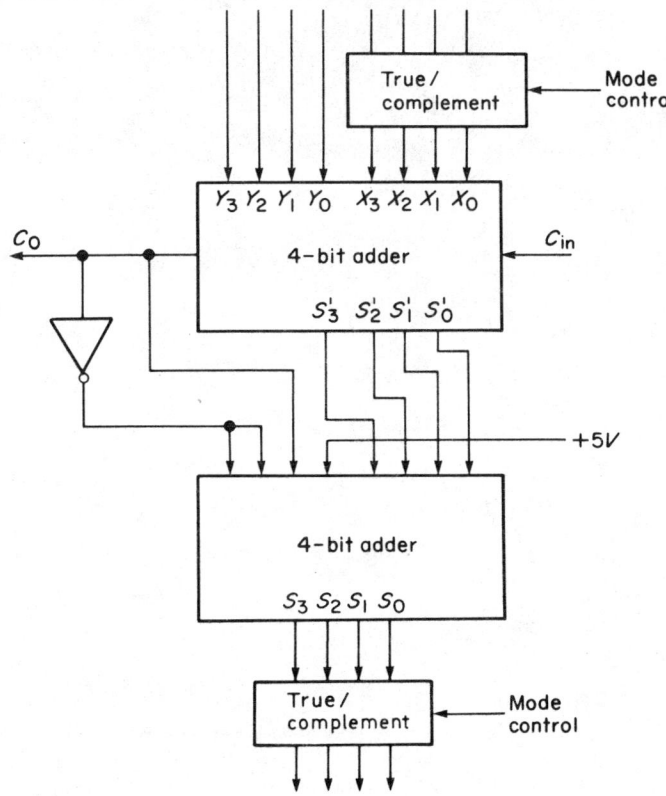

Chapter 11

11.1 (a) (i) $f = \overline{CD} + \overline{A}\overline{D} + B\overline{D} + \overline{A}B\overline{C} + ABC + ACD + A\overline{B}D + A\overline{B}\overline{C}$
(ii) $f = (A + B + \overline{D})(A + \overline{C} + \overline{D})(\overline{A} + \overline{B} + C + \overline{D})(\overline{A} + B + \overline{C} + D)$
(b) (i) $f = AB + B\overline{C} + B\overline{D} + ACD + \overline{B}CD$, (ii) $f = (B + C)(B + D)(A + \overline{B} + \overline{C} + \overline{D})$.
It is left to the student to satisfy the fan-in requirements

11.2 (a) Static 1-hazards $A = B = C = 1$ and $A = 1, B = 1, C = 0$; Static 0-hazards $B = C = 0, D = 1$.
(b) Static 1-hazards $A = 0, C = D = 1$ and $B = D = 0$; Static 0-hazards $A = B = C = 0$.

11.4 (a) Dynamic hazard $A = 1, B = C = 0, C$ changing from $0 \to 1$.
(b) Dynamic hazard $A = 0, B = 0, C = 1, C$ changing from $1 \to 0$.

Bibliography

Bannister, B.R. and Whitehead, D.G., Fundamentals of digital systems, McGraw-Hill, 1973

Blakeslee, T.R., Digital design with standard MSI and LSI, Wiley, 1975

Boyce, J.C., Digital logic and switching circuits: operation and analysis, Prentice-Hall; 1975

Caldwell, S.H., Switching circuits and logical design, Wiley, 1967

Clare, C.R., Designing logic systems using state machines, McGraw-Hill, 1973

Dietmeyer, D.L., Logic design of digital systems, Boston, Allyn & Bacon, 1978

Doktor, F. and Steinhauer, J., Digital Electronics, Macmillan, 1973

Fletcher, W.I., An engineering approach to digital design, Prentice-Hall, 1980

Friedman, A.D., Logical design of digital systems, Woodland Hills, Computer Science Press, 1975

Friedman, A.D. and Menon, P.R., Theory and design of switching circuits Woodland Hills, Computer Science Press, 1975

Givonie, D.D., Introduction to switching circuit theory, McGraw-Hill, 1976

Gothmann, W.H., Digital Electronics: an introduction to theory and practice, Prentice-Hall, 1977

Greenfield, J.D., Practical digital design using ICs, Wiley, 1977

Hill, F.J. and Petersen, G.R., Introduction to switching theory and logical design, Wiley, 1974

Humphrey, W.H., Switching circuits with computer applications, McGraw-Hill, 1958

Bibliography

Hurley, R.B., Transistor Logic Circuits, Wiley, 1963
Hurst, S.L., The logical processing of digital signals, Crane, 1978
Johnson, D.E., Digital circuits and microcomputers, Prentice-Hall, 1979
Kohavi, Z., Switching theory and finite automatic theory, McGraw-Hill, 1978
Kostopoulos, G.K., Digital engineering, Wiley, 1975
Lee, S.C., Digital circuits and logic design, Prentice-Hall, 1976
Lee, S.C., Modern switching theory and digital design, Prentice-Hall, 1978
Lenk, J.D., Handbook of Logic Circuits, Resten, 1972
Lenk, J.D., Logic designer's manual, Resten, 1977
Levine, M.E., Digital theory and practice using integrated circuits, Prentice-Hall, 1978
Lewin, D., Logical design of switching circuits, Neben, 1976
Lind, L.F. and Nelson, J.C.C., Analysis and design of sequential digital Systems, Macmillan, 1977
McCluskey, E.J., Introduction to the theory of switching circuits, McGraw-Hill, 1965
Maley, G.A. and Earle, J., Logic design of transistor digital computers, Prentice-Hall, 1963
Malvino, A.P. and Leach, D.P., Digital principles and applications, McGraw-Hill, 1975
Mano, M., Digital logic and computer design, Prentice-Hall, 1979
Marcus, M.P., Switching circuits for engineers, Prentice-Hall, 1967
Morris, N.M., Digital electronic circuits and systems, Macmillan, 1974
Morris, N.M., Logic circuits, McGraw-Hill, 1976
Nagle, H.T., Carroll, B.D. and Irwin, J.D., An introduction to computer logic, Prentice-Hall, 1975
Oberman, R.M.M., Digital circuits for binary arithmetic, Macmillan, 1979
Oberman, R.M.M. Disciplines in combinational and sequential circuit design, McGraw-Hill, 1970
Poratt, D.J. and Barna, A., Introduction to digital techniques, Wiley, 1979
Rhyne, V.T., Fundamentals of digital systems design, Prentice-Hall, 1973
Roth, C.H., Fundamentals of logic design, West, 1975
Triebel, W.A., Integrated digital electronics, Prentice-Hall, 1979
Unger, S.H., Asynchronous sequential switching circuits, Wiley, 1969
Wickes, W.E., Logic design with integrated circuits, Wiley, 1968
Wood, P.E., Switching theory, McGraw-Hill, 1968
Zissos, D., Logic design algorithms, OUP, 1972
Zissos, D., Problems and solutions in logic design, OUP, 1976

Index

1's complement 67
2's complement 66–69
Absorption theorem
   redundancy of 9
   switch contact circuits of 10
Accumulator 297
Adders
   carry look-ahead 61–63
   four-bit parallel 60, 62
   full 58
   half 57
   MSI chips 252
   NBCD 257
ADD instruction 293
Addressing techniques for ROM 240
AND function 2, 3
   inversion of 11
   NAND gates 37
   NOR gates 42
AND gate 3
AND/OR circuits 240
AND ROM 246
A-O-I module 46
Assembly programming language 300
Association theorem 12
Asychronous binary counter 119
   resettable 123
Auxiliary carry flag 299

Bi-directional bus system 297
Binary/hexadecimal equivalences
   294, 295

Binary multiplication 70, 281
Binary multiplier 282, 283
   programme-controlled 288
Binary, reflected 74
Binary representation 67, 68
Binary ring counter 145
Binary sequence generator 140
   pseudo-random 153
Bipolar transistor 1, 238
Block diagram (microprocessors) 296
Boolean algebra 1–15
   Addition function (OR) 4
   AND function 3
   consensus theorem 13
   factorising 13, 40
   function generator – PLA 248
   function generator – ROM 239
   function, two variables 19
   implementation using NAND
      gates 40
   inversion function 5, 27
   multiplication function 3
   simplification of 25
   truth tables 3, 4

Caldwell's merging procedure 166,
   167, 179, 199
Canonical form 18
Carry-look-ahead adder 60–62
'Can't happen' terms 28
Characteristic equation of flip-flops
   85, 89, 95

Index

Chips
  cascading of 127
  connections 125, 126
  TTL IC 124
Circuit equations 168
Circuit implementation 173
CLEAR instruction 293
Clock-driven sequential circuits 158–194
  (see also sequential circuits)
Clock signals 108, 121
Code conversion 71–74
Collector dot 48
Combinational logic design 56–82
  circuit 56
  daisy-chaining 79
  full-adder 58
  half-adder 57
Combinational networks hazards 264, 265
  design of hazard free 269
Combination detector 167
  internal state diagram 165
  J and K input 169
Commutation theorem 12
Complementary circuits 11
Complementation theorem 5
  determination of 10
Connection matrix of
  PLA 247, 248
  ROM 238–240
Consensus theorem 13
  definition of 13
  truth table 13
Control bus system 297
CONTROL instructions (microprocessors) 291
Counters 101–128
  asynchronous binary 119
  asynchronous resettable 123
  chips 127
  decade binary down 113
  decade binary up 111, 112
  decade Gray code 'up' 113–115
  integrated circuit 124
  hexadecimal 251, 252
  parallel connection of 105
  programme 290
  scale-of-two 101
  scale-of-four 103
  scale-of-five 106

Counters (cont.)
  scale-of-eight 103
  scale-of-ten asynchronous up 122
  scale-of-sixteen 118
  series connection of 105
  synchronous down 106
  TTL IC chips 124
Cycles, event-driven circuits 201

Daisy-chain technique 79, 80
Data selector see multiplexer
De Bruijn diagram 142
Decade counter
  binary down 113
  binary up 112
  design of 137–139
Decimal addition using MSI adders 254
Decimal ring counter 143, 145
Decoder applications 232–234
  Intel 8205 3-to-8 235
Decrement instructions 293
De Morgan's theorem 12, 36
Demultiplexers 231, 232
D flip-flops 95, 96
Diminished radix complement 67
Distribution theorem 12
'Don't care' terms 28, 73
Double-rail input shift registers 132
Down-counters 106, 113
Dualisation process 6
Dynamic hazards 274–277

Electronic gates (Boolean equations) 6, 7,
Essential hazards 277, 278
Event-driven circuits 195–219
  circuit cycles 200, 201
  essential hazards 277, 278
  four-state machine 207, 208
  internal state diagram 197
  pump problem 203
  race conditions 201–203
  ROM's 243
  secondary variables 200–202, 207
  sequence detector 209–211
  slings 212
  three-state machine 202
Exclusive-OR function
  cascading 52
  gate 49, 52

Exclusive-OR function (*cont.*)
  implementation 52
  K-map plots 51, 53
  rules of 54
  truth table for 50
Expansion gates 45

Feedback equations, maximum length sequences 152
Feedback functions 139, 143, 152
Feedback loop 86
Feedback shift register 135
Five-variable functions 26
Flip-flops 83–97
  characteristic equation of 85, 89, 95
  D 95, 96
  excitation map 213
  JK 91–95, 283
  latching action of 96
  race conditions 84, 92, 93
  SR 87–91, 213
  state diagrams 84
  state tables 85
  T 83–87
Floppy discs 297
Flow diagrams 285–288
Four-bit parallel adder 60
Full-adder 58
  implementation of 59
Full subtractor 64
  truth table for 65
Function simplification 16–34
  Boolean functions 25

Gates
  AND 3, 262
  AND-OR 47
  delays 261
  electronic 6, 7,
  exclusive-OR 49, 52
  expansion 45
  inverse functions 27
  JK flip-flop 93
  miscellaneous 46
  NAND 36–38, 261
  NOR 42, 269, 270
  number of 16
  open-collector 47, 48
  tri-state 49

Generators
  binary sequence 140
  MLS self-starting 154, 155
  sequence 171
Gray code converter 74–76
Gray code counter 113–117

Half-adder 57
HALT instruction 292
Hazard free networks 269
Hazards 261–279
  detection of 272, 273
  dynamic 274–277
  essential 277, 278
  static 264–274
Hexadecimal counter 250, 251
Hexadecimal number system 293, 294

Idempotency theorem 7, 8
Implication tables 181–184
Input ports 295, 296
Inputs
  equation of 108
  number of 45, 46
Instruction register 291
Instructions 291–293
Intel 8085 295, 300
Intel 8205 3-to-8 decoder 235
Interface unit 296, 297
Internal state diagram 164, 165
  sequence detector 172, 211
Interrupt signals 80
Interrupt sorters 76
  implementation of 78
  truth table for 77
Intersection theorem 8
Invalid code detector 187, 189
Inversion function 5, 27, 31, 32
  double 6
  gates 27
Inverter 6
  exclusive-OR gate as 53
  NAND gate as 37
  NOR gate as 41
  tri-state 49

JK flip-flop 91–95
  master/slave connection 94, 160
  scale of two counter 102
  sequential circuits 158, 159, 168
  steering table for 102

Index

Johnson counter, see twisted ring counter
JUMP instruction 292

Karnaugh (K) maps 16–34
  canonical forms 18
  counters 103, 104, 111
  exclusive NOR functions 51
  exclusive OR functions 51, 53
  feedback function 139, 143, 155
  flip-flop input signals 117
  four-input multiplexer 227
  full adder 59
  full subtractor 65
  hexadecimal counter 251
  NAND function 36
  NOR function 41
  OR functions 37
  race-free assignment 207
  reflected binary 75
  SR flip-flop 88
  static hazards 276
  twisted ring counter 148

Latching action of flip-flop 96
Latching circuit 123, 124, 127
LOAD COUNT REGISTER instructions 292
LOAD MULTIPLICAND instruction 290, 292
LOAD MULTIPLIER instruction 291
Logic function generator 222, 225
LSI chips 71, 280

Machine code language 299
'Master/slave' connection 94, 95
Maximum-length sequence 151
  feedback functions for 152
Mealy State machines 174
Memory 289
  elements 83–97
Microprocessors 280–299
  accumulator 297
  binary multiplier 282–285
  block diagram for 296, 297
  comparison with multiplier 295

Microprocessors (*cont.*)
  CPU 8085 298
  flow charts 285–288
  instructions register 291
  memory 289
  programme counter 290, 297
  programming of 299
  word length 289
Minuend 64
MLS self-starting generator 154, 155
Mnemonic programme form 300
Moore State machine 174
MSI adders 252, 253
  decimal addition 254
MSI chips 252, 280
MSI circuits 220
  multiplexers 220–231
  PLA's 246–252
  ROM's 237–245
Multiplexer 220–231
  eight-input 228
  four-input 225–227
  logic function generator 222, 225
Multiplexing levels 229–231
Multiplicand register 283

NAND function
  definition of 35
  K map of 36
  truth table for 36
NAND gates 38, 40, 269, 270
  carry look-ahead adder 62–64
  input 45
NAND sequential equation 89
NBCD adder 257
NBCD code 29, 30, 71–73
NOR function
  definition of 41
  gates 42, 269, 270
  K map of 41
  truth table for 42

Open-collector gate 47, 48
Operational instruction 291
OR feedback, exclusive 150
OR function 4
  inversion of 11
  NAND gates 37
  NOR gates 42

Index

OR gate 5
OR-ROM 246
Output ports 295, 296
Output signal (Z) 159, 161

Parallel-in shift registers 132
Parity flag 299
Partitioning 179, 180
Preset facility (SR flip-flop) 91
Prime implicants of simplified functions 25, 26
Prime implicants of simplified functions 25, 26
Programmable logic array (PLA) 246
   connection matrix of 247, 248
   sequential circuit design 249
Programme-controlled multiplier 288
Programme counter 285, 290
Pseudo-random binary sequence generator 153
P/S expressions 16, 17
   NOR gates 43, 45
   plotting of 30
P term 17, 25
Pull-up resistor 48
Pulsed synchronous circuits 176
   internal date diagram 177

Quine-McCluskey Method 16

Race circuit 200-203
   conditions, JK flip-flop 92, 93
   conditions, T flip-flop 84
   critical 202
   free assignment for 3-state machine 202
   free assignment for 4-state machine 206
Radix complement 66, 68
Random access memory (RAM) 289
Read-only memory (ROM) 237
   address decoder with 64 bit 237-239
   addressing techniques for 239
   sequential circuit design 241
   memory sections 289
Redundancy theorem 9, 14
Reflected binary 74

Registers (microprocessors) 290, 297, 299
'Ribble-through' counter, see asynchronous binary counter
Ribble-through, maximum 120
Ring counter 143-147
   binary 145
   decimal 143, 145
   state table for 144
   three-stage 146
   twisted 147-150

Secondary variables 200-202, 207
Sequence detector 209-211
Sequence generator 140
   circuit equations 173
   design of 171
   internal state diagram 172
Sequential circuits 101
   JK flip-flop 153, 159, 168
Sequential circuits, clock driven 158-194
   analysis of 160
   block diagram of 163
   design procedure for 163
   equation development 168
   implementation 171
   internal diagrams 164, 165
   output equation 161
   PLA's 249
   ROM's 241
   state assignment 184, 185
   state reduction 166
   state table of 159, 161
   timing diagram 159
Sequential circuits, Even tdrive 195-219
Serial-in shift registers 131
Shift instruction 293
Shift pulse 283
Shift registers 131-156
   applications 132
   bi-directional shifting 134
   counters 134
   eight-state sequence 142
   exclusive-OR feedback 150, 151
   four-bit 133
   MLS self-starting generator 154
   ring counter 143-150
   sequence generator 140
   universal state diagram for 135

Sign flag 299
Single-rail input shift register 132
Single switch contact 2
Slings 212
S/P expression 16
  NAND gates 38
  NOR gates 43
Spike generation 262, 263
Stack pointer 297
State diagrams
  counters 102–116
  external 85
  internal 164, 165
  JK flip-flop 92
  SR flip-flop 88
  T flip-flop 84
State reduction 179, 180
  implication table 181–184
  partitioning 180, 181
State table 85
  minimal 166
Static hazards 264–274
Steering table 102
S-term 17
STORE PRODUCT instruction 292
Subtrahend 64
Switch contact circuit 10
Switch, logic of 2

T flip-flop 83–87
  instability in 86
  state diagram of 84

Timing diagrams 95, 97, 121, 125
Traffic controller 243–245
TRANSFER instruction 291
Tri-state gate 49
Truth table 2
  collector dotting operation 48
  daisy-chaining 79
  full-adder 59
  full-subtractor 65
  half-adder 57
  interrupt sorter 78
TTL IC chips 124
Turn-off equations, flip-flop 108, 170
Turn-on equations, flip-flop 108, 170
Twisted ring counter 147–150
  K-map for 148
  undesired count sequences 149

Unidirectional address bus 297
Universal state diagram (shift registers) 135, 137
Union theorem (Boolean algebra) 8
Unused inputs 42
Up-counters 101–106

Waveform information 164
Word length 289

Zero flag 299
Zissos Method 41